AN ILLUSTRATED EXAMINATION OF

# ADVANCED TECHNOLOGY WARFARE

AN ILLUSTRATED EXAMINATION OF

# ADVANCED TECHNOLOGY WARFARE

A detailed study of the latest weapons and techniques for warfare today and into the 21st century

Col. Richard S. Friedman
Lt. Col. David Miller • Bill Gunston • Doug Richardson
David Hobbs • Max Walmer

a Salamander book

Published by Salamander Books Limited
LONDON

# A SALAMANDER BOOK

Published by Salamander Books Ltd.,
Salamander House,
27 Old Gloucester Street,
London WC1N 3AF,
United Kingdom.



**ISBN 0 86101 255 0**

Distributed in the UK by
Hodder & Stoughton Services,
P.O. Box 6, Mill Road,
Dunton Green, Sevenoaks, Kent TN3 2XX.



All correspondence concerning the content of this volume should be addressed to Salamander Books Ltd.

## CREDITS

**Editor:** Ray Bonds
**Designer:** Mark Holt

**Diagrams:** TIGA
**Colour aircraft artwork:** Stephen Seymour
**Jacket:** David Palmer

**Colour and monochrome reproduction:**
Melbourne Graphics,
Rodney Howe Ltd.,
York House Graphics Ltd.,
David Bruce Graphics Ltd.

**Filmset:** SX Composing Ltd.

**Printed** in Italy by G. Canale e C SpA, Turin.

## ACKNOWLEDGEMENTS

Literally hundreds of individuals, industrial organizations, research groups, and various official departments of defence throughout the world have contributed information and illustrations during the preparation of this volume. The publishers wish to thank them all wholeheartedly for their great assistance.

# THE AUTHORS

**COLONEL (Retd) RICHARD S. FRIEDMAN,** Military Intelligence, US Army, has served in a variety of assignments, most recently with the US Army General Staff in Washington, as Defense and Army Attache in Budapest, Hungary, and, prior to his retirement at the end of 1984, as Deputy Assistant Director, Inteligence, on the International Military Staff of NATO Headquarters in Brussels. He is a consultant and writer on international affairs, and is co-author of the Salamander book "The Intelligence War".

**BILL GUNSTON** is a former RAF pilot and flying instructor. Since leaving the Service he has acted as an advisor to several aviation companies and become one of the most internationally respected authors and broadcasters on aviation and scientific subjects. He is an Assistant Compiler of "Jane's All The World's Aircraft", former Technical Editor of "Flight International" and is the author of numerous Salamander books, including the most recent "Warplanes of the Future".

**DAVID HOBBS** is Director of the North Atlantic Assembly's Scientific and Technical Committee, based in Brussels. Trained as a physicist, he was a researcher at the prestigious Center for Defence Studies at Aberdeen University. He pays close attention to both the military and civil exploitation of space and writes extensively on defence technology.

**LIEUTENANT-COLONEL DAVID MILLER** is a serving officer in the British Army. He has served in Singapore, Malaysia, Germany and the Falkland Islands, and has filled several staff posts in Army headquarters, besides commanding a regiment of the Royal Corps of Signals in the United Kingdom. In addition, he has contributed numerous articles to technical defence journals, and is co-author of Salamander's "The Intelligence War", "The Vietnam War" and "The Balance of Military Power".

**DOUG RICHARDSON** is a defence journalist specializing in the fields of aviation, guided missiles and electronics. After an electronic R&D career he served as Defence Editor of the internationally respected aerospace journal "Flight International", and as Editor of "Military Technology and Economics" and "Defence Materiel". He is author and co-author of numerous Salamander books.

**MAX WALMER** is a defence journalist who has contributed numerous articles to international defence journals. He has specialized in "special warfare" aspects and is the author of Salamander's much acclaimed "Guide to Modern Elite Forces".

# CONTENTS



*The views of Col. Friedman do not purport to reflect the positions of the Department of the Army or the Department of Defense of the United States.

# INTRODUCTION

THERE CAN be few people in the West who are not aware of the extraordinary technological explosion that is taking place all around them. Pocket computers, unheard of thirty years ago, are now commonplace; hand-held cordless telephones are beginning to become fashionable in people's homes; there are talking computers in private motor-cars; word-processors abound in offices; factories are full of robots; in the skies supersonic travel in Concorde is now a matter of simple routine for British Airways and Air France; and flights of America's Space Shuttle are becoming so regular it has been described as a "Spaceliner" – it is even taking non-professional astronauts as passengers, including a US Senator! This advanced technology is greeted as a wonder when it arrives, but within months, perhaps even weeks in some cases, it is considered commonplace.

If this is the case in everyday civilian life then it is even more so in the military field where technological advances are leading to developments which are quite literally turning science fiction into science fact.

The purpose of this book is to investigate the defence-related technological developments that are taking place, not just on the battlefields of the world and within the armed forces which face each other over many disputed borders, but also in the laboratories. Indeed, in some cases it could be argued that the scientists and "boffins" are fighting – and perhaps deciding – World War Three in their laboratories today. However, it must be made clear that high technology is manifest not simply in the hardware – the ships, aircraft and tanks – which epitomises military might, but also in the materials from which the equipment is made, the power source which makes it work and, increasingly, the electronics which provide its command and control functions.

Sometimes such advances can be totally hidden from the public eye and in many ways are quite unglamorous, yet their

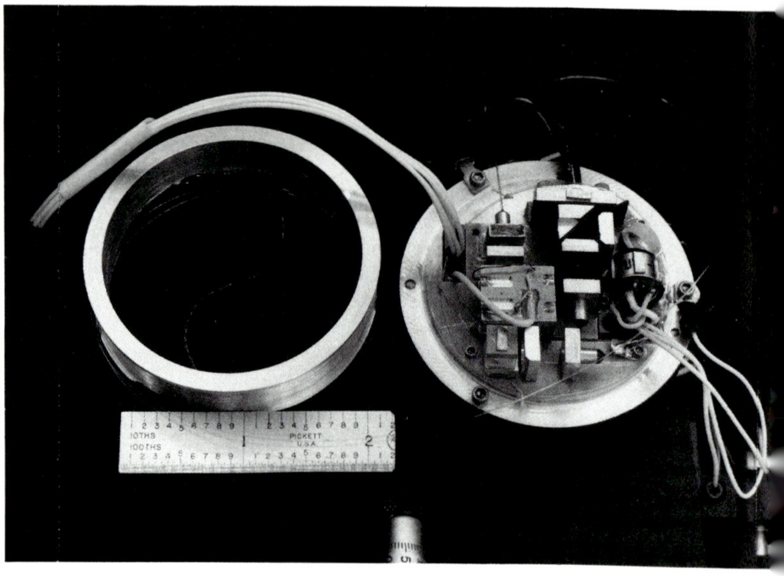

**Above: Many significant advances in military technology are hidden from public gaze. For example, miles of heavy, fault-prone electrical wiring can now be replaced by fibre-optics which carry enormous amounts of information at high speed, saving both space and weight.**

**Below: Aerodynamicists are now looking at daring new concepts to improve aircraft performance, especially in combat. The F-16 needs a computer to enable it to stay in the air, while this Grumman X-29 has forward-swept wings and canard control surfaces: inherently unstable, but highly manoeuvrable.**

**Rapid technological advances are changing the very shapes, scope and speed of warfare, literally turning science fiction into science fact. Miniaturization, computer-aided controls and intense materials research are propagating more deadly weapons and countermeasures in the air, at sea, on the ground and, increasingly, in the "final frontier" – space.**

**Fibre-optic control for tactical missiles**
Wire-guided tactical missiles (for example, anti-tank guided weapons) are widely used as the wires give a secure and unjammable means of communication between controller and missile. Such systems depend, however, on a clear line-of-sight between controller and target, which obviously imposes limitations on their use. But fibre-optics offer the possibility of firing and controlling weapons from safety behind obstacles; a video camera in the missile's nose passes real-time pictures to a controller, who would then direct the missile down on to the target, without seeing it directly himself. Such a system could be secure, cheap and hard to counter. American firms like McDonnell Douglas are leading the way.

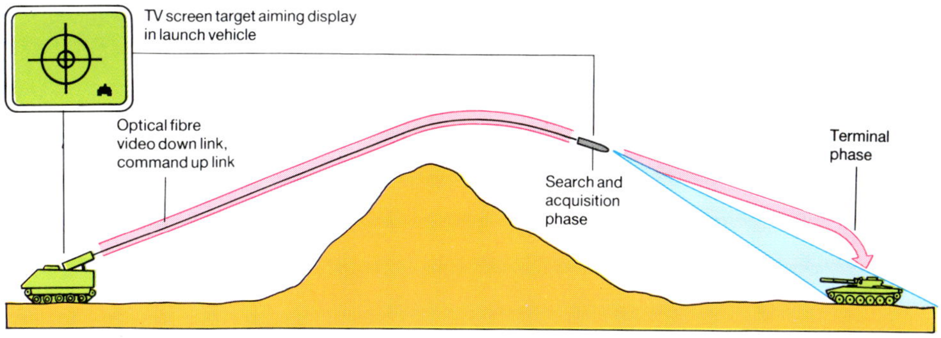


**Below: The effective life of new technology is getting shorter. Solid-state laser rangefinders entered service just a few years ago, but they are already outdated by this carbon-dioxide laser, which penetrates battlefield smoke and dust far more effectively than its predecessor does.**

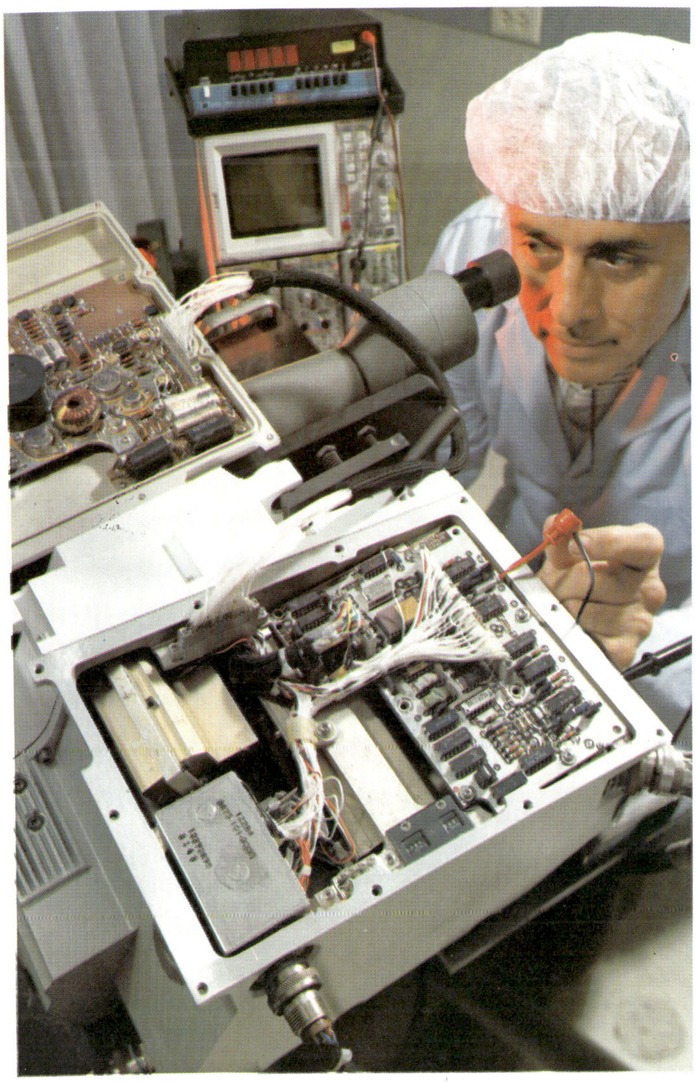

effect is profound. For example, as equipments have become more complicated and more dependent upon electrical and electronic control, they have required ever-increasing amounts of wiring – in general terms, two wires for each circuit. Such wiring is expensive, heavy, frequently a fire hazard, and very demanding in maintenance resources. Today, however, such wiring is being increasingly replaced by multiplexers, which stack up circuits electronically, and then pass them over one pair of wires, or even over a fibre-optic cable.

In air warfare new shapes are taking to the skies with novel wing designs, new engines and armaments with totally new capabilities. Perhaps the most significant of the known developments, however, is "stealth" – a combination of various technological advances that results in an aircraft which might ultimately be "invisible" to radar, electronic, visual and aural detection. If this technology really is as good as is claimed then it will, of course, lead to a great dilemma, in that it will be impossible to tell whether or not your opponent also has such an aircraft because, by definition, you will not be able to detect it!

In land warfare, technology is leading to major advances in many areas, with one of the key fields being that of the armour/anti-armour confrontation, where there is an interesting example of the point-/counter-point competition in military technology, which is making a direct and important contribution to the ever-escalating cost of defence hardware.

In the 1950s and 1960s one of the major defences against the tank was the hollow-charge warhead: this had excellent penetration capabilities which, most importantly, were not dependent upon the speed of the projectile. This meant that man-portable, accurate and reasonably lethal anti-tank weapons could be designed for the infantry; many missile systems were designed on this basis, such as the US TOW, the British Swingfire and the Soviet Swatter, Sagger and Spandrel. However, the British Chobham armour, a ceramic sandwich material of highly classified composition, apparently provides excellent protection against such warheads and thus virtually negates a whole generation of anti-tank weapons. The other major type of warhead, and one not defeated by Chobham armour, is the kinetic energy round, but "reactive armour" (such as the Israeli "Blazer") has now been developed which may reduce the effect even of this. Thus, vast sums are now having to be spent by both Warsaw Pact and NATO nations to develop an anti-tank warhead which will defeat both these new types of armour and restore the effectiveness of the defence once again.

At sea arguments rage about the effectiveness of surface ships and their value in the primary maritime tactical battle: that of

anti-submarine warfare (ASW). Ships have in many ways changed remarkably little in the past 40 years in fundamental technology, or in size and shape, although internally there have been great advances, especially in electronics, and command and control. In naval armaments the primary weapon is now the missile, while the gun, although in no way obsolete, has nevertheless been firmly relegated to second place, except perhaps as an anti-missile weapon.

Perhaps one of the more significant developments at sea, if not the most widely understood, has been the rapid advance in submarine technology. Unlike that of surface ships, the speed of submarines has increased by a factor of at least four in the past 40 years, and many submarines can now outpace their surface hunters. Most significant of all, however, has been the emergence of the submarine as a strategic weapon system with the ability to carry out direct attacks against the opponent's land mass, using ballistic missiles, and cruise missiles, as well.

The effectiveness, and unique value, of the strategic missile submarine lies in its ability to hide itself in the depths of the ocean. However, vast sums have been, and will continue to be, devoted to ASW technology, and in particular to finding new means of locating and tracking submerged submarines. Should there be a breakthrough – and there seems to be no reason why there should not – then the SSBN's invulnerability as a second-strike deterrent will be seriously eroded, even if not totally negated. Such a technological breakthrough may be difficult and expensive, but experience shows that given the incentive and resources scientists can now produce the solution to almost any problem. In such a case the effort to counter a ballistic missile attack needs to swing from deterrence to an effective defence.

Just such considerations are behind President Reagan's Strategic Defense Initiative (SDI), where it is proposed that an effective non-nuclear defence should provide the United States (and perhaps her allies, too) a virtually invulnerable umbrella against incoming ICBMs and SLBMs. For many people SDI is seen as an unnecessary escalation of the arms race and a negation of the US strategists' own arguments about deterrence. However, if it is seen as one possible outcome of an appreciation that SSBNs are about to lose their invulnerability, then it is a logical and reasonable step – one that could lead to a dramatic change in the balance of strategic power.

One of the more significant areas of technological advance has been in sensors, which enable man to detect, locate, classify and track enemy devices beyond the range available through his own senses. Radar, discovered just before World War II, is now capable of operating effectively far beyond the horizon, and of providing such an accurate image of a target that the very class of ship can be determined, or even, by scanning from one aircraft to another astern, determining what type of engine is being used (by make) and thus aiding positive identification. Optical sensors are also developing fast; indeed, the limiting factor tends to be the control and aiming technology rather than that of the optical device itself. As is shown in this book, it is now feasible to aim a laser at a space-borne mirror and use the reflected beam to help modify the outgoing beam in such a way that it corrects itself; such a beam can then be used to communicate with submerged submarines thousands of miles away.

The computer revolution, while significant in civil life, has been revolutionary in the military sphere. As sensors have multiplied in numbers and capability, as weapons systems have

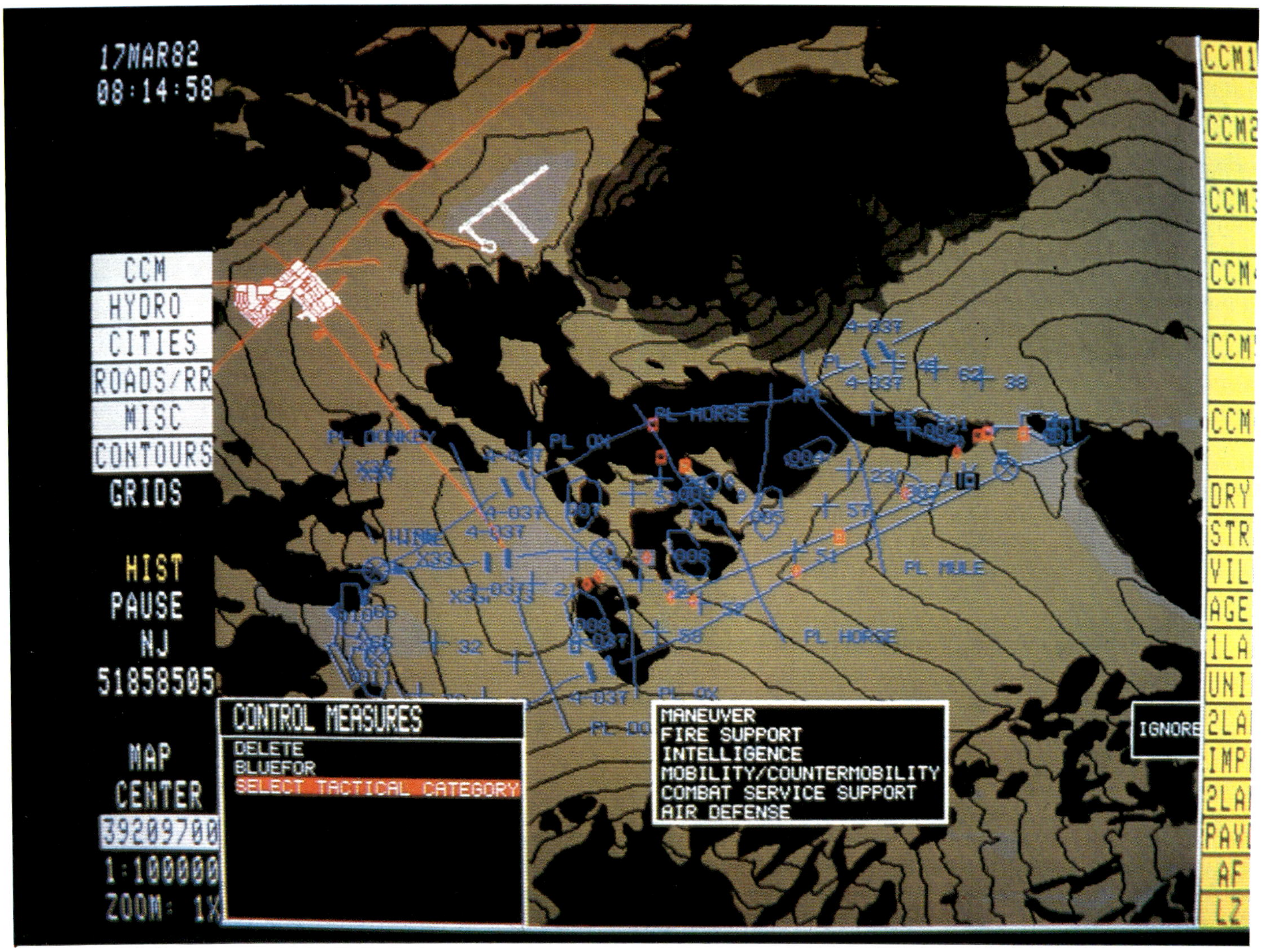


**Left: British Challenger main battle tank and MCV80 armoured personnel carrier, the very epitomy of modern military might. Technologically very advanced, they are conceptually little different from MBTs and APCs used thirty years ago. In twenty years time they could be as out-of-date as the dinosaur.**

**Above: As the land battle has become ever faster and more complex command-and-control methods have had to be brought up-to-date to enable commanders to maintain control over events. Displays such as this depend upon reliable, secure computers and communications systems.**

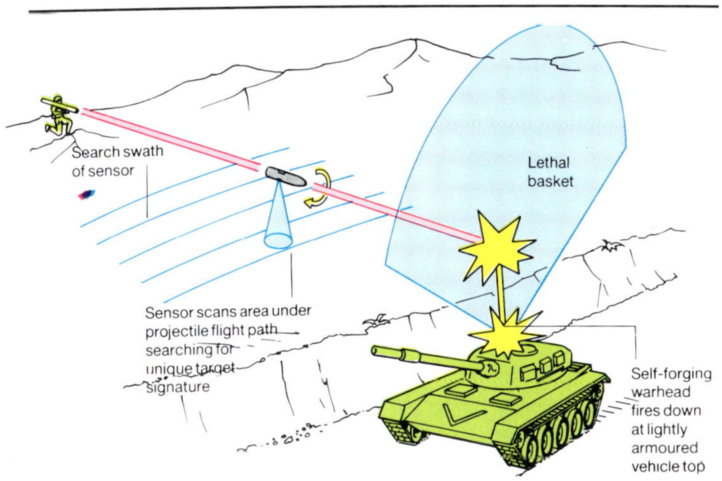


**Top attack of armour**
The massive numerical imbalance in Warsaw Pact armour and APCs, coupled with advances in armour technology, has led to many new methods of attack. One such is the Aerojet Electro Systems Smart Top Attack Fire and Forget (STAFF) system, which could be used for projectiles fired from artillery, tanks or, as shown here, shoulder-fired rocket launchers. The STAFF projectile and its sensor spin rapidly in flight, searching for an armoured target. When detected, a hyper-velocity self-forging fragment warhead is fired downwards to attack the tank or APC through the roof, its least armoured and thus most vulnerable sector. AFV designers are rapidly up-armouring their vehicles to counter this new threat.

become increasingly electronically controlled, and as the military forces have become ever faster moving, so the requirement for effective command, control and communications (the so-called $C^3$) has become ever more pressing. The solution has become available through the development of very powerful digital computers and communications systems, and this, allied with the concurrent dramatic reduction in size of electronic devices, enables scientists to produce equipment which helps commanders to keep control of the fluid battle situation.

Another revolution is under way in metallurgy and non-metal materials. Well-established materials such as steels and aluminium alloys are now being used less frequently as technology enables new materials to be used in production processes. Thus, for example, most of the new tank armours are based on some form of ceramic sandwich. Titanium has long been known as a metal that is very strong, very light and (extremely important in some applications) non-magnetic. The West has not been able to develop a production process which would enable titanium to be welded without some 200 passes of the machine; obviously an expensive, time-consuming process. But the Soviet Union has enabled the weld to be made in just six passes, and thus they are able to construct submarines of this excellent material. Another remarkable material is the Swiss-based Dupont company's Kevlar, a very strong and very light carbon-fibre material, and which has found many military applications from infantryman's body armour to aircraft components.

A material that has had a marked effect on military equipment is plastics, which appears in an almost infinite number of varieties. Combining light weight with considerable strength and durability, allied to ease of manufacture, plastics seemed to be ideal for many military uses until the South Atlantic War of 1982 showed that many plastics products suffer some serious limitations as both fire and gaseous hazards. This is an example of a not-uncommon occurrence, where a major advance in one field produces a significant step back in another.

Advancing technology also has both a positive and a negative impact on another of the most significant problems in defence – that of cost. In some areas technology has without a doubt brought the cost of military equipment down. As with domestic devices such as radios and "hi-fi", home computers and wrist watches, much in the military electronics field is very cheap compared with 20 years ago. However, while the relative cost of electronic hardware may have decreased in many instances, the cost of software has escalated rapidly, to the point where it far outweighs that of hardware. Even worse, it also seems to continue escalating throughout the life of the equipment.

In major equipments, such as ships, tanks and aircraft, the development and manufacturing costs have reached a point where few but the largest and wealthiest nations will be able to afford the most sophisticated equipment that scientists can provide. The cost of a main battle tank (MBT), for example, has increased from £250,000 for a Centurion in the mid-1950s to £1,250,000 for a Challenger today. Technology can help in some ways, by providing cheaper and less manpower-intensive manufacturing processes, or cheaper components, but all too frequently it also seems to escalate the costs by encouraging the user to ask for yet more "capability" to be built into a particular platform.

**Below: The US is devoting great resources to laser-weapon programmes, one of which has been the Airborne Laser Laboratory (ALL) mounted in an NKC-135. This drone was destroyed when the laser burned through its skin, destroying vital components. Subsequently, the ALL downed five Sidewinder AAMs in five separate engagements.**

Below: Microminiaturization, such as in this tiny, but sophisticated electronic component of the Trident I SLBM (left), enables designers to produce ever more capable weapons systems which are nevertheless smaller, lighter and more reliable. However, this is achieved only at a great increase in development and capital costs, to the point where many nations can no longer afford many of the systems that technology offers.

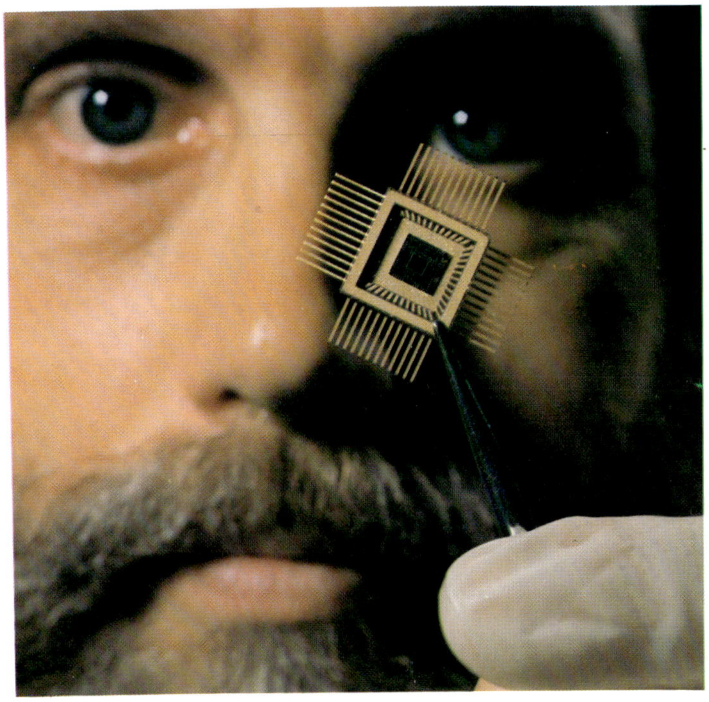

An area of increasing concern to strategists is that of fuel. Apart from a relatively few nuclear-powered ships, all military equipment is dependent upon fossil fuels for its power. The problem is two-fold: not only is the world's supply obviously finite, but much of the current resources are controlled by a relatively small number of producers. Efforts are being made in some nations to find new fuels, but such efforts seem to be concentrated into particular fields. For example, fuel-cell research for submarine power is being undertaken with a view to finding an effective alternative to nuclear-power to free diesel-powered submarines from having to go regularly to the surface – rather than for the simple purpose of replacing oil as a fuel.

The final word, however, must be on the subject of space. In the popular television programme "Star Trek" space is described as "the final frontier" and this seems to be a remarkably accurate description of the situation. Military space commands have been set up in the USA and the USSR where generals are as responsible for the space "combat zone" as other commanders are responsible for a continent or an ocean. The Soviets put far more military than civilian missions into space, and the USA, after some very well-publicized civil flights, recently made a very highly classified military flight with their Space Shuttle. There have been a number of serious proposals for space interceptors, and actual anti-satellite tests have been carried out. Many of the current satellites have military roles.

All this tends to suggest it is inevitable that space will become an area of military conflict, where high technology will rule, and the "final frontier" may well become the "final battlefield".

Below: In a popular TV series, space is described as "the final frontier", and it could well be that this turns out to be an accurate description, not just for peaceful exploration, but for warfare, as well. Already there is great military use of space by both superpowers – mainly for "spying" and communications. A third of America's planned Shuttle missions are military.

Above: An Earth-generated laser beam being reflected toward a high-altitude target by a space-based mirror, a concept tested on the June 1985 US Space Shuttle mission. This is one of many ideas being researched in President Reagan's Strategic Defence Initiative (SDI), popularly called "Star Wars".

# ELECTRONIC WARFARE

An ever-increasing proportion of the cost, complexity and performance of modern weapon systems is directly due to electronics. Back in 1950, the only electronic system which the crew of the Soviet T-54 tank had to contend with was the radio (although a simple gun-stabilisation system was to follow on the improved T-54A and T-55 models) and the contemporary MiG-15 had little more avionics that a World War II fighter.

The crew of a present-day tank such as the US Army's M1 Abrams have to deal with laser-rangefinders, fire-control computers, thermal-imaging systems, and radiation detectors, with frequency-hopping radios likely to follow. The pilot and radar operator of the F-14 Tomcat have to cope with track-while-scan radar, long-range electro-optical systems, built-in countermeasures, and also with aiming and launching the 100-mile (160km) range Phoenix missile.

Such is the pace of electronic technology that the personal computer purchased in a Western store can exceed the complexity of many military systems, and the US Government is already applying export restrictions to the more complex business models.

There is inevitably a penalty to be paid for the use of advanced electronic technology in warfare. Combat effectiveness may be increased, but massive "black box" payloads often reduce reliability, while cost and development time rise astronomically compared with earlier weaponry.

The US F-86 Sabre fighter, whose most complex item of avionics was a radar gunsight, first flew in October 1947 and was in service by December 1948; its contemporary, the Soviet MiG-15, went through the same process in less than a year. Typical first-flight-to-service-entry times for radar-equipped all-weather interceptors such as the F-4 Phantom II was three years, but later programmes such as the Tornado and Soviet MiG-29 Fulcrum have involved times from first flight to service entry of around six years.

**Above: The gunner's controls in the US Army's new M1 Abrams tank are associated with a complex fire-control system which maximises the chance of first-round hits against combat targets.**

**Right: By the year 2000, soldiers may be equipped with voice-commanded "expert" computer systems able to help them locate and attack ground and airborne targets by day or night.**

**Below: The AWG-9 fire-control system, whose electronics consoles are visible in the rear (NFO's, left) and front (pilot's right) cockpits of the Grumman F-14 Tomcat, give the aircraft the ability to pick off targets at ranges of up to 100nm (180km) using long-range AIM-54 Phoenix missiles.**

Despite such penalties (which may be reduced by the latest solid-state electronic technology) the trend to ever-growing complexity and increased firepower will continue. In 1984, the British R&D company Scicon even unveiled its concept of the "Infantryman 2000" electronics-based combat kit likely to be worn by soldiers in the year 2000. At first sight, the masked figure on the company's exhibition stand might have seemed like an "extra" from the film "Star Wars", but this futuristic figure and its equipment were based on current research concepts.

Given the projected degree of miniaturisation for computers of 2000AD vintage, the soldier would be able to carry around 100Mbyte of memory with the capability of rapid number processing and also symbolic processing for expert systems. All input commands to the soldier's weapons and equipment would be by voice, including special word strings reserved for the activation of weapons. Communication to command centres or specialised weapon systems would be via the computer, which would provide time-compression, frequency-hopping, encryption and error compensation.

"Infantryman 2000's" helmet would be equipped with a throat microphone for voice input commands, and also feature an image intensifier and thermal imager for target acquisition, and a gyro-stabilised

**Today's giants of the modern defence industry are the companies which develop and manufacture the supporting electronic systems – sensors, data-processing equipment, communications equipment and even specialised electronic warfare systems used to disrupt the operation of these new tools of the "wizard" war.**

laser target designator. Sound sensors would provide full 360 degree coverage, and incorporate a sound-analysis facility able to identify targets by examining their acoustic signature. Information from the sensors and computer could be overlaid onto the external view by means of a head-up display built into the helmet, or viewed as a separate high-resolution picture.

The soldier's multi-purpose personal weapon would be fired using sighting facilities built into the helmet display system, while two vertically-launched missiles built into the back pack would offer anti-aircraft and anti-AFV firepower. The total effect may seem like science-fiction, but if Scicon's projections are correct, the soldiers who would wear such futuristic kit have already been born.

## Miniaturisation of electronics

The impact of modern electronic technology is often difficult to visualise, but John Frichtel, manager of the American company General Electric's APG-67 radar programme, claims that this unit – designed for use in the Northrop F-20 Tigershark – is a good example of 1980s technology. This miniature set weighs 270lb (122.5kg) and occupies a volume of only 3.3 cubic ft (0.93 cubic metres). "This radar in 1960 technology would have weighed 25,000lb (11,340kg)", says Frichtel – a weight roughly equal to that of the complete aircraft when ready for take-off! Mean time between failure is greater than 200 hours. Future radars could raise the latter figure to 500 or even 1,000 hours.

Much of the performance of future Western weapons will be the result of the current Very High Speed Integrated Circuits (VHSIC) programme – a US tri-service effort to develop two generations of integrated circuits (ICs) with very high data processing capacity for a wide range of military systems. Conventional electronic systems are made of a number of interconnected ICs, but VHSIC technology should allow entire systems to be packed in a single chip, with corresponding savings in size, and weight. The end product will be ICs so complex that just one could take the place of over 50 conventional components.

Military electronic systems made from such components will have a ten-fold increase in processing capability compared with earlier equipments; they will require less electrical power, and will be more reliable. VHSIC-based equipment should also be less expensive than its

**Above: The largest wooden structure in the world is this platform at Kirtland AFB. The antenna wires seen in this photo are used to test the resistance of aircraft to electromagnetic pulses.**

**Below: This Soviet printed-circuit card uses 133 Series integrated circuits. These components are the equivalent of the Texas Instruments 5400 Series, says the US Department of Defense.**

present-day counterparts. Phase 2 of the VHSIC programme will develop second-generation technology chips, with the use of submicrometer geometries allowing 100,000 or more gates to be placed on a single chip.

Each of the US services has specific target systems for early use of VHSIC technology. Initial applications will be in digital signal processors for radar, ASW, communications, missile guidance, electronic warfare and optical sensor systems.

In the early 1980s, the US Government displayed an increasing concern over the access which the Soviet Union had to Western electronics technology, and the use which Intelligence officials claimed that the Soviets were making of such technology in creating advanced weaponry and in modernising its defence industry. Certainly, VHSIC will be the target of both industrial and military espionage, and the US Government has placed strict controls on the export of such technology.

The exact degree of use which the Soviet Union makes of Western technology is difficult to assess, and has been the subject of wild rumour and occasional "scare" stories. However, there is sufficient hard evidence to cause concern. In several instances, items of Soviet hardware which have fallen into NATO hands have been found to contain integrated circuits based on Western originals. A large shrouded exhibit carried into a classified US briefing in the early 1980s is reported to have been a Soviet sonobuoy containing integrated circuits based on Texas Instruments designs. According to one report in the late 1970s based on defector testimony, at least one IBM computer was at that time installed at the Soviet anti-ballistic missile (ABM) test site at Shary Shagan.

Probably the best unclassified "hard" evidence of Soviet copying came several years ago when a US technical journal obtained samples of Soviet integrated circuits, and published photographs of the internal structure – the pattern etched into the tiny silicon wafer. One US semiconductor manufacturer promptly released a similar photograph of one of its own products as graphic evidence that the Soviet product was a direct copy.

## Vulnerability to EMP

Growing reliance on electronics and computers has raised new areas of vulnerability in modern defence equipment. The additional complexity is certainly resulting in "smarter" and more effective systems, but has also created new and unconventional threats.

In addition to the heat, blast and radiation effects from nuclear explosions, there is also in many cases an intense pulse of radio energy caused by gamma photons from high-altitude nuclear explosions, liberating large numbers of electrons from atoms of atmospheric oxygen or nitrogen. This electro-magnetic pulse (EMP) of radio energy is of very short duration, but of intense power. A high-altitude nuclear blast of about one megaton yield would generate an electromagnetic field of several tens of kilovolts per metre. Although the pulse would last only some 200ns, the instantaneous power level would be around 500 thousand million megawatts, sufficient to wreak havoc with electronic systems, particularly those using solid-state electronics. Electrical fields induced within semiconductor devices can alter their characteristics or even cause complete malfunction.

Above: This subsidence marks the site of the Huron King underground nuclear test. Such trials prove the design and yield of nuclear weapons, but cannot confirm blast or EMP effects.

Below: EMP-shielded building under construction for the British MoD. When completed, the steel walls, roof and floor will form a giant metal box which will screen the interior.

The threat of radiation damage must be faced by the designers of advanced military systems, and procurement agencies are increasingly specifying that new equipment and even some existing systems be "radiation hardened". Sensitive electronics may be mounted within a metal box which is electrically "sealed" – a good technique for upgrading existing equipment – or designers can use modern components designed to have a good resistance to EMP.

Another threat now being recognised is the inadvertent "leakage" of data from computer systems. The internal circuitry within computer equipment operates at radio frequencies, and tends to radiate electrical noise. This is at a very low level, but could be intercepted for espionage purposes – the computer is virtually broadcasting the information it is handling.

To curb the military possibilities of this leakage, the US National Security Agency (NSA) has set up a programme known as Tempest. In order to be Tempest-rated, equipment must be designed to suppress unwanted radiation – partly a matter of arranging good electrical screening or shielding within the equipment. To eliminate the risk of signals "leaking" from interconnecting cables, transmission of data between items of computer equipment can be carried out by using fibre optics instead of conventional wiring.

## "Hackers"

The cinema film "War Games" released in 1983 highlighted another potential threat to military computer systems – the risk that unauthorised individuals or organisations may attempt to gain access to the machine and its data via the telephone. Many computers exchange data via normal telephone lines by using interface units known as modems (MOdulator-DEModulators). Breaking into such systems – an activity known as "hacking" – has become a hobby to some home computer enthusiasts. Once having gained access to the computer, most "hackers" are content simply to read confidential data, perhaps leaving the electronic equivalent of a "Kilroy was here" graffiti in electronic form as evidence of their prowess. But a small minority attempt to destroy data files, or even to disrupt the normal running of the computer they have accessed.

The degree to which military computer networks have been penetrated by hackers is probably exaggerated, but some systems have been broken. The most common "victim " is probably the Arpanet network which the US DoD uses to link many of its research contractors.

So far, there have been no recorded instances of hacking skills being applied to espionage, or of data obtained by hacking having been passed to an unfriendly power, but the possibility is clearly troubling the US DoD which certainly takes the threat seriously. Additional security devices are being added to existing telecommunications systems, while newly-issued design specifications for the latter now define suitable "anti-hacker" measures.

The part of the electromagnetic spectrum used for military purposes is largely that extending from Low Frequency (LF) transmission (starting at 20kHz) up to 18GHz (the top end of the centimetric radar band). Recent trends involve exploiting the extreme ends of spectrum – Very Low Frequencies (VLF) and millimetre waves. (Reference to centimetres and millimetres may seem odd at first sight, but these apply to the wavelength of the signals involved. Engineers tend to specify transmissions in terms of frequency rather than wavelength, but the latter is often used in radar work, since it gives a direct clue to the physical size of the components used for signal transmission.)

Extremely Low Frequency (ELF) transmissions – below 30kHz – are used largely for communication with strategic submarines, while Low Frequency (LF) transmissions – 30 to 300kHz – are used for low-quality but reliable long-range communications. Medium Frequency (MF) – 0.3 to 3MHz – is old "medium waves", as used largely for broadcasting, while High Frequency (HF) – 3 to 30MHz – is what used to be called "short wave". HF signals can travel long distances after being reflected by the Earth's ionosphere. Until the arrival of satellite communications, HF was the only method of long-range radio communication.

Very High Frequency (VHF) and Ultra-High Frequency (UHF) span the spectrum from 30 to 150MHz and 150 to 400MHz, respectively. These frequencies are not reflected by the ionosphere, but are ideal for short-range work and for long-range search radars.

At still higher frequencies, radio waves can no longer be passed down conventional cables, but must be led from one part of a electronic system to another using waveguides, which are metal pipes often of rectangular cross-section. Antennas (aerials) tend to feature the solid-metal or mesh reflectors used to create a narrow beam.

The frequencies widely used by radar range from 1 to 20GHz. For convenience, this area of the spectrum is divided into bands designated from A to M. An older system of band designations originally

**Battleground for EW**
One of the unseen battles takes place in the electromagnetic spectrum, which is becoming ever more crowded as an increasing number of people want to use some sort of transmitting device. The frequency of a transmission is measured by the number of oscillations per second, the unit being the "Herz". Thus, one Kiloherz (1KHz) is one thousand cycles per second, one Megaherz (1MHz) one million cycles, and so on. The lower end of the scale is the "audio" band which extends from about 20Hz to 20KHz, although most people's hearing is actually in the range 80Hz to 15KHz, despite what the "Hi-Fi" salesmen may say! The second characteristic of these transmissions is "wavelength", which is the distance between one wave and the same point on the next wave. As frequency rises the wavelength gets shorter and the radio path becomes more of a straight line. Radar bands have letter designations; the wartime system was revised in 1972 and is now as shown. The electromagnetic spectrum extends from low frequencies associated with audio, through the radio and radar spectrum, into the infra-red, visual and ultra-violet, and then into the realms of X-rays, gamma rays, and other radiations from nuclear weapons. Most military applications use the radio, radar, infra-red, visual and ultra-violet portions of the spectrum. Radars tend to operate in the region between 1GHz and 40GHz, shown separately on an expanded scale (right). Some radar bands such as D, E, F, I and the lower part of J are crowded with signals. Soviet designers cleverly spread the operation of their weapons over a wide frequency range, thus reducing mutual interference and increasing the range of frequencies which NATO ECM systems must cover. Western designers seem less bold, so that congestion and mutual interference sometimes results. The E-3A Sentry and Rapier surveillance/acquisition radars operate at similar frequencies, so an interference problem was noted when the first AWACS aircraft visited Europe.

| BAND | SURVEILLANCE/ TRACKING | AIRCRAFT | MISSILE CONTROL |
|---|---|---|---|
| D | Martello 3-D surv (UK)<br>FPS-117 3-D surv (US) | | SA-3 command link (Sov) |
| E | FPS-6 heightfinder (US)<br>P-50 "Barlock" GCI (Sov) | | Crotale target acquisition (France) |
| F | RAT-31S surv (Italy)<br>RAN-10S surv (Italy) | APY-1 EW (NATO E-3) | Aegis SPY-1 multi-role (UK)<br>Rapier target acquisition (US)<br>SA-2 missile control (Sov) |
| G | | | SA-6 target acquisition (Sov) |
| H | | "Skip Spin" AI (Sov Su-15 "Flagon") | |
| I | Rasit battlefield surv (France)<br>PPS-6 battlefield surv (US)<br>TPN-18A GCA (US) | "Puff Ball" search (Sov M-4 "Bison")<br>APG-66/-68 multimode (NATO F-16) | SA-6 target tracking (Sov)<br>SA-3 target tracking (Sov)<br>Crotale command link (Fr) |
| J | Cymbeline anti-mortar (UK)<br>"Gun Dish" ZSU-23-4 fire-control (Sov)<br>MPQ-4 anti-mortar (US) | APQ-99 TF/ground map (NATO RF-4)<br>"Jay Bird" AI (Sov MiG-21 "Fishbed")<br>APQ-113 multimode (US F-111) | Seawolf target tracking (UK)<br>Rapier command link (UK)<br>Seawolf command link (UK) |
| K | | | |

AI: Airborne intercept
FAD: Fleet air defence
GCA: Ground-controlled approach
GCI: Ground-controlled intercept
surv: Surveillance
TF: Terrain-following

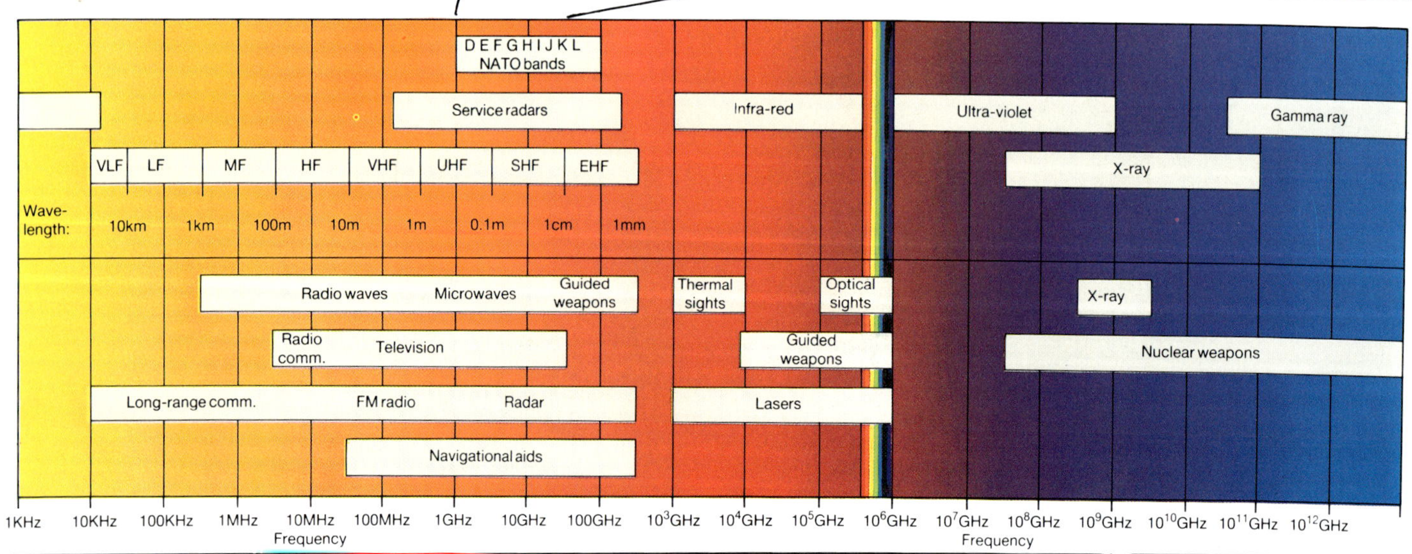

devised as a security measure during World War II is still in limited use.

Super-High Frequency (SHF) – 3 to 30GHz – transmissions are less affected by jamming and the effects of nuclear blasts than conventional UHF signals, a feature which explains the installation of SHF satellite communications (satcom) links on the US Air Force's E-4B airborne command posts.

The region between 30 and 300GHz – known as EHF (Extremely-High Frequency) is that of millimetre waves, a part of the spectrum which may be of vital importance to future weapon systems. These extremely high frequencies allow high-gain/high resolution antennas of modest dimensions to be created. The signal beam can be made very narrow, and thus difficult to detect and monitor. Sidelobes – spurious secondary beams which reduce system performance and act as an "Achilles Heel" which enemy ECM designers will attempt to exploit – are minimised.

One potential problem is atmospheric attenuation caused by absorption of the signal by atmospheric oxygen or water vapour. For applications requiring long range, millimetre-wave equipment must operate at frequencies where these effects are minimal – regions (referred to as "windows") centred around 35, 94, 140 and 220GHz. These high frequencies allow millimetre-wave transmissions to carry a large amount of data. Any one of the "windows" described above offers a greater bandwidth than all the conventional radio/radar spectrum put together.

One key application of millimetre-wave technology is in the field of radiometry. Millimetre waves are almost a half-way house between the world of radar and that of infra-red. In the same way that all objects radiate heat, they also radiate millimetre waves. By exploiting such millimetric radiation, designers can create passive seekers or even dual mode active/passive millimetre-wave seekers. Other patterns of dual-mode sensor may combine active or passive millimetric modes with laser or IR homing.

**Above: The small circular antenna on this Norden millimetre-wave tactical radio generates a much narrower beam than would be possible at conventional microwave frequencies.**

Although the theoretical upper limit of the millimetric portion of the spectrum is 300GHz, a practical upper limit for the use of conventional components such as transmission lines and waveguides is around 200GHz. Beyond this point, signals must be treated in a manner similar to infra-red or light, rather than by radio/radar techniques.

All objects radiate infra-red energy. The hotter they are, the greater the energy emitted. The infra-red band covers the frequency range from 1.5 microns to 14 microns. The peak of the emission pattern of the emitted energy depends on its temperature – the hotter the object, the shorter the wavelength.

Not all of the IR spectrum is militarily useful, however, since here again water vapour, carbon dioxide and other constituents of the atmosphere absorb energy in many regions of this part of the spectrum. As a result, IR system designers tend to use one of two "windows" in the spectrum – 2 to 3 microns and 8 to 13 microns – at which such losses are minimal.

**Below: The superstructure of new CG-47 Ticonderoga class US Navy cruisers carries flat planar-array antennas whose beams are steered electronically. These allow the AEGIS missile system to engage multiple targets.**

Radar remains the primary long-range search sensor for targets in space, in the air, on land and on the surface of the sea. It is also used for mapping and navigation, and for the guidance of interceptors, missiles and other weapons.

The basic forms of radar are pulse and CW (continuous-wave) types. Both detect targets by transmitting a powerful signal and searching for reflected "returns" from targets. Pulse radar sets measure range by observing the interval of time between the transmission of a pulse and reception of the return signal, while CW sets impress upon the signal a modulation which will allow timing circuits to identify the time at which a returned signal was originally transmitted.

Recent development such as the use of phased-array (flat-plate) antennas, digital signal processing, pulse Doppler and pulse-compression techniques, and a trend toward use of higher frequencies have not significantly altered these basic principles, but have greatly improved radar performance, particularly in the face of radar clutter and electronic countermeasures (ECM).

To provide long-range early-warning of bomber attack on the continental USA, the Department of Defense is engaged in a massive programme to upgrade the radar coverage. The northern approaches to the USA are already protected by the DEW (Distant Early Warning) Line of radars sited in northern Alaska, Canada and Western Greenland. Now obsolescent, this is being upgraded with new long-range radars, while development of shorter-range gap-filler radars is almost complete. The obsolete CADIN-Pinetree radar network in southern Canada is being phased out as an economy measure.

First line monitoring of the Soviet ICBM fields and potential SLBM launch areas is provided by America's early-warning satellites. Two ground-based radar systems monitor the likely flight paths of hostile missiles – BMEWS (Ballistic Missile Early-Warning System) is intended to detect ICBM launches, while Pave Paws radars and the PARCS radar in North Dakota watch for SLBM launches.

**Below: A technician operates a Pave Paws console at the USAF 6th Missile Warning Station at Otis AFB, Massachusetts, part of the US SLBM-detection network.**

**Below: The Pave Paws network monitors probable directions of SLBM attack upon the Continental USA. This electronically-steerable antenna is at Beale AFB, California.**

Two Pave Paws phased-array radars are already in service – one on the east coast of the USA, the other on the west. These are being joined by two more located in Texas and Georgia. When the latter are operational, the SLBM-warning network will be complete, allowing the US DoD to close down obsolescent FPS-85 and FSS-7 radars currently operating in Florida.

The DEW line may guard the northern approaches to the USA, but there are no convenient land masses to the west, east and south on which similar radars may be located. To provide all-round coverage, the US DoD plans to deploy eight Over-The-Horizon (OTH) radars intended to cover the eastern, southern and western approaches. Each set will cover a 60 degree arc. Priority is being given to the east coast, but funding has already been requested for the first radars to be installed on the west coast.

OTH-B detects aircraft by sending high-frequency signals from its transmit antenna to the ionosphere, which in turn refracts the signals back towards the Earth's surface. Aircraft passing through areas where the beam is returned to the surface in this way may be detected when the reflected energy follows the reverse path back to the ionosphere and, by refraction, back to the receive antenna. Despite the complexity of this operating mode, the beam may be steered in order to cover airspace from 500

**Right: The earlier and obsolescent FPS-85 anti-SLBM radar located in Florida will be phased out once the Pave Paws sites in Georgia and Texas have become operational.**

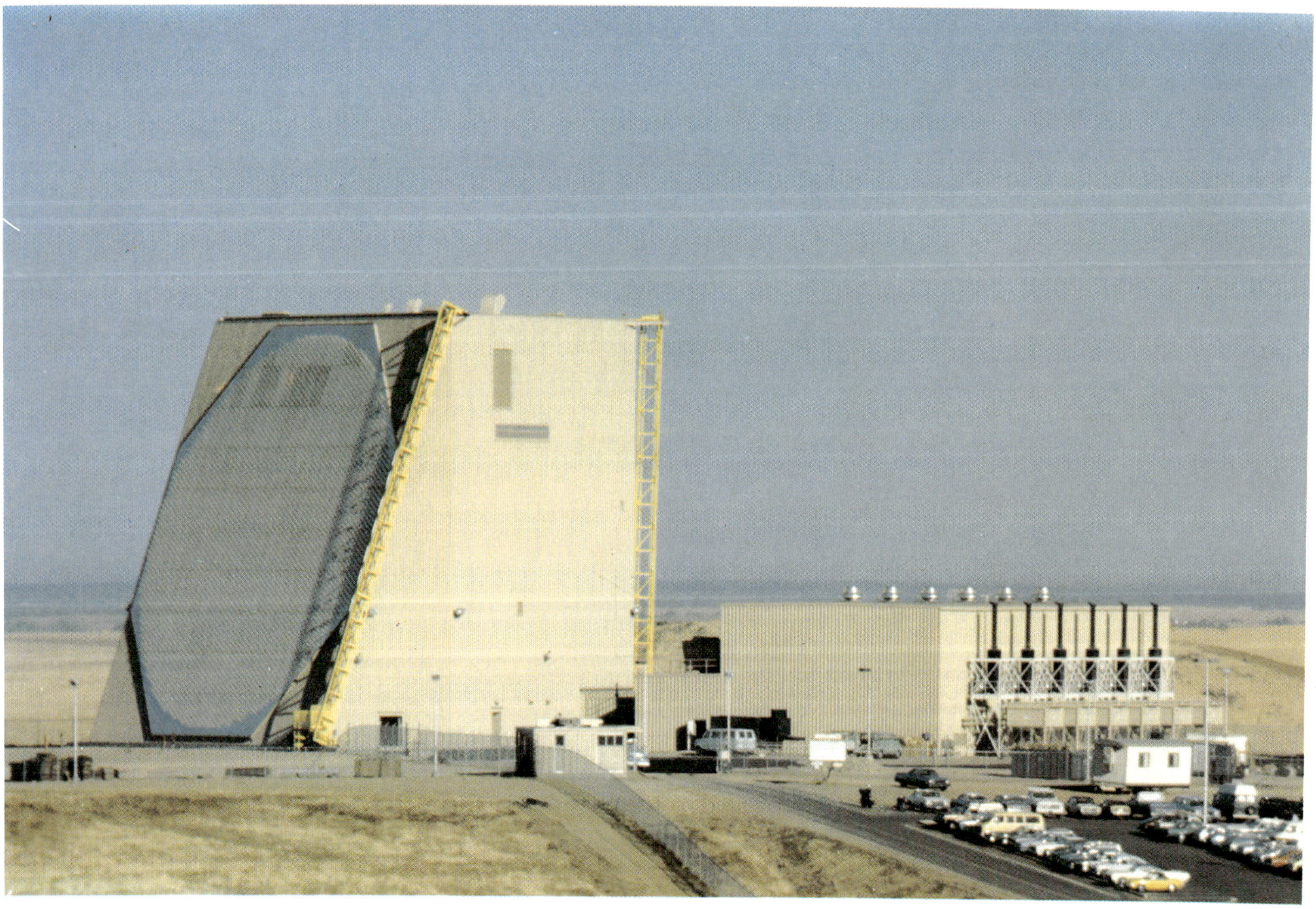

**Detecting the long-range air threat**
Over-the-horizon (OTH) radars such as the US OTH-B use HF signals. These are refracted from the ionosphere, returning to Earth at great distances from the radar. The reflected echo returns to the radar by a similar route. Parallel beams are used to scan several sectors in azimuth, while range can be varied by altering the operating frequency so that the signal is refracted by a greater or lesser amount.

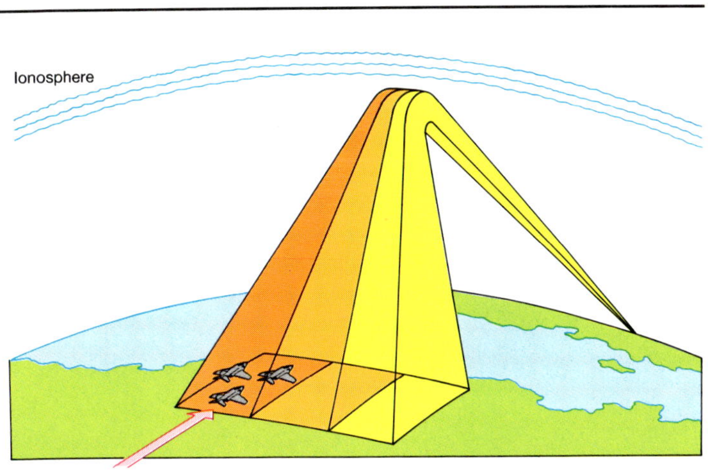

to 1,800 miles (about 800 to 2,900km) offshore.

A tactical OTH radar is now under development for the US Navy. These new radars will have a range of up to 1,800 miles (2,900km), and be located in geographic locations which will allow them to monitor the routes likely to be used by hostile bombers flying towards USN task forces. The set is designed to allow rapid deployment to prepared sites, a feature which will permit the US to establish emergency long-range coverage of areas not normally within the OTH radar "umbrella".

Radar is the ideal long-range surveillance sensor, and has been mounted on land, sea, air and space platforms. Downward-looking air and space systems are the most difficult to design, since the radar returns from the target can be swamped by unwanted "clutter" signals from the land or sea. The Soviet Union lags behind the West in this technology, and is only now deploying reliable "look-down shoot-down" systems on its fighters, while British attempts to develop the Nimrod as an equivalent to the Boeing E-3A Sentry AWACS airborne early-warning aircraft have bogged down (at least temporarily) in a series of radar problems which limit the aircraft's surveillance performance.

For maritime-patrol aircraft, radar remains the primary long-range sensor. Optical and infra-red viewing systems are not suitable for such applications, since they are badly affected by poor visibility, Electronic Support Measures (ESM) may be nullified by targets operating in radio/radar silence, while IFF (Identification Friend or Foe) systems cannot discriminate

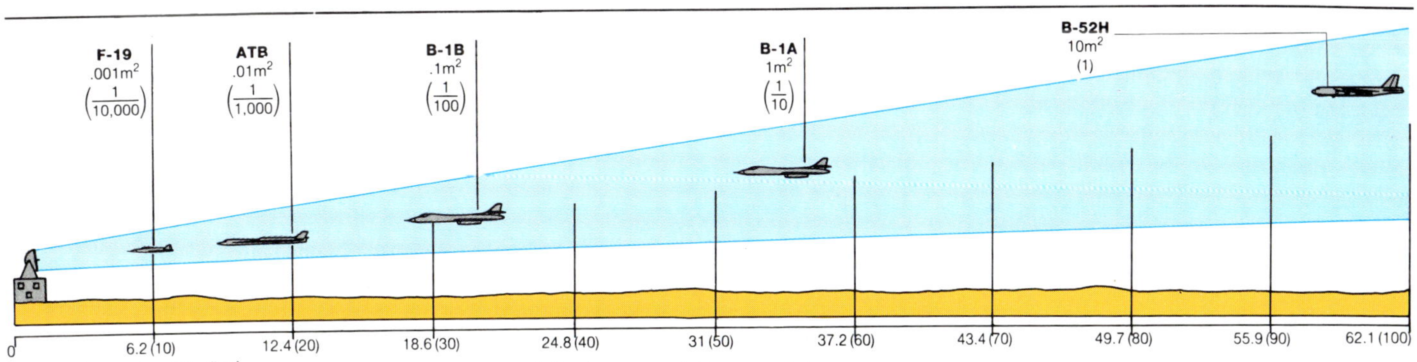


**Stealth in action**
Radar is just one of many aspects stealth technologists consider. The effect of reducing the radar cross-section of an aircraft by a factor of 100 does not reduce the radar detection range by a similar amount. Assuming that the radar shown at the far left can detect a B-52 (with a 107sqft/10sqm target area) at almost 62 mile (100km) range, the predicted detection range for smaller targets is shown. From right to left these are the B-1A (10.7sqft/1sqm), B-1B (0.1sqm), the new Advanced Technology Bomber (0.10sqft/0.01sqm) and the F-19 Stealth Fighter (0.010sqft/0.001sqm). The radar cross-sections quoted are obviously "guesstimates" – actual values are classified. The important point is that the values of radar signature reduction mentioned above do not and cannot make aircraft "invisible" to all hostile radars, but simply reduce the detection range to values which greatly inhibit detection. Given good pre-flight planning, the aircraft can be routed well away from known radar sites. Radar invisibility will require target areas several orders of magnitude below the levels discussed here.

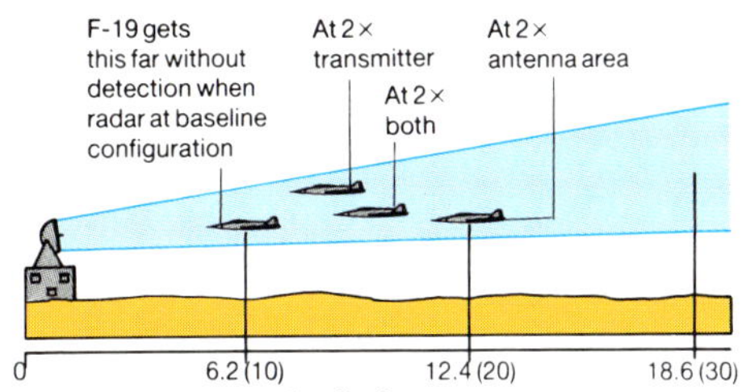


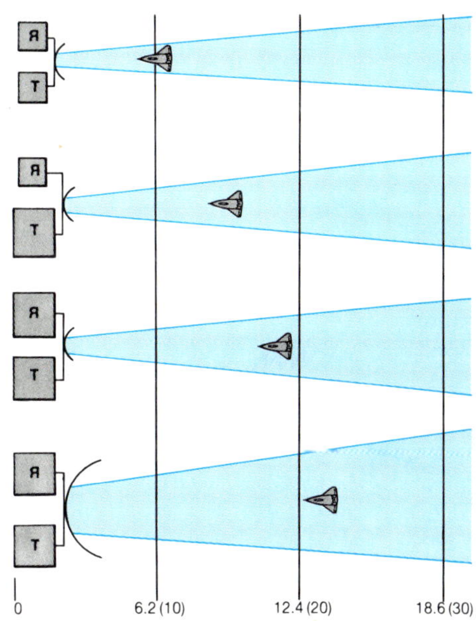


**"Quick fix" impossible**
Given the hypothetical radar and "F-19" described above, the enemy will naturally try to restore radar performance by various up-grades. Three possibilities are shown at the right and in plan view above. Starting from the baseline configuration (top right), transmitter power may be doubled (upper centre), receiver sensitivity may also be doubled (lower centre), then antenna area may be doubled (bottom right). Even if these upgrades prove possible the resulting performance gains are modest. Detection ranges rise from 5.4nm (10km) to 6.3nm (11.8km), then to 7.5nm (14km), and finally to 10.8nm (20km). Radar performance remains inadequate for tactical use.

Above: The Signaal Goalkeeper naval anti-missile gun incorporates a narrow-beam millimetre-wave radar to track sea-skimming targets with minimal radar "clutter" from the sea surface.

Left: The narrow beamwidth of millimetre-wave radars is ideal for tracking low-flying targets. This Krauss Maffei 5PFZ-CA1 anti-aircraft tank has a millimetric radar on the turret front.

between enemy ships, neutral vessels or "friendlies" whose IFF transponder is unserviceable.

One advanced maritime surveillance radar of this type is the Thorn EMI Searchwater, which entered RAF service in 1979 aboard the Nimrod MR.2 maritime-patrol aircraft. This incorporates several features intended to help discriminate between genuine target returns and the effects of clutter. The narrow beamwidth of the antenna plus the use of pulse-compression techniques keep the size of the illuminated patch of sea surface to a minimum, while the use of frequency agility allows the radar to "sample" the radar behaviour of the target at several radar frequencies.

Although the main radar display is a PPI (Plan-Position Indicator) screen, radar data may be examined in A-scope (amplitude/range) and B-scope (range/azimuth) form. Trials have shown that under the correct conditions, the A-scope imagery bears a distinct resemblance to the target ship's silhouette.

New air-to-ground radar technology also offers improved methods of detecting and attacking attacks by massed armoured formations. In order to provide a reliable means of locating and tracking moving targets at extended range, providing the data needed by units tasked with attacking targets deep behind enemy lines, the earlier US Army Battlefield Data System (BDS) and USAF Pave Mover programmes were merged in 1984, creating the joint STARS airborne radar programme. Once located and identified by STARS radar, targets can be attacked by ordnance such as terminally-guided sub-munitions delivered to the target area by aircraft or missiles.

Recent advances in radar technology are naturally being applied to new and more effective systems currently on trials or still on the drawing board. For example, USAF Project 2314 (Tactical Air Surveillance) has evaluated technology for future tactical air surveillance tasks, particularly the near real-time positive identification of hostile, friendly and neutral aircraft. Development of an experimental tactical air surveillance radar started in the late 1970s. By the end of the decade a signal processor suitable for use in a jam-resistant air surveillance set was on bench test, and a wideband transmitter had been delivered.

Early tests used a two-dimensional antenna, while the design of a three-dimensional digitally-coded unit was still under way. Following trials of an advanced tracker module interfaced with a TPS-43 test radar, two experimental phased-array antennas were designed, built and tested in severe jamming and high clutter environments.

After these tests, advanced-development sets were ordered in the early 1980s. A laboratory demonstration of the basic surveillance network with an integrated aircraft identification capability – using an advanced tracking system, a laboratory radar and an AN/TPS-43E radar – has now been carried out. Development is continuing, and recent trials interfaced this system with data from the E-3A AWACS and the US Army's air-defence system.

## Millimetre-wave systems

The advantages of millimetre waves have already been exploited by a number of radar systems. Sets such as the British Marconi Blindfire and the Dutch dual-frequency I/K-band Signaal Flycatcher were early applications of such high frequencies to the problem of tracking low-level aircraft targets in the presence of ground clutter. Narrow beamwidth and relative absence of sidelobes greatly reduce the multi-path effects (the equivalent of "ghost" multiple images on a TV screen) which confuse conventional radars when attempting to track targets flying at elevations of below a few degrees.

This feature is particularly useful for tracking sea-skimming missiles such as Exocet. The Marconi Type 911 tracker for the Lightweight Seawolf missile is based on the Blindfire set, while trackers such as the Thomson-CSF Canopus and BAe Dynamics Sea Archer fire-control systems, and the tracking radars on the Signaal Goalkeeper CIWS (Close-In Weapon System), all operate at millimetric frequencies.

Much of the current US millimetre-wave effort is being carried out by the Army, which sees this technology as being well-suited to the often-poor weather in Western Europe. Millimetre-wave radars could give the crews of AFVs such as the M1 tank the ability to operate in smoke or fog.

An early US millimetric radar was the 94GHz Martin Marietta STARTLE (Surveillance and Target-Acquisition Radar for Tank Location and Engagement). Constructed in 1980, this was integrated with the thermal sight and fire-control system on an M60A3 tank, and subjected to field tests. These showed that as visibility deteriorated due to weather or smoke-screens, the performance of thermal imaging systems steadily deteriorated, while that of the radar remained constant.

Interdiction and close support operations are difficult by night and in adverse weather. One possible solution is an air-to-surface radar in advanced development for the USAF by Norden Systems under the Millimeter Wave (MMW) Technology project. Rather than attempt the development of an all-new millimetre-wave fire-

**Above: The threat posed by anti-radar missiles resulted in low-signature radars whose transmissions are difficult to detect. This is the millimetre-wave Ericsson Eagle Ka-band radar.**

**Below: The use of electro-optically guided weapons has forced the development of new forms of camouflage such as this FFV system designed to lay coloured foam to cover visible and IR tracks.**

**Below: Modern camouflage systems must be able to screen targets at visual and infra-red wavelengths. This demonstration by the British company Bridgeport Aviation Products shows how a parked Land-Rover (left) may be screened even from the attention of thermal imaging systems (right).**

control radar, this adapts an existing set developed for the Aquila RPV (remotely piloted vehicle). The system consists of a millimetre-wave radar sensor, a signal processor and a HUD, and will be able to detect moving and stationary targets, using signal processing to select high-density groups of moving targets for display on the HUD.

A joint USAF/DARPA project to develop a "stealth" fire-control radar for airborne applications is reported to be pursuing millimetre-wave and conventional I/J-band frequencies. Few details are available, but it will almost certainly use spread-spectrum techniques intended to disguise its signals and make them difficult to detect and identify.

Soviet designers are believed to be placing much effort into millimetre-wave weapons. At least one of their next-generation anti-tank missile uses a millimetric seeker, while tank and aircraft-mounted millimetre-wave radars were reported in the late 1970s.

## Seeing through camouflage

Foliage and camouflage can defeat many types of sensor. At short range, infra-red (IR) systems offer a solution, but for all-weather, long-range reconnaissance, long wavelength radar is more promising, and could complement electro-optical systems such as forward-looking infra-red (FLIR) sensors. The ability of long-wavelength radar signals to penetrate foliage was first demonstrated by a Pentagon R&D programme in the late 1970s. Technology for an operational camouflage-penetrating radar will probably be based on the results of the earlier Integrated Multiple Frequency Radar (IMFRAD) and Concealed Target Detection (CONTAD) programmes.

Despite the search for alternatives, sonar remains the only reliable area-search sensor for the location and tracking of submarines. Unlike radar, sonar is of limited effectiveness, since sea water does not permit sound waves to travel in a straight line, but refracts them in an unpredictable manner.

Passive sonar systems attempt to detect submarines by listening to the sounds generated by the vessel's machinery and propellers, and the disturbed flow of water around the hull. As submarine targets become quieter, detection range drops, and cannot be restored simply by increasing the amplification of the sonar receiver. The wanted signal must be picked out from amid a mass of background noise generated by other vessels, sea creatures, underwater currents, one's own ship and also the inevitable electronic "noise" (low-level spurious signals) generated by the electronics of the sonar receiver.

The amount of unwanted signal may be minimised by listening to only a small sector, while discrimination of the wanted signal may be eased by a time-history-versus-bearing display, but in the long run there remains a threshold level beneath which sounds cannot be reliably detected.

Passive sonar yields only bearing information. In order to obtain range, the sonar platform must move to another position in order to create a baseline for triangulation. The target itself is moving, and may change course. If a change of heading such as a zig-zag manoeuvre is not detected and allowed for, the derived range will be wrong. As range decreases, the usefulness of triangulation by passive sonar is greatly reduced.

This problem is particularly acute in submarine-versus-submarine actions. At the very time that good target data is required the attacker is reduced to receiving only bearing data. In order to attack, it must either switch to active sonar or rely on firing homing torpedoes down the target bearing. Active sonars transmit a powerful sonic pulse, and listen for return signals reflected by the target. Once the attacker starts "pinging", its transmissions will alert its intended victim, which may counter by launching a long-range anti-submarine weapon of its own.

Methods of avoiding this problem have been postulated, but none yet shows signs of being practical. Use of a remotely-positioned sonar transmitter would hide the true bearing of the attacking submarine, but would still warn the victim. Some form of "covert" sonar with frequency-hopping transmissions is the likely long-term answer.

Today's ultimate passive sonar is the US Navy's Sound Surveillance System (SOSUS) – a series of fixed-site sonars positioned around the world to detect and track submarines operating off the coasts of the United States, or transiting choke-points in the North Atlantic, Baltic, and near the Azores. The effectiveness of the system is classified, but is known to be sufficient to detect and roughly locate most patterns of submarine. Schemes to modernize the system's underwater and shore-based components started in the mid-1970s. Most are secret - contracts for SOSUS improvements are hidden in the US budget as "oceanographic research" or "oceanographic systems".

**Above: An operator working at the switch-free display console of a Dutch Signaal PHS sonar controls the set by touching box-shaped markings on the screen with the light-pen.**

Fixed surveillance systems need to be backed up with more flexible systems able to plug gaps in the coverage caused by failure or combat damage, or able to monitor areas of sea not under permanent underwater surveillance. Tagos SURTASS (Surveillance Towed-Array Sonar System) vessels first entered service in late 1984, and a total of 18 are planned. By the end of the decade, SURTASS should be supplemented by the Rapidly Deployable Surveillance System (RDSS) which is now under development. This is a sensor system which may be dropped by maritime-patrol aircraft to give extended time-urgent coverage of areas of special interest.

Long-range detection of targets is best tackled using towed-array sonars – the most recent breakthrough in anti-submarine warfare (ASW) sensor technology. This type of equipment first saw service aboard submarines, but larger and more elaborate systems have since been deployed aboard surface vessels. The sensor elements consist of long lines of neutrally-buoyant line towed behind the sonar-equipped vessel and carrying a series of low-frequency hydrophones. Towed sonars are passive systems, and the beam-

**Sonar for submarines**
Active and passive sonar form the primary underwater sensors of submarines, the various acoustic arrays being distributed around the hull and fin of the vessel.

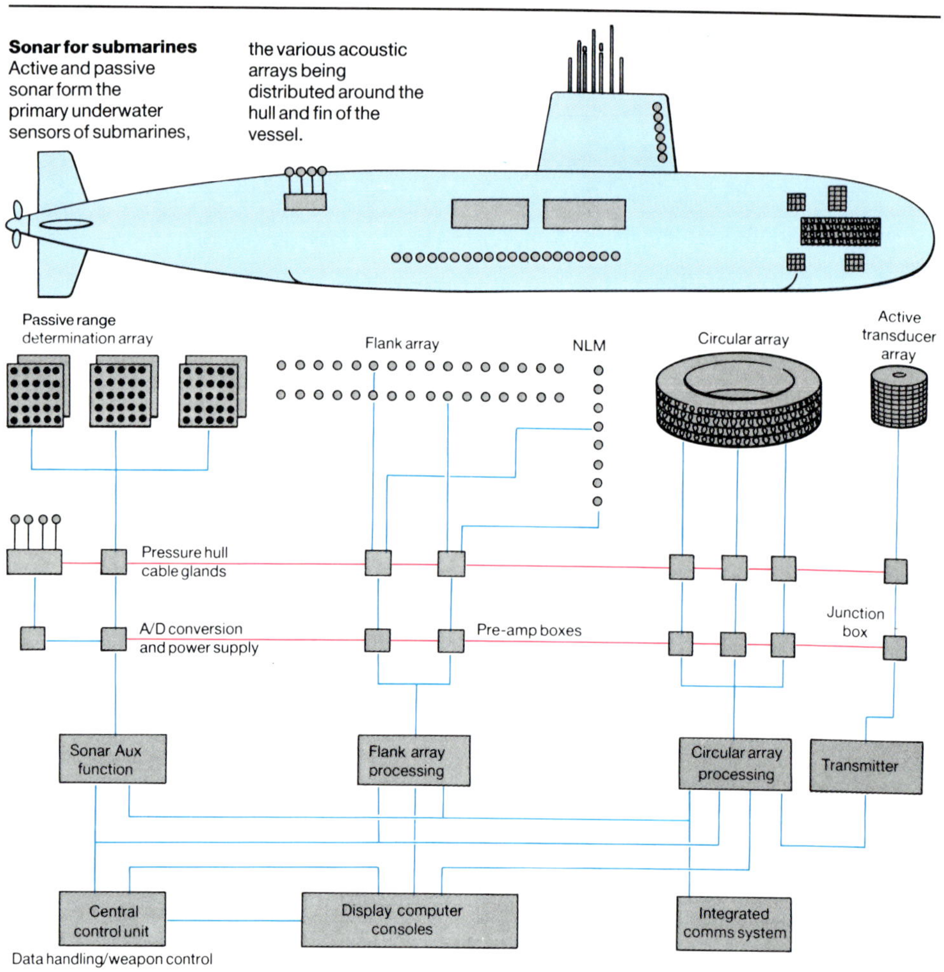


**Submarine sensors**
Navies are secretive about their submarine-mounted sonars so few details of modern systems are available. This diagram illustrates the SIASS-2 version of the Signaal Submarine Integrated Attack and Surveillance Sonar System. This acts as an integrated part of the company's Submarine Sensor Weapon and Command System (SEWACO). It can observe noise-radiation targets at long and medium range, automatically track multiple targets, analyse target motion (determining position from bearing-only passive data), operate in active mode, and act as an intelligence-gathering system.

width of the array is dependent on operating frequency, array length, and the angle at which sounds reach the array.

Fixed-wing maritime-patrol aircraft such as the P-3 Orion, Nimrod and Atlantique drop sonobuoys in order to get sensors into the water. On landing in the water, the sonobuoy raises its antenna, and releases a sensor which falls to the end of a submerged cable whose length may often be preset before the unit is dropped in order to obtain the best performance in the water conditions and depth. The unit's operating life is fixed or in some cases preselected before release, the buoy scuttling automatically once this time limit has expired.

Sonobuoys may be directional or omnidirectional and, like sonars, may operate on active or passive principles. The most sophisticated units such as Dowty Electronics CAMBS (Command Active Multibeam Sonobuoy) are active and directional. CAMBS can vary its sensor depth on radio command, giving the aircraft which releases it much of the flexibility normally associated with dunking sonars.

Protection for naval task forces is provided by surface warships and sonar-equipped ASW helicopters. The increased range of the missiles and torpedoes now being carried by Soviet submarines has forced Western navies to supplement conventional hull-mounted active sonars and towed variable-depth sonars with towed passive-arrays similar to those used on submarines. The current US Navy towed-array sonar is the SQR-18 fitted to Knox and Perry class frigates, but in the late 1980s vessels such as Ticonderoga class cruisers, Spruance and Kidd class destroyers, and the planned DDG-51 class destroyers will be fitted with the advanced SQR-19 now in development.

Latest technology in the "dipping" sonar field may be seen in the Marconi Avionics/Plessey Marine HISOS 1 (Helicopter Integrated SOnics System) intended to equip the Westland Sea King Mk5, and also likely to be installed aboard the British Navy's planned fleet of EH.101 helicopters. The "wet" end of the system is the Plessey Marine Cormorant. Dunked into the water as a compact cylindrical unit, this deploys beneath the surface to extend its array of sensors. The individual sections unfold under the power of electro-mechanical actuators to create a large effective aperture for high resolution.

Each of the array's five "arms" contains three tubes each equipped with six wide-band hydrophones. When the set is used in the active mode, three free-flooding transducer arrays provide the power. All signals are passed to the parent aircraft by means of a multiplex system feeding a single co-axial cable strengthened by a Kevlar aramid webbing layer.

Quieter and deeper-diving submarines which the Soviet Navy is known to be developing have created the requirement for advanced acoustic sensors to re-equip US Navy attack submarines and ASW aircraft. The increased sonar performance needed will probably require the use of multiplexing and fibre-optic data transmission, plus other completely new techniques.

The next class of USN SSNs will probably be fitted with the Advanced Conformal Submarine Acoustic Sensor, a transducer system which will take advantage of the improved locations made possible by a new hull design. These vessels will also carry the Submarine Advanced Combat System (SUBACS), a system which will combine improved sonar, fire-control and weapons, applying the latest computer technology to maintain the current US Navy lead in target detection and tracking.

At least two new patterns of sonobuoy are planned for use by the USN. Following tests of horizontal line arrays sized for sonobuoy applications, development of air-drop models for large-area ASW commenced. In the early 1980s, work started on the Advanced Localization Sonobuoy intended to provide a shorter-range sensor.

**Left: The "wet end" of the Marconi/Plessey HISOS helicopter-mounted sonar is the Plessey Cormorant dunking sensor. This is compact, and may be carried in the extended position at high airspeeds as the aircraft moves to its next search location. Once in the water, Cormorant extends its sensitive arrays (below) into a configuration which will give optimum resolution.**

Electro-optically (EO) guided weapons first saw service with the US forces in the Vietnam War, and the latest generation – systems such as the Maverick TV-guided missile and Walleye "smart" bomb – are widely deployed. These carry a miniature TV sensor in the nose, and may be locked onto a suitable point of high contrast on the target. Similar EO seekers are also used as target trackers.

Like image-intensification night sights and viewing systems (which magnify the tiny amount of light available from the sky on even the darkest night), they are passive, so do not warn the enemy that they are being used. But of course the main problem is that of visibility, particularly in Western Europe's often-foul weather conditions.

EO-guided weapons now in development and entering service tend for this reason to use imaging infra-red technology (described below), although one novel future application of electro-optics is the seeker of the planned man-portable anti-aircraft missile being developed by the Japan Self-Defence Force Technical Institute. Rather than use IR homing, the Japanese weapon will lock on to the visual image of the target, allowing all-aspect engagements.

The two main military applications of infra-red technology are missile seeker heads, and night-vision systems. IR system designers tend to use one of two "windows" in the spectrum: 2 to 3 microns and 8 to 13 microns. Objects heated to around 300 Kelvin (K) have emission peaks in the 8 to 14 micron band; a rise in temperature to around 700 to 800 K is needed to create a peak at 2 to 3 microns.

Infra-red homing systems fall into two categories – imaging and non-imaging.

**Above: A Maverick-armed fixed-wing aircraft can expect to knock out up to three tanks per sortie. With simpler weapons such as cluster bombs, the kill rate is well below one per sortie.**

**Below: The high resolution of electro-optical seekers results in high accuracy. Miss distance of TV-guided Maverick is only a few feet, as this test shot against a bunker demonstrates.**

Widely used in heat-seeking anti-aircraft missiles, non-imaging systems are relatively-simple sensors which can be locked on to a source of heat energy. Normally made in the form of a small telescope, these are designed to focus the incoming infra-red energy onto a sensor – effectively the IR equivalent of a photocell, consisting of one or more detecting elements. The best IR sensitive materials available at the present time are lead sulphide, lead antimonide, cadmium mercury telluride, and indium antimonide. Each exhibits the greatest sensitivity at different parts of the spectrum. The designer of a 2 to 3 micron seeker may use lead sulphide, while cadmium mercury telluride is better at 8 to 15 microns.

Lead sulphide works well at room temperature, but most detectors give better results when cooled. Cadmium mercury telluride works best at temperatures of around 70 K, for example. Some missiles such as Israel's Raphael Shafrir have uncooled seeker heads to minimise the time delay between deciding to use a missile and the moment it is ready for launch, but most design teams accept the delays caused by pre-cooling the seeker in order to obtain the highest performance.

**Above: TV-guided seekers are sensitive to patterns of light intensity or to contours. Once locked on to part of a TV image (in this case the Ninh Binh road/rail bridge near Thanh Hoa in North Vietnam) a Walleye guided bomb continues to home. As the weapon nears its target (centre and above right), the seeker remains locked on to the end of the main span. A post-strike reconnaissance photo (below) shows that the weapon impacted accurately.**

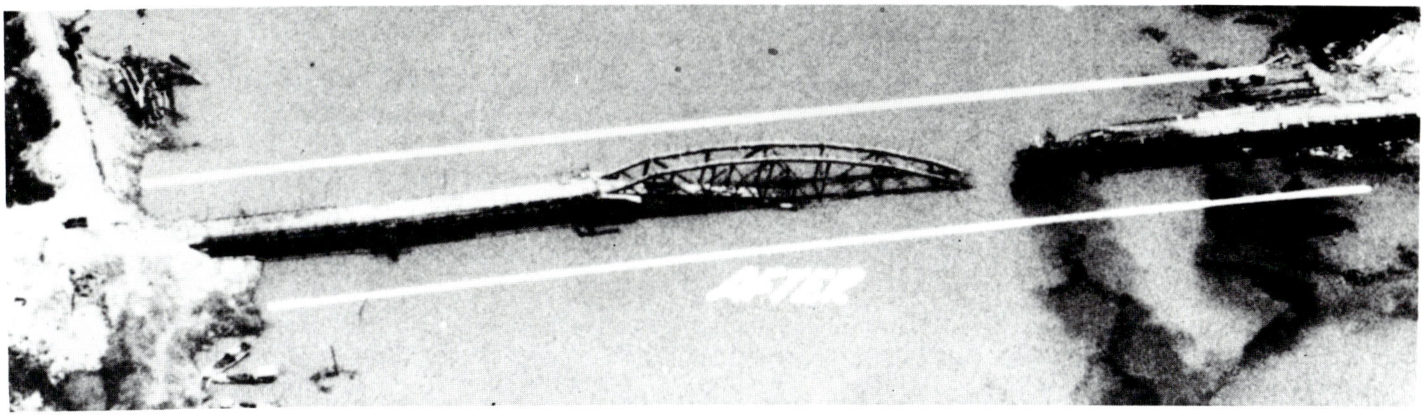

The first infra-red guided weapons such as the AIM-9 Sidewinder air-to-air missile entered service in the early 1950s, and had seekers with a fixed forward-looking gaze. In combat, the pilot of the launch aircraft had to manoeuvre so as to bring the intended victim into the seeker field of view – a narrow cone only a few degrees wide. Seekers were of low sensitivity, so could respond only to the hot rear section of the target's engines, flying virtually "up the jetpipe" of their victims. As a result, they could be fired only from immediately behind the target. If the victim realised that he was under attack, he could pull a tight turn so that the missile seeker head could no longer "see" the jet pipe.

During the 1960s and 1970s, missile designers concentrated on the creation of weapons which were easier to use. With the development of more sensitive IR detectors, current weapons such as the Matra Magic and the AIM-9L version of Sidewinder can respond to the lower intensity heat for the most if not all of the target's airframe, so can be launched at virtually any angle. These missiles also make target acquisition easier. Magic is programmed to carry out a pre-launch search procedure over a selected area, while the more recent Magic 2 and the AIM-9L have seeker heads which can be slewed in response to signals from the parent aircraft's radar or other sensors.

Non-imaging seekers can cope only with a single hot source against a fairly uniform background. Such simple technology lacks the sophistication needed to analyse a complex infra-red image. As a result, they are useful only for anti-aircraft and anti-ship missions – instances when the target may be expected to be much hotter than the background. When simple IR-guided missiles such as the Soviet AA-2 Atoll and early-model Sidewinders attempt to lock on to targets flying lower than the launch aircraft and seen against a relatively hot background such as desert terrain, their seekers are easily confused, and often fail to locate the target. During one series of low-level IR missile trials carried out in the USA in the late 1970s, the AIM-9 Sidewinder was dubbed "Groundwinder" by test personnel in reference to its propensity to lock on to terrain hot spots.

**Below: Although outwardly similar to the earlier Magic 1, Matra's Magic 2 which entered service on the French Air Force's Mirage 2000 fighters embodies the latest technology in its guidance system, including a multi-element cell homing head and a digital autopilot with a built-in microprocessor. The new seeker is "several score times" more sensitive than that of Magic 1.**

The "fire-and-forget" properties of IR guidance made it attractive for air-to-ground use over the battlefield, but this requires the use of more complex seekers based on imaging infra-red (IIR) technology. These seekers are virtually infra-red TV cameras which build up a "heat image" of the target, then rely on sophisticated signal processing to lock on to a designated part of the image. Unlike TV systems, IIR sensors works equally well in total darkness, and are better than visual systems in coping with haze and smoke. But since fog and cloud consist of suspended water vapour which attenuates IR, these conditions remain troublesome.

IIR technology has also been applied to

night-vision systems. Primitive night-vision systems relying on the use of infra-red searchlights have been in service for decades, but these simply converted the infra-red scene lit by the searchlight. Once the enemy had deployed IR viewing systems, using these searchlight-based equipments was little more subtle than shining a conventional searchlight.

Modern thermal imaging systems are passive, relying on the heat emitted by objects in the field of view. Now widely deployed as surveillance systems, navigation aids and weapon sights, they have proved reliable and efficient, but have triggered off R&D efforts into IR-opaque smokes capable of screening military targets – a simple form of IR countermeasure.

Among the most exotic platforms for night-vision systems are the customised fixed-wing aircraft and helicopters devised by the USAF to support Special Operations Forces. For example, the Lockheed MC-130H Combat Talon II version of the Hercules transport is equipped with high-accuracy navigation systems, plus terrain-following and self-protection systems which enable it to penetrate enemy air-space at night and at low altitude in order to drop combat personnel and equipment behind enemy lines. A total of 35 will be in service by FY91.

Another, the HH-60A Nighthawk, a specialised version of the UH-60 Blackhawk, is also fitted with avionics for precision navigation at night or in bad weather. Some will be deployed with the Special Operations Forces, others with combat-rescue formations. The HH-60E has simpler avionics, and will be used for combat rescue in more favourable weather conditions.

**Above: The MIRA night-vision sight jointly developed by MBB and SNIAS gives the widely-deployed Euromissile Milan man-portable anti-tank missile round-the-clock combat effectiveness.**

**Below: The wartime task of night-vision-equipped MC-130 version of the Hercules would be to land special forces. The aircraft would also probably be used in anti-terrorist operations.**

**Below: The Hughes Detecting and Ranging Sets (a laser/infra-red sensor) will allow the US Navy's Grumman A-6E Intruders to locate and attack targets shrouded by haze or smoke.**

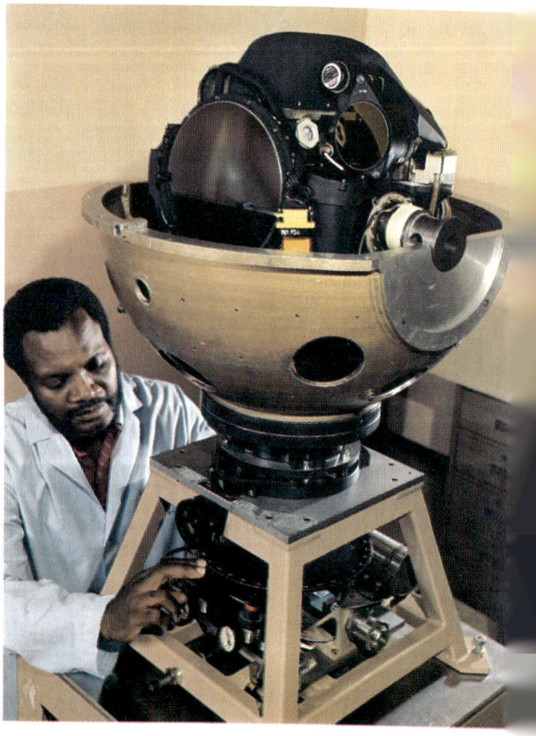

**Above: The infra-red image being displayed by this Rank Pullin thermal imager shows an obsolescent Conqueror tank. IR images of current British AFVs are classified.**

In the late 1980s, the F-16 Fighting Falcon fighter and A-10 Thunderbolt II attack aircraft are scheduled to carry the pod-mounted Martin Marietta LANTIRN (Low-Altitude Navigation Targeting Infra-Red Night) system, which will provide terrain-following radar and FLIR imagery for navigation. In the attack phase, the pod sensors will automatically acquire, identify and categorise tank targets, for instance, passing target information to the aircraft's fire-control system so that targets may be attacked using AGM-65 Maverick missiles or "designated" for attack using laser-guided munitions.

The equipment is contained in two pods, one for navigation and the other for target acquisition/tracking sensors. The navigation pod carries a Ku-band terrain-following radar, and a wide field-of-view FLIR. The wide azimuth coverage of the radar – a product of sophisticated signal processing techniques – allows the pilot to fly high-rate turns at low level in order to avoid or confuse the defences, while the 28deg×21deg field of view of the HUD allows FLIR imagery to be superimposed on the external view for night or bad-weather attacks.

The most advanced IR system currently under development in the Western world is probably Teal Ruby. In an attempt to devise a reliable method of detecting low-flying bombers and cruise missiles, the US Defense Advanced Research Projects Agency (DARPA) plans to orbit this experimental system on a DoD spacecraft designated P-80-1. If successful, this project could lead to operational orbiting hardware by the late 1980s.

**Below: The experimental Teal Ruby spacecraft is designed to test advanced infra-red sensors. The amount of cryogenic coolant carried will limit spacecraft lifetime to around a year.**

The sensors will be mosaic focal plane arrays – the most advanced IR technology now leaving the laboratory, and one which shows great promise for future "intelligent" guided missiles. Instead of using a single infra-red detector element to scan sequentially a large field of view, these new focal plane arrays are rectangular mosaics of individual IR-sensing elements. Since all parts of the IR "view" are under simultaneous observation instead of scan-based regular sampling, sensors based on focal plane arrays are more sensitive and responsive than earlier scanned patterns. The operating frequency will be optimised to detect manned bombers and cruise missiles, but tests will also be carried out at wavelengths suitable for ballistic missile launch detection, and for the tracking of ground targets.

Detection of small targets from orbit must be done against the normal level of IR emission from the terrain background. As we have seen, this poses a problem for many current-generation IR sensors, but Teal Ruby will attempt to circumvent this problem by means of infra-red spectroscopy. The on-board sensors will simultaneously measure the thermal emission of bomber or cruise missile targets, and that of the terrain, building up a large reference collection of IR signatures. Complex signal processing equipment carried on the P-80-1 spacecraft will filter out background clutter, and attempt real-time detection and tracking of targets. Information on high-priority targets will be transmitted to a ground station using a secure laser-communications link, while data on secondary targets will be stored and replayed back to ground stations at regular intervals.

Teal Ruby is due to fly in late 1985 or early 1986 during the first launch of the US Space Shuttle from Vandenberg AFB, California. Once established in a 460 mile (740km) high orbit, and checked out, it will begin trials against a series of pre-arranged air targets. Background thermal data will also be gathered for a wide range of global areas, priority being given to regions of critical importance to US defence plans. The sensors are cryogenically cooled, so spacecraft lifetime will be limited.

In its swift journey in time from a laboratory curiosity of the 1950s to an essential component of the 1980s Compact Disk record player, the laser has found a number of military applications which utilise its light weight, intense power output, and narrow highly-directional beam. The narrowness of a laser beam can be exploited in point-to-point communications links. These can even be made hand-held for possible covert communications purposes, but the popular image of lasers as high-powered devices able to burn through steel plates makes user acceptance difficult in some quarters. Manufacturers of hand-held laser equipments often meet what some call the "Goldfinger" effect – memories of how movie hero James Bond was threatened with emasculation and eventual physical bisection by Goldfinger's laser. Informal demonstrations of a hand-held laser communications link in the corridor of one European defence ministry were momentarily halted when a senior military officer ordered, "Don't switch that thing on until my b***s are safely out of range!"

The earliest military applications of lasers have been for ranging. Some may be hand-held units, or tripod mounted equipments for artillery sighting and surveying. Others include target-designating systems mounted in the noses of fighters. Most modern main battle tanks now rely on laser ranging rather than optical rangefinders or ranging machine guns. Such lasers often form the central component of a complete fire-control system.

From simple ranging, the next obvious step was the development of similar equipments for target marking. Reflected laser energy could be detected by a suitable seeker mounted in either a manned aircraft (such as the SEPECAT Jaguars of the RAF, and the Soviet MiG-27 "Flogger"), or in the noses of guided bombs or missiles.

Despite the limitations imposed by weather, semi-active laser guidance is now widely used in "smart" weaponry. The best-known laser-guided bomb (LGB) is the Texas Instruments Paveway which smashed the Thanh Hoa rail bridge in North Vietnam in 1972. During the invasion of South Vietnam which eventually led to the collapse of that nation, Paveway-equipped F-4 Phantoms, working in conjunction with laser-equipped Rockwell OV-10 Bronco forward air control (FAC) aircraft, attacked North Vietnamese armour, proving that LGBs were also viable anti-tank weapons.

Paveway also saw action in the closing stages of the 1982 Falklands War when the British forces' Ferranti Laser Target Marker and Ranger units located on Two Sisters were used to designate targets on Tumbledown on the final day of combat. Four Paveway LGBs were "tossed" from a stand-off range of about 4 miles (6.5km) by Harrier GR.3s. Two attacks were mounted, each with a single aircraft running in from the south-west, releasing an LGB at a 30deg angle during a 3g pull-up. The bombs climbed over Mount Harriet, reaching a maximum height of around 1,500ft (460m) before falling back toward the target. Errors in target-designation procedure caused the illuminating laser to be turned on too soon during two of the attacks, and the bombs therefore missed the target, but the other two scored direct hits on Argentinian positions. Some observers credit the shock of these pinpoint attacks as playing a major role in the Argentinian decision to surrender.

**Above: If friendly forces can mark tactical targets with laser energy, laser-guided weapons can quickly score direct hits. This is a test of the Texas Instruments Paveway laser-guided bomb (LGB).**

**Above: US artillerymen load a Copperhead laser-guided projectile into a 155mm howitzer during trials at White Sands. The nose-mounted seeker is designed to survive the shock of gun launch.**

**Above: Although TV and imaging infra-red seekers were considered for the Hellfire anti-tank missile (seen here being test-fired from an AH-1), semi-active laser guidance proved the best solution.**

Other types of LGB are now under development or in service. The latest Texas Instruments production model is Paveway III, designed for release at low level, while Matra's LGB for the French Air Force and export customers completed development trials in 1984 and is now in service.

Laser-guided missiles include the Rockwell AGM-114A Hellfire in production for the AH-64 Apache helicopter, the Martin Marietta Copperhead 6in (155mm) "smart" artillery shell, and the US Navy's 5in (130mm) guided projectile.

Older weapons may sometimes be given a new lease of life by the installation of laser seekers. Aerospatiale, for instance, is now producing the AS.30 Laser variant of what was originally a command-guided air-to-surface missile, while Bendix has proposed mating the standard 127mm Zuni US Navy/USMC unguided rocket with a low-cost laser seeker created by reworking obsolescent AIM-9B Sidewinder seekers currently in storage.

Little is known about Soviet laser-guided weapons. Improved models of LGB are reported to be entering service to replace older weapons of this type, while MiG-27, Su-17 and Su-24 attack aircraft are armed with the 6.2 mile (10km) range AS-10 laser-homing missile.

In the early stages of laser development, national design teams adopted independent standards of operating frequency and modulation pattern. Although close, these were often incompatible. During the Falklands War one RAF Harrier GR.3 attempted to use its nose-mounted laser rangefinder to designate the Port Stanley runway while another delivered a Paveway. Engineers had warned that the pulse codes used by the two systems were not theoretically compatible, so were not surprised when the LGB seeker failed to lock on.

In theory, NATO could have ended up with a series of incompatible systems, but some effort was put into standardisation, backed up by extensive testing. For example, during firing trials of the Hellfire missile from a Westland Lynx helicopter at a Norwegian range, British and US laser designators were used.

**Above and below: A Copperhead laser-guided trials round makes its final approach to an M47 tank target, then shatters (below) on the turret. Despite such pinpoint accuracy, Copperhead remains a controversial weapon – some critics doubt whether the round would have sufficient time to guide after exiting the low cloud cover often found in Western Europe.**

**Above: After launch, Bofors' aptly-named RBS70 Rayrider lightweight surface-to-air missile flies along a laser beam radiated by the launcher – a new and ECM-resistant method of missile guidance.**

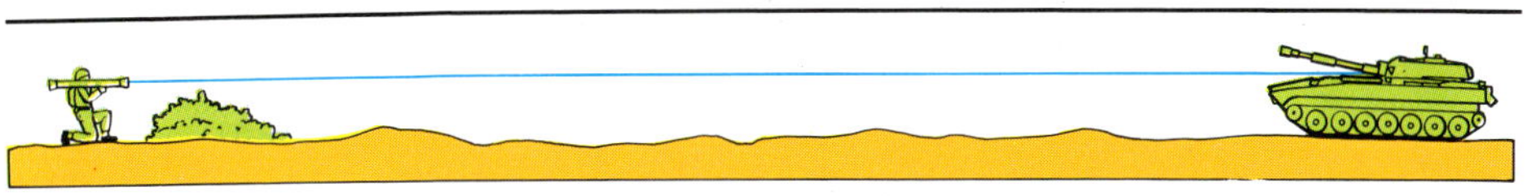

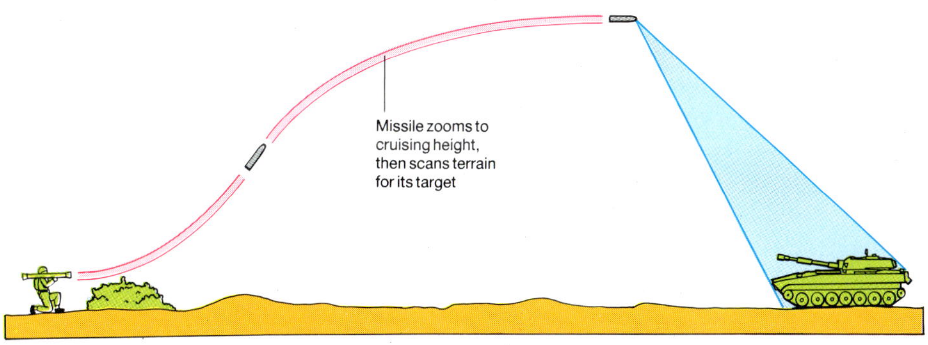


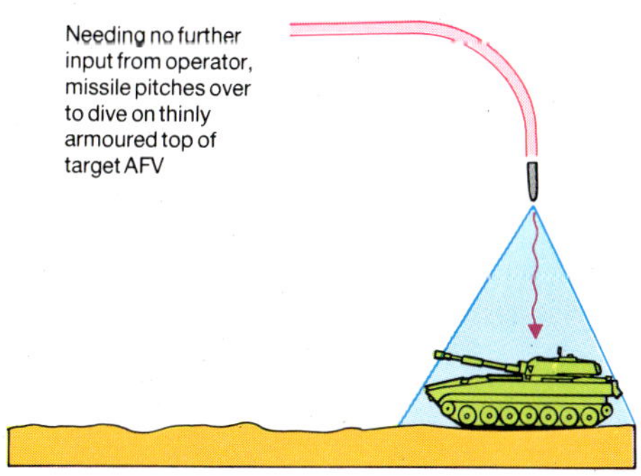


**Tomorrow's tank killer?** One possible scheme for a next generation man-portable anti-tank missile involves the infantryman, operating from the benefit of cover, using a laser to measure target range (top). After launch, the weapon climbs to a cruising altitude as it approaches the target (above), then begins to scan the terrain with a passive millimetre-wave sensor. The missile is of the fire-and-forget type, not requiring any post-launch input from the operator. This is a tremendous advantage, since he may be tired, afraid or under great pressure so with this system he can select other targets or move to another area. Once the missile has detected the target, it can pitch over into a near vertical dive, attacking the target through the relatively thin roof armour.

One little-used technique which makes good use of a laser's narrow beam is laser beam-riding, the optical equivalent of the radar beam-riding used by some early guided missiles. One of the few users is Sweden's Bofors, which adopted such a scheme for its RBS70 man-portable SAM, aptly known as Rayrider. This guidance system is difficult to jam, since the missile-mounted sensors used to detect the guidance beam have a narrow field of view, and face backwards towards the launcher.

Since lasers can be pulsed, their application to radar was logical. Being optical, such systems may be built with high angular resolution but small size. Range resolution is also good, since the output from a pulsed laser is of very short duration. One of the first applications of laser radar is the British Aerospace Dynamics Laserfire – a private-venture lightweight missile system based on the Rapier SAM. Targets detected by the Laserfire radar are engaged by a laser tracker based on a Ferranti Type 629 high repetition rate neodymium-YAG laser. Bearing and range data from this unit is passed to the system computer.

The Hyper-Velocity Missile being developed by Vought under a USAF contract uses a track-while-guide fire-control system based on a laser radar known as the Multi-functional Infrared Coherent Optical Scanner (MICOS). This can detect and classify multiple moving targets, transmitting commands to a formation of several ripple-fired missiles by sensing data down the laser beam.

Recent investigations in the USA and Soviet Union have concentrated on the possibility of using high-power lasers as direct-energy weapons. Some experimental work is being carried out on anti-aircraft systems, but the most likely application of direct energy would be in ABM or ASAT weaponry discussed in the Space Warfare section. For the shorter term, high-power lasers could find application in electro-optical countermeasures systems designed to dazzle or even damage optical target tracking and homing devices, or the eyesight of missile and gun aimers.

Communications and computer technology are combining to give military commanders a flexibility in command and control which would have seemed like science fiction a decade or so ago. In place of the cumbersome telephone equipments and unreliable radio links of the past, today's commander can rely on complex but reliable systems able to pass speech and data around the combat zone, into the civil or military telephone systems or even via communications satellites back to higher command authorities on the other side of the globe.

Even the front-line soldier is profiting from this new technology. Integrated-circuit electronics and microprocessors may not be able to alter the fundamental characteristics of the various radio-frequency bands, but they can make a battlefield radio link as reliable as the household telephone. Microchip technology makes complex jam-resistance operating techniques a possibility, and allows the creation of "user-friendly" tactical radios which may be used by personnel lacking the traditional skills of a radio operator.

## Communications security

The increasing use of signal Intelligence (sigint) and EW techniques has forced military users to protect communications traffic from eavesdropping or even detection. Latest-generation military radio equipments make use of the latest solid-state microelectronic technology. Most have provision for the use of encryption and decryption equipment to ensure good communications security (COMSEC), while others rely on frequency hopping to avoid detection and jamming.

In a frequency-hopping radio the output frequency of the transmitter and the tuning of the receiver is switched rapidly over a range of frequencies. The brief time spent on any one frequency is sufficiently short to make detection of the transmission difficult, while the fact that succeeding frequencies are chosen by a pseudo-random law makes the pattern impossible for an unauthorised listener to predict. Detailed information on hopping rate is normally classified, but most sets operate at medium

**Above: Emerson's MSQ-103A Teampack Dragoon vehicle-mounted ESM system is intended for use at divisional level. Its receiving systems monitor the RF spectrum from 500MHz to 40GHz.**

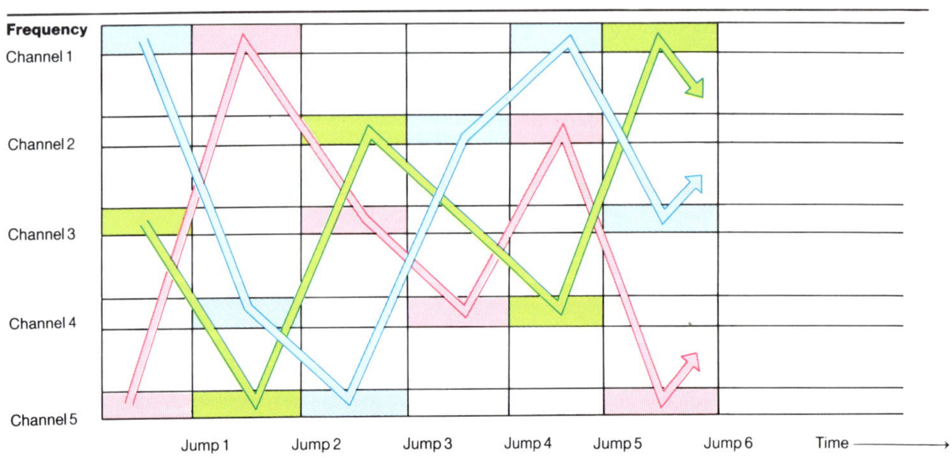


**Frequency hopping – the agile radio link**
In order to avoid interception of signals by hostile ESM systems, modern tactical radio equipments rely on frequency hopping. All the sets on a net jump at regular intervals of time – usually several hundred times a second – from one channel to another in a predetermined but psuedo-random sequence. In the scheme shown there are three nets over five channels without mutual interference.

**Below: Plessey's ICE (Interference Cancelling Equipment) can reject interference from a hostile jammer, or even from a friendly jammer being used to screen communications from enemy ESM.**

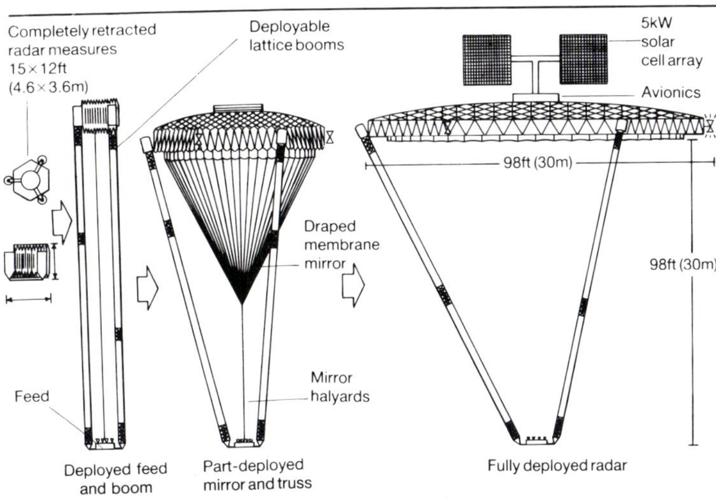


**Building the "dish"**
Creating Rhyolite-sized dish antennas in space involves extending telescopic struts, then erecting a large circular structure made from a deployable tetrahedral truss members, while a lanyard holds an aluminised membrane clear of the expanding structure. Once the structure is fully deployed, electrostatic charges would suck the membrane back into the circular assembly, creating a paraboloidal shape.

**Rhyolite: America's space-based "super spy"**
Like all point-to-point microwave links, the Soviet Union's highly secure communications links radiated a tiny amount of radio energy in the form of unwanted sidelobes (minor beams radiating at angles away from the main beam). In a good antenna design, the energy in any sidelobe is a thousandth or less than that in the main beam. Conventional sigint satellites move in low orbits, and being fast moving must rely on low-gain antennas of wide beamwidth. These lacked the gain needed to make use of sidelobe energy, so point-to-point links deep within Soviet territory seemed secure from interception. By orbiting the Rhyolite spacecraft – virtually a small radio telescope – in the 22,300 mile (35,887km) geostationary orbit – US spacecraft designers were able to steer a high-gain beam Earthwards, eavesdropping for the first time on faint signals such as sidelobe energy. Signals were then passed to Earth stations for analysis by specialised sigint personnel of the US National Security Agency and UK GCHQ. Traitors in both organisations eventually betrayed this highly-classified espionage operation.

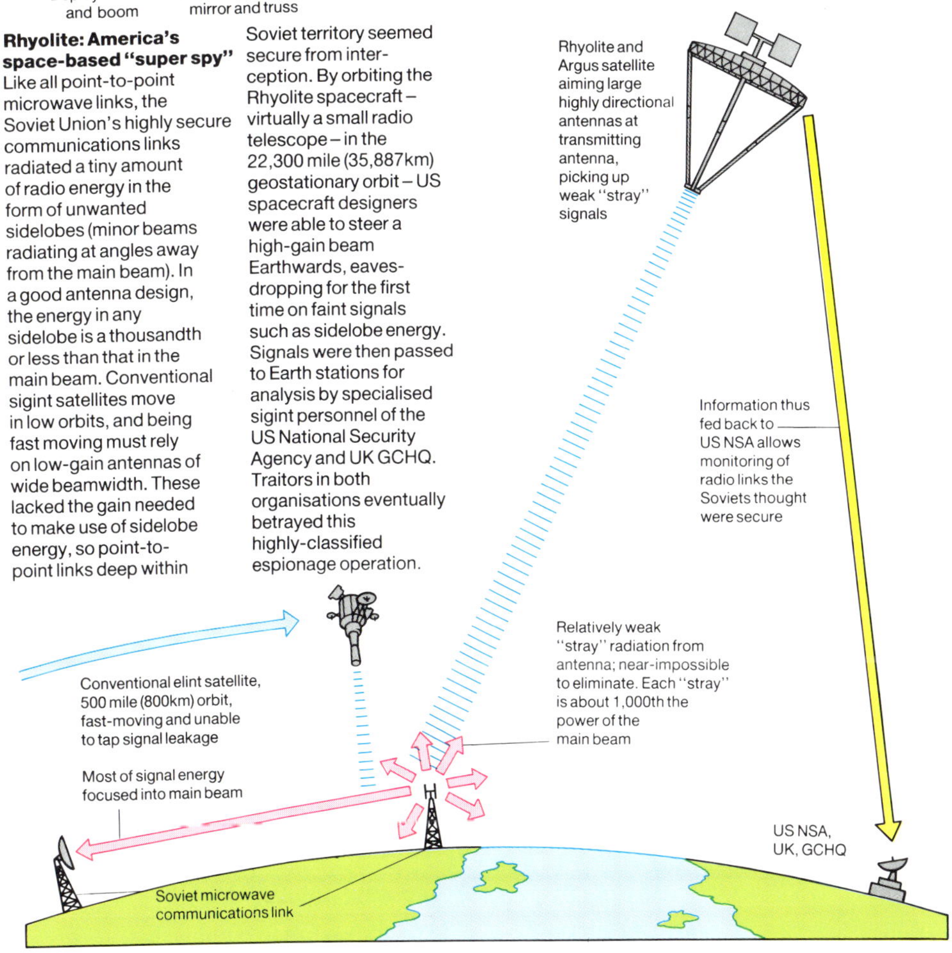

speeds – several hundreds per second – although some companies (such as Rockwell-Collins) favour high speeds involving several thousands of hops per second.

For basic communications, the US Army is beginning to rely on the SINCGARS-V (SINgle-Channel Ground And AiRborne System-VHF), a frequency-hopping radio intended to replace the 20-year-old PRC-77 radio system currently used by combat battalions and companies. Deployment began in FY85. It is a massive programme, with at least a quarter of a million sets being built for the US forces, plus others for export.

The SINCGARS programme has moved slowly, with ITT Avionics finally being selected in July 1983, defeating the rival Cincinnati Electronics. But while the selection and evaluation process was still under way foreign electronics companies noted that some customers were not prepared to wait. Despite the advantages of buying US-compatible equipment, it was clear that some nations which perceived a sophisticated EW threat would prefer an off-the-shelf frequency hopper available immediately rather than the prospect of being able to procure a US-standard set at a later date.

The UK was quick to offer systems such as Jaguar (Racal), Scimitar-V (Marconi Space & Defence) and System 4000 (Plessey), while other equipments are offered by nations including Australia, France, Israel, South Africa, and Sweden. This proliferation of rival systems could produce inter-operability problems, particularly for NATO. Even if a system such as Jaguar, Scimitar or System 4000 is deployed for normal tactical communications, a smaller number of SINCGARS-compatible equipments will still be needed for communications with flanking or rear-area formations equipped to the US standard.

USAF efforts to develop a secure, jam-resistant radio system received a setback when the US Congress ordered that development of the planned Seek Talk system be ended. The short-term solution is the modification of existing tactical UHF sets with the Have Quick system. Several longer-term alternatives have been explored, the preferred solution being an enhanced version of the Joint Tactical Information Distribution System (JTIDS). The basic version of this secure jam-resistant digital data and voice communications system is already in service aboard the Boeing E-3A Sentry AWACS and at selected ground sites in the USA and Western Europe. It will be deployed throughout the 1980s and into the 1990s.

Typical of the latest-generation secure long-range data links is the US Army's new Ground Mobile Force project, which will provide reliable jam-resistant satellite-communications terminals able to link field headquarters, giving commanders a means of transmitting orders and intelli-

**Left: Add-on encryption units can be used to protect radio transmissions from interception. The unit shown here linked to a tactical radio by a modem is made by Racal-Datacom.**

gence information over long distances. Several hundred sets will be deployed with the Army and USMC.

## Flexible communications systems

Many armed forces are currently equipping themselves with systems which have the flexibility and responsiveness perceived necessary in future battle conditions. Most combine the latest in computer and communications technology. Ptarmigan is one of the British Army's new advanced communications networks. Intended for tactical trunk communications, it is a digital communications network which combines the addressability of the telephone with the mobility of radio links. Production Ptarmigan units have already been deployed for field trials, and the system became operational in 1985.

Ptarmigan consists of a network of nodes, each linked by at least three multichannel radio relays to other nodes. The system functions rather like a telephone network – users have individual identities (rather like telephone numbers) and may be either fixed or mobile. Messages are routed around the network, avoiding any area of damage. If a route exists from the caller to the addressee, the system will find it. During a recent British Army exercise in West Germany, one defence correspondent put the system to the test by telephoning his editor in London from the seat of a fast-moving military vehicle. Ptarmigan routed his call from the vehicle, through the military radio network, and into the West German phone system.

Wavell is the British Army's data-processing and distribution system. Conventional database systems hold their data in a central mainframe computer which users can access via remote terminals, but a system with such a vulnerable central heart is unacceptable for military use. In Wavell, the data is distributed around a network of vehicle-mounted computers, which are linked via trunk communications. Data fed in at one location is available to all, so loss of any individual vehicle does not dramatically degrade the system. This fast distribution of data also frees commanders and their staffs from the laborious traditional tasks of requesting data and compiling status reports.

**Above: If tactical communications are to remain effective in the face of enemy ECM and physical attacks on signals units, obsolete equipment must be replaced by modern multi-node systems such as the British Army's Ptarmigan.**

Ptarmigan technology could find its way into the US Tri-Tac communications programme, one component of which will be the Mobile Subscriber Equipment (MSE) – a common-user telephone system for US Army static and mobile subscribers. The US Army does not want the delays, costs and risks of a new development programme, so plans to buy an "off-the-shelf" system. The obvious candidates would seem to be the British Ptarmigan or French RITA (Reseau Integre de Transmissions Automatique) systems.

The Pentagon also wants to be able to transmit large amounts of data, and several projects are currently under development with this in mind. To give US military commanders information on the tactical disposition of friendly and hostile forces, the Maneuver Control System – a network of small computers adapted for military use – is being developed. A prototype system has already been fielded in Europe to allow users to participate directly in refining the design of the definitive system. The flow of data between command and control, Intelligence, air defence, fire-support, and EW systems will be handled from FY88 onwards by the US Army Data Distribution System (ADDS) – a secure, jam-resistant digital communications system.

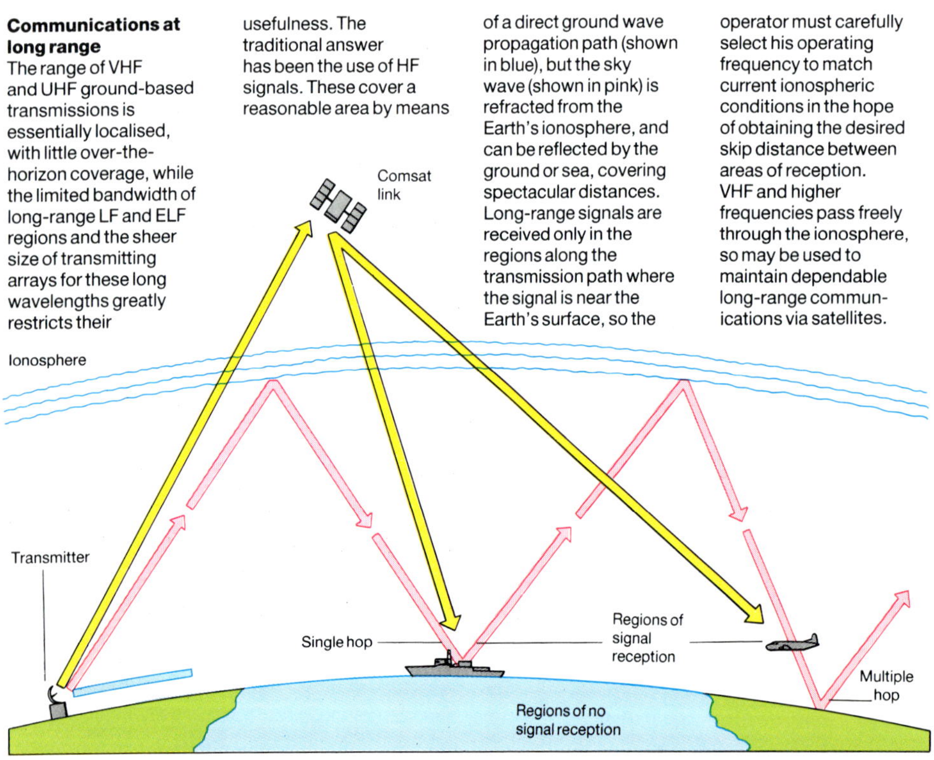


## Processing Intelligence

Effective and timely Intelligence-gathering in mobile warfare makes massive demands on computing power. Conventional computers obey instructions one at a time, but many defence tasks require specialised hardware designed to work in more novel ways, tackling several elements of a task simultaneously. A typical example is the Britton Lee Intelligent Database Machine (IDM), a functionally designed high-speed processor specifically developed for fast database processing. A conventional computer used for database work reads data into its memory, then carries out a search. With the IDM, electronic "pages" of data from the disk store can be searched while they are being transferred to memory.

In the future, "expert systems" – programmes which are able to absorb and re-use the knowledge of human experts – may be used to simplify complex tasks to the point where these may carried out by unskilled operators. The full impact of such technology will not be available until the widely-predicted "fifth generation" of

Above: The infantryman's kit of the future, being studied by British R&D company Scicon (see photo on p13) assumes that the helmet can display infra-red imagery and an aiming reticle.

Above: Computerised "expert systems" carried by future soldiers could aid target recognition, says Scicon. Here the system compares the profile of a distant vehicle sighted by the soldier with built-in threat data. By rotating the shape of known enemy vehicles until a "best match" is obtained, the system tentatively identifies the target and indicates its degree of certainty.

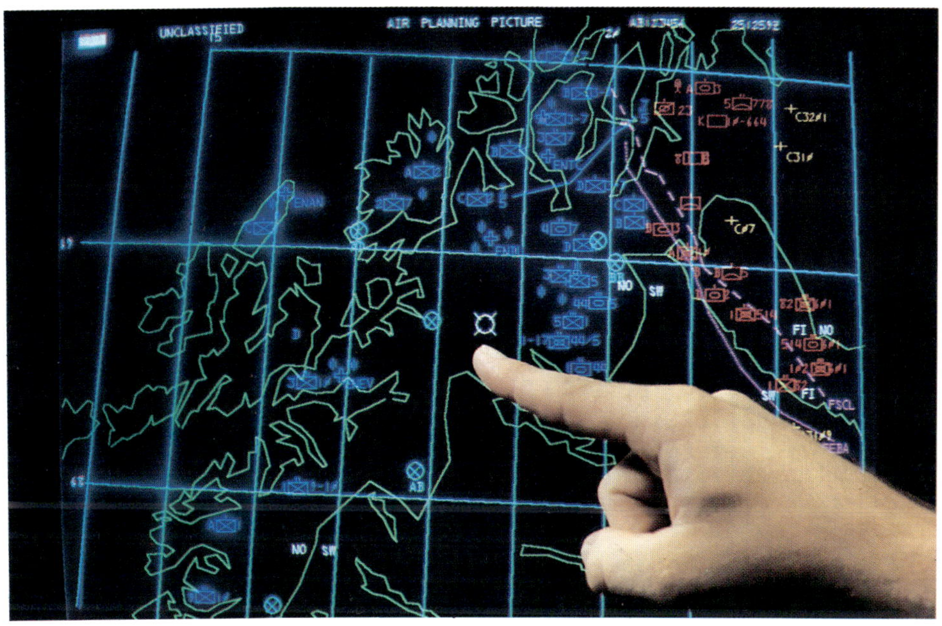


Above: High-resolution colour graphics generated by computer provide a simple means of presenting data to an operator, while touch-sensitive screens allow quick and unambiguous feedback of commands to the system.

Below: One of the most interesting of artificial intelligence programme studies is this Martin Marietta Autonomous Land Vehicle, a robot which could carry out missions in environments hazardous to humans.

computers is developed, but the first steps toward the creation of such systems for military applications have already been taken.

To take one example, a privately-funded research study carried out by Britain's Scicon has investigated methods of applying expert systems to battlefield Intelligence-gathering. The goal of such work is to create systems capable of accepting information from multiple sources – data whose validity and meaning inevitably degrade with time – then speedily collating this into a database representing the currently-perceived Intelligence "picture".

A demonstration system intended to help photo-analysts discriminate between MBTs, APCs, self-propelled guns and other battlefield targets has already been tested. Faced with an indistinct photograph of what might be an MBT, the user replies to a series of questions asked by the system. As this process of question and answer progresses, the CRT display indicates the likely identity of the target, complete with a probability rating. The computer may be asked at any time to justify its line of questioning or its conclusion.

## Enlisting the aid of "robots"

Recent developments in technologies such as artificial Intelligence, computer science and microelectronics, have resulted in the US DARPA Strategic Computing programme – a scheme to create "machine intelligence technology" for application in future defence systems such as autonomous vehicles (robots), battle-management systems and a "pilot's associate" (an advanced information summary and display system). The latter would almost constitute a robot co-pilot, bringing the era of human pilot and 'droid co-pilot depicted in the movie "Star Wars" out of the realm of science fiction and into the laboratory.

Speech and video input, and understanding of natural language may simplify human interaction with computers, allowing new alternatives in computer design. Multiple computations may be performed in parallel, allowing large improvements in machine performance.

In the days before navigation aids (navaids), the pilot of a ship or aircraft had to rely on dead reckoning, a simple mathematic calculation of probable position. This simple technique is by no means dead in the electronics era – new technology can be applied even to simple navigation aids such as the ship's log. The AGI Microlog is based on the well-known principle that a voltage is induced in a conductor by a moving magnetic field. By moving the magnetic field with the ship, and using the sea water as the conductor, an accurate measurement of ship's speed can be made.

A combination of a simple odometer (distance-travelled indicator) and a heading sensor forms the basis of many navigation aids for ground vehicles such as tanks. Odometers normally measure distance by counting rotations of the vehicle's wheels, or of a shaft geared to the wheel. As the tyre on a wheeled vehicle wears, its diameter reduces. Once 5mm of rubber has been worn away from a tyre 50cm in radius, the circumference, and thus the distance travelled per revolution, will have fallen by one per cent. Tyre size must therefore be measured at regular intervals and the odometer re-calibrated. Heading may be measured using either a directional gyro or a magnetic compass, the former offering high accuracy, the latter low cost.

Radio navigation aids such as Decca and Loran are intended for general long-range navigation, while TACAN is specifically designed for military aircraft use. All are relatively inexpensive, and reliable in normal use. Being based on radio, they rely on ground stations which may be jammed or attacked. The civil VOR/DME (VHF Omni-Range/Distance-Measuring Equipment) operates on VHF, but the military TACAN system operates at UHF. This higher frequency allows the use of a smaller antenna, and is less affected by weather and the terrain around the beacon site.

New technology is giving radio-navigation a fresh lease of life. Systems such as the Hughes Precision Location Reporting System (PLRS) couple UHF operating frequencies and Frequency Time-Division Multiple-Access (FTDMA) techniques to create a navaid capable of monitoring the positions of up to 400 vehicles, aircraft or other units. A single Master Unit assigns transmission time slots to all of the User Units under its control. It also assigns to each User Unit a list of the User Units which it must listen for and report, and those which it should use as a radio relay should a direct Master/User radio link not be possible.

When its time-slot comes round, each User Unit transmits back to the Master Unit. While awaiting its turn, it listens for the pulses from other User Units, and compiles a list of the times at which such signals were heard. By comparing this data with a list of the time slots assigned to all the individual User Units, the Master unit is able to build up an overall picture of the battlefield, and the location of all its User Units.

Navstar/GPS (Ground Positioning System) is the latest US satellite-based navaid. Based on a constellation of 18 satellites, plus high-accuracy atomic clocks carried in each spacecraft and ground station, this provides two levels of accuracy – one for general users, and a more precise version based on encrypted signals for the use of the US Armed Forces and their allies. In order to obtain a fix, a ground station receives the signals from one of the satellites "visible" from its current location. The signal is coded with its exact time of transmission. By noting the difference between this and the time at which the signal was received, the ground station can deduce its range from the satellite. After taking fixes from three spacecraft, the ground station is able to deduce its own position.

Inertial navigation systems (INS) are costly, but completely self-contained,

**Above: Mobile warfare presents an appreciable risk that units, disoriented by darkness, bad weather, unfamiliar terrain, and the stress of battle, will accidentally engage other friendly forces. Hughes' solution to this problem is the PLRS (Position Location Reporting System). A single master station, linked by radio to up to 370 manpack, vehicle-mounted or airborne user units, can measure the position of the latter using multilocation techniques.**

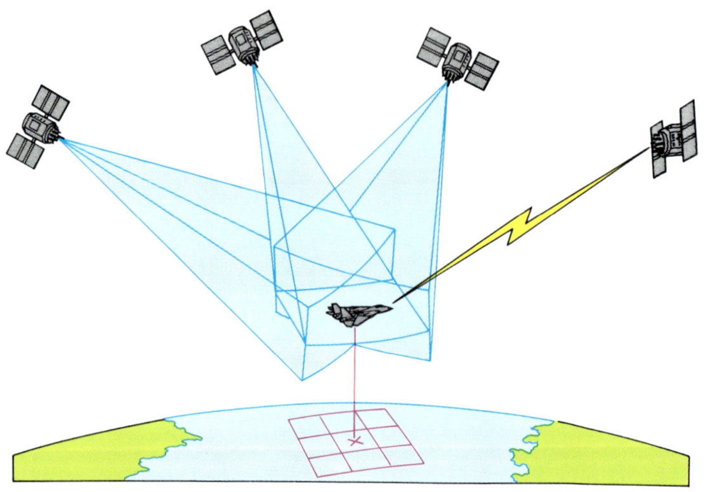

**NavStar in action**
Each NavStar Global Positioning Satellite continuously transmits its identity, current position in space, and current time according to its on-board clock. By noting when this message is received, an aircraft or other platform can tell how long the signal took to arrive, and thus its own distance from the satellite. Measurements from three spacecraft give a positional fix, while a fourth indicates any error in the user's own reference clock. Thus, no ship, tank or aircraft need be "lost".

using accelerometers to detect motion. Initially developed for strategic-missile applications, they quickly found service aboard nuclear submarines, then manned aircraft. They are now offered for simpler navigation applications such as surface ships and AFVs. The tendency to drift with passing time is a problem, but this can be solved by regular positional fixes and INS upgrades. A more serious problem until recently has been that of cost – the precision gyros and accelerometers required are very expensive.

In the past, the use of naval INS has largely been confined to high-value vessels such as nuclear submarines and aircraft carriers, but this shows signs of changing. By using inertial-navigation technology originally developed for aircraft, the Navigation Systems department of Ferranti has developed an INS likely to be about one third less costly than earlier units, making it an economic proposition for smaller warships such as frigates and destroyers. Since naval applications are very much less critical (by an order of magnitude) than aircraft systems, the designers were able to use conical gas-bearing gyros rather than the more traditional and expensive ball-bearing units used in aerospace applications.

**Above: A USAF Navstar Global Positioning Satellite undergoes simulated mission testing in the space simulation chamber of the Arnold Engineering Center, before being successfully orbited.**

**Below: A GPS terminal indicates its current latitude, longitude, and height above mean sea level. It can also indicate velocity, groundspeed, and time/distance to the next waypoint.**

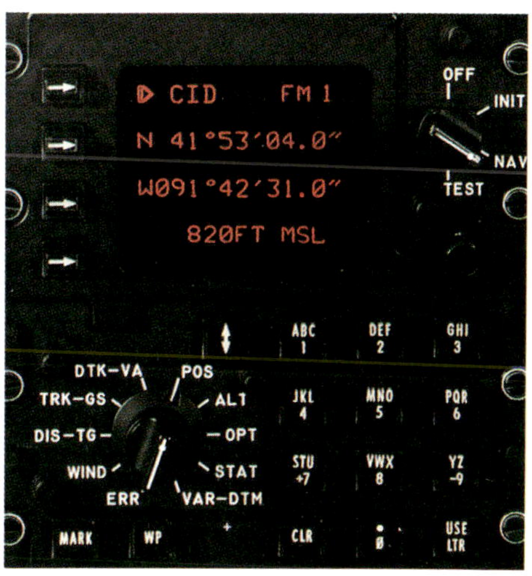


The resulting Ferranti Inertial Navigation System (FINS) consists of three main components – the inertial platform unit (IPU), electronic unit, and remote control and display unit. Apart from data from sensors such as the ship's log – used to provide damping – the system needs only 350W of power from a 24 to 28V DC supply. No cooling air is required, so the unit may easily be retrofitted to vessels already in service. FINS is capable of determining ship position to within a 1.7nm (3.1km) CEP (circular error probability) after 24 hours of operation following alignment, and can give other navigational information such as steering error and distance/time to waypoints.

Other developments in INS technology have led to the introduction of relatively simple strap-down sensors in place of the traditional but expensive gimballed platforms used in conventional systems, plus the use of specially shaped glass blocks to create robust inertial sensors based on beams of laser light. These techniques do not eliminate drift error, but back-up secondary systems such as Navstar/GPS or the Tercom radar-based guidance system devised for use on US cruise missiles are overcoming this problem.

Obvious targets for attack by electronic warfare (EW) include the enemy's surveillance, target-acquisition and tracking systems, plus the guidance systems of missiles and "smart" weapons. EW may also be used to good effect against speech, data and missile-guidance communications links, while even a simple warning system able to detect threats can be a useful tactical aid.

The most basic item of electronic countermeasures (ECM) equipment is therefore the radar-warning receiver (RWR). This is designed to detect the signals from hostile radars, and to signal to a fighter pilot, for instance, that he is under radar observation. Most patterns of RWR give an approximate indication of the threat bearing, frequency and signal characteristics, while some identify the threat using a built-in catalogue of threat parameters.

If a surface or airborne radar manages to lock-on to the aircraft, this will register on the RWR display. Pilots attempting to break the lock of threat signals by deploying chaff or flying very low will be able to determine whether their tactic was successful by watching the RWR output to see if the hostile signal is still present.

Infra-red warning receivers can be used to detect the launching of tactical missiles, enabling ECM systems to be turned on or evasive tactics to be carried out. This is particularly valuable if IR jammers are to be used, since the life of the lamp sources used in many systems is limited. Early IR receivers suffered a high false-alarm rate, but the newer designs are reported to be more reliable.

The simplest anti-radar countermeasure is chaff – small strips of conducting material whose length is selected to make them good reflectors of radar energy. The first type used during World War II was made from aluminium foil, but modern chaff is made from thin glass or plastic fibres coated with a metal film. Once dispersed into a compact cloud, a small chaff package about twice the size of a packet of cigarettes will match the echoing area of a jet fighter.

The trick of stuffing chaff behind the air brakes or inside the wheel wells of an aircraft then momentarily opening these in combat is a useful wartime improvisation for air arms not equipped with proper dispensers, but is hardly the ideal technique.

**Above: Ferranti's CREST radar simulator displays the effect created when a chaff corridor is laid by EW aircraft (stepped vertical line at left of image).**

**Above: Chaff filaments (glass fibre coated with aluminium) are so fine that several hundred can share the eye of a needle with a length of cotton thread.**

Chaff is often packaged in cartridges containing pre-cut lengths, then ejected from aircraft-mounted dispensers or ship-mounted launchers. Rockets can also be used to carry chaff ahead of an aircraft, or well way from a ship. Some airborne chaff systems consist of pods fitted with drums of continuous chaff material and cutting mechanisms. Chaff may thus be customised in combat to match the exact frequency of the threat.

Jamming systems for use against radars operate in noise or deception modes. Noise jamming was the first technique to be used during the infancy of EW. Essentially a "brute-force" technique, it aims to swamp the receiver of the target system with unwanted radio-frequency noise, so that the genuine and wanted signal cannot be distinguished from the background.

If the exact operating frequency of the target is known, the jammer may be tuned to the same frequency, concentrating its power against the enemy. This is known as spot jamming. If the exact threat frequency is not known, or if the threat is frequency-agile, the output of the jammer can be spread over a range of frequencies likely to contain the threat. This technique - known as barrage jamming – dilutes the effect of the jammer, since the equipment's limited output power must be spread over the range of frequencies to be covered.

Noise jamming may be simple to implement, since very little needs to be known about the characteristics of the threats, but its effectiveness against sophisticated opposition is minimal. Much more effective are deception techniques which manipulate the known weaknesses of radar technology in order to confuse the system under attack. By receiving the threat signal, processing it in some way, then re-transmitting it, the threat system can be persuaded to accept the "doctored" signal as

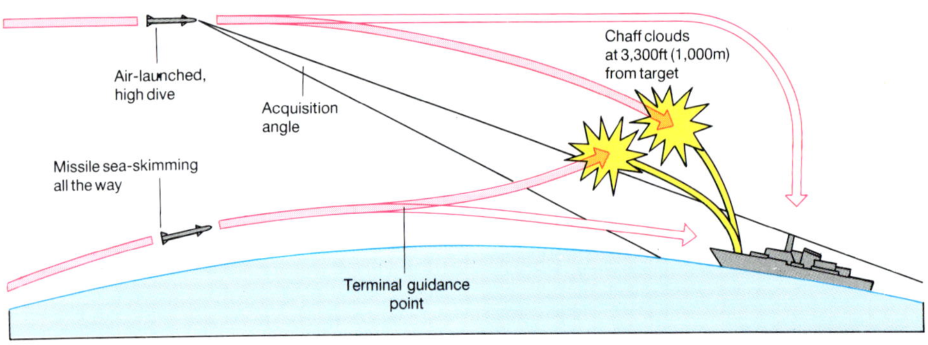


**Chaff as a naval decoy** Anti-ship missiles which operate at high-altitude or in sea-skimming mode may be decoyed by the subtle use of naval chaff rockets. In the case of a sea-skimming missile, chaff may be shot to an altitude of 3,000ft (1,000m) or more while the target ship is still below the missile's radar horizon, presenting a tempting target. High-altitude missiles designed for diving attacks – a type of weapon widely deployed by the Soviet Union – may be lured into a premature dive by a similar chaff cloud.

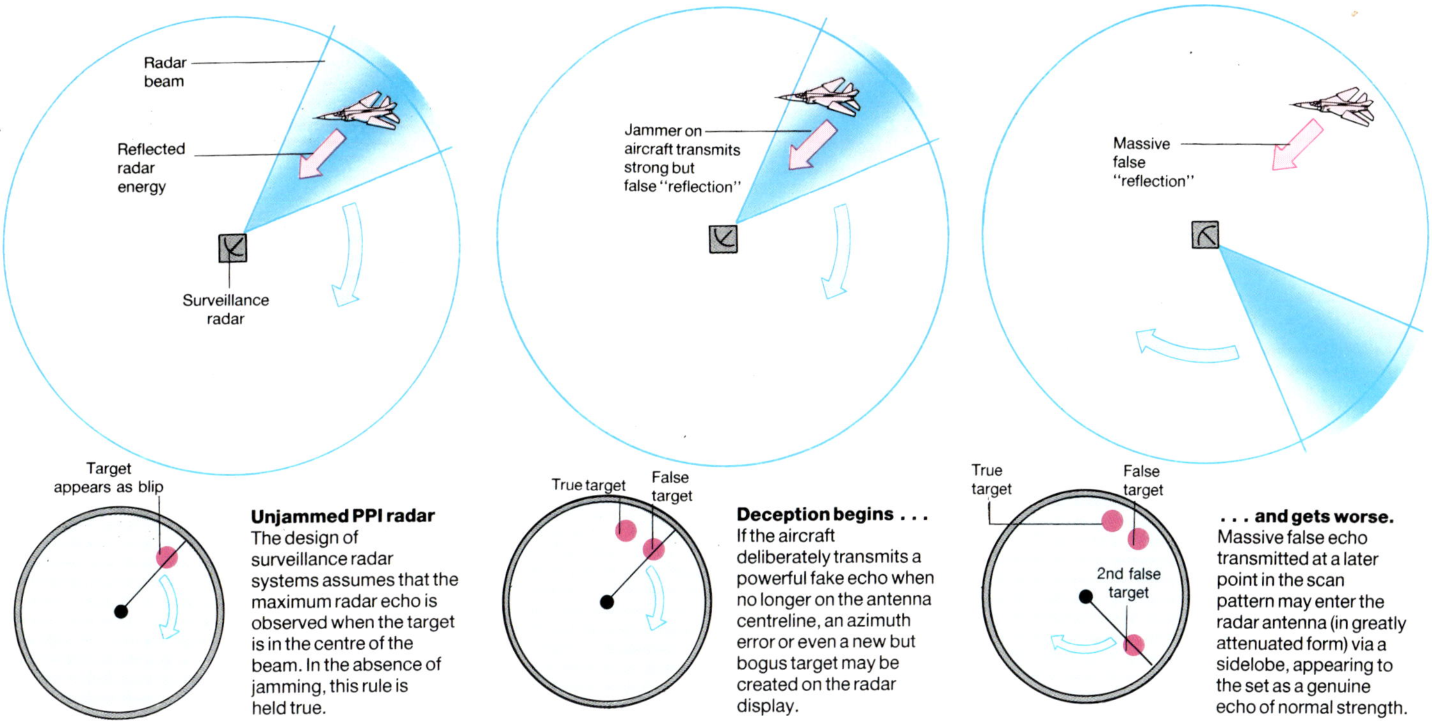


**Unjammed PPI radar** The design of surveillance radar systems assumes that the maximum radar echo is observed when the target is in the centre of the beam. In the absence of jamming, this rule is held true.

**Deception begins . . .** If the aircraft deliberately transmits a powerful fake echo when no longer on the antenna centreline, an azimuth error or even a new but bogus target may be created on the radar display.

**. . . and gets worse.** Massive false echo transmitted at a later point in the scan pattern may enter the radar antenna (in greatly attenuated form) via a sidelobe, appearing to the set as a genuine echo of normal strength.

genuine. By choosing the manner in which the signal is processed before transmission, the ECM designer can feed false range or bearing information to an enemy radar, or even cause multiple false targets to be detected.

This relatively subtle approach does require accurate data on the characteristics of the threat system, and also the techniques which the latter uses. If the deception technique is not properly matched to the threat, the jamming will be useless.

Most deception jammers use a built-in "set-on" receiver to detect threat signals, and to measure the frequency to which the jammer transmitters must be tuned. The more sophisticated systems compare the parameters of the detected signals with threat data stored in a built-in threat library. Once those representing the greatest threat to the aircraft have been identified, the jammer assigns the power available from its various jamming transmitters accordingly. This technique is known as power-management. The type of jamming to be used will be chosen to match the threat, and on the most advanced jammers, the transmitters will be turned off momentarily at regular intervals, allowing the receiver to monitor the hostile transmission and assess the effect of the jamming. If the latter is not proving effective, a different jamming method may be automatically selected.

**Above: The ALQ-99E jammer carried by the USAF's EF-111A EW aircraft can "white out" a circular PPI display with hundreds of false targets.**

**Below: The Westinghouse ALQ-131 (seen here under test on a YF-16) is the USAF's standard jamming pod. Improved models are now being deployed.**

Obsolescence is a problem with any jamming system, and particularly with deception systems. The hardware is often specifically built to deal with identified threats. A new threat can involve modifications to existing hardware, or even completely new sub-systems. For this reason the design of last-generation systems such as the United States" ALQ-131 is modular in order to facilitate upgrading.

In developing the Westinghouse ALQ-165 Advanced Self-Protection Jammer (ASPJ), the USAF and USN decided to adopt a software-controlled system whose computer could match countermeasure techniques to threats in accordance with a combination of established threat data and real-time information obtained from the system's built-in receiver/signal processor. The operational requirement called for ASPJ to cover several octaves of frequency, handling multiple threat signals against a dense background of other signals. The system can simultaneously jam many threats using its power-managed dual-mode transmitters. These have travelling-wave tubes (TWTs) which operate in parallel, generating pulse or CW signals as demanded by the system software.

## Countering IR-guided weapons

The simplest countermeasure against IR-guided weapons involves the release of flares by the target aircraft in an attempt to decoy the missile with a false target. This can partly be countered by measuring the level of IR energy at two different wavelengths in order to estimate roughly where the peak level of IR output lies. To produce a large amount of IR energy from a small package, the flare must burn at a high temperature, while the genuine target is a large mass of metal (such as the aircraft aft fuselage and jetpipe) at a somewhat lower temperature. Since the flare (being a higher-temperature source) will have its peak response at a higher part of the IR spectrum, the seeker head can, by means of filters and signal processing, distinguish the real target from the false.

Like chaff, flares are often released from aircraft-mounted dispensers, which in many cases can also carry expendable radar jammers. Reliability of these dispensers is apparently not as high as users would like, but the problem cannot be serious; funding for a US programme to improve such equipment was cancelled in the early 1980s. Flares and expendable radar jammers for naval use are often launched by rockets, as is chaff.

The rotating-scan operating principle used by most non-imaging IR guided weapons may be simple and cheap, but the use of a regularly rotating reticle is a weakness exploited by designers of countermeasures. Since a flickering IR image is seen by the missile as evidence that the seeker is not pointed directly at the target, IR countermeasure systems deliberately generate a false source of flickering IR energy which the missile could interpret as an aiming error. If accepted by the missile seeker, this false error will result in a spurious steering command being sent to the seeker head, driving it away from the target bearing.

Typical US IR countermeasure systems include the Loral ALQ-123 (pod-mounted on the A-6 and A-7), Northrop AAQ-8 (pod-

**Above: An F-16 releases pyrotechnic flares – a simple but often effective counter to many types of heat-seeking missile. Many modern missiles sample the incoming infra-red energy at two wavelengths, and are less easily fooled.**

**Below: The Martin-Marietta TADS/PNVS turret houses the sensors and laser designator of the AH-64 Apache attack helicopter. Deployment of this effective system will have spurred the development of Soviet electro-optical EW equipment.**

mounted on the F-4), and the Northrop AAQ-4 internally-mounted system hastily applied to C-5A Galaxy transports during the Vietnam War. Many IR countermeasures systems such as the Sanders ALQ-144 and -147 (pod-mounted on helicopters and OV-1D FAC aircraft, respectively) use a continuously-operating source of radiation, and rely on a mechanical shutter system to provide the necessary modulation, but the AAQ-4 and -8 mentioned earlier use electronic modulation, and the -123 probably does also.

Northrop has also developed the 53lb (24kg) Modularised Infrared Transmitting Set (MIRTS) intended to provide 180deg countermeasures coverage for fixed-wing aircraft and helicopters. It is reported to use a sapphire lamp capable of reliable multi-band output covering both the atmospheric "windows" used by IR systems.

Several new US IR countermeasures projects have been identified, including the Advanced IRCM programme. These are under development to meet the threat posed by the latest generation of IR seekers, trackers and FLIR systems. Reports that a new guidance system resistant to IR countermeasures is being developed for the Chaparral IR-homing point-defence SAM is evidence of the growing deployment of IR countermeasures by the Soviet Air Force, while US Navy plans to improve aircraft decoy techniques suggest that similar improvements are expected in Soviet IR-guided SAMs.

## Countering EO- and laser-guided weapons

Visually or electro-optically guided surface-to-air and air-to-air weapons may also be attacked by EW. Detection systems can warn aircrew that they are under attack, while decoys and jammers may be used to confuse tracking and homing devices. Most of this work is classified, but a number of programmes can be identified. Again, American technologists appear to lead the field. Their Advanced Laser Warning System will provide a complete coverage of the optical spectrum. A warning system for the Detection of Laser Emitters (DOLE) has also been developed and flight tested.

The growing deployment of EO-guided and millimetre-wave sensors and weapons has forced EW designers to expand the coverage of their products. In 1982/83, the US DOLRAM (Detection of Laser, Radar and Millimeter Waves) programme saw the flight testing of an integrated warning suite incorporating a laser warning receiver, angle-of-arrival receiver and a millimetre wave receiver.

**Above: Xerox developed the ALQ-157 infra-red jammer to protect tactical helicopters from attack by IR weapons. Shown here on a CH-47 Chinook, the equipment may also be carried by the CH-46 and CH-53.**

Under project F34-375, an Integrated Tactical Electronic Warfare System (ITEWS) is being developed for use aboard US Navy aircraft. This advanced system will include electro-optical, infrared and ultraviolet technology for target detection and classification, and may also incorporate facilities for targeting weapons against detected threats.

An advanced EO countermeasures pod, originally codenamed Compass Hammer, is being developed by both the USAF and USN. Flight trials of experimental pods started in the early 1980s. The Coronet Prince programme now in advanced development applies technology developed under Compass Hammer to an operational system intended to enter production in 1990. Coronet Prince is usually described as an optical countermeasures pod, but has also been described as an "optical threat acquisition and cueing system". Working under Compass Hammer, Westinghouse and Martin Marietta developed rival systems carrying the unofficial designations ALQ-179 and ALQ-180, one of which could be adopted in modified form by the Coronet Prince programme.

The Expendable Laser Jammer (ELJ) is about to commence flight trials, while development of at least three other EO countermeasures systems has been reported. One is for an air-to-air application, another is a counter to projected electro-optical threats, while a third is intended to cope with laser-guided weapons.

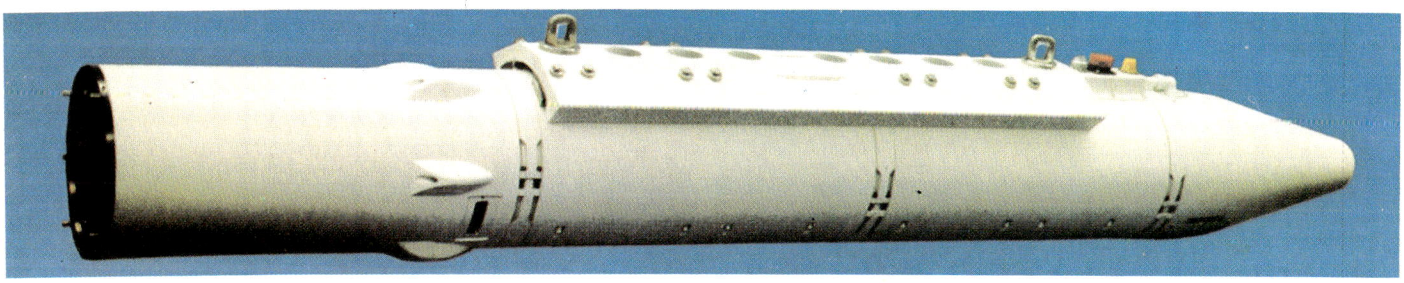

**Above: Northrop's AAQ-8 is a pod-mounted IR jammer for use on tactical aircraft. Unlike earlier systems, it uses an electronically modulated IR source.**

**Below: Transmitter and control unit of the Sanders ALQ-144 infra-red jammer. This uses an electrically-heated IR source.**

**Below: More than 1,000 ALQ-144 have been ordered. The unit can be mounted above or below the helicopter (in this case a Bell UH-1N).**

Given the widespread use of thermal imaging systems, it is hardly surprising that specialised smokes are being developed to "blind" these. Most of the announced work in this new field is taking place in the UK, the British Army having decided to fit its AFVs with an anti-IR system.

Schermuly's Multi-Band Screen operates by firing a vertical "fountain" of hot particles from a series of cartridges. These are designed to land outlet-side-up after launch from current AFV grenade launchers. The Royal Ordnance Factory's VIRSS (Visual and Infra-Red Screening System) is capable of providing screening over a 120 degree sector. Unlike the Schermuly system, this does not generate the screen from a source on the ground, but develops a smoke cloud able to act both as a visual and an IR barrier. An experimental Chieftain MBT fit used in trials features six launchers on each side of the vehicle. Each consists of 20 individual launch tubes of small calibre.

The Guard decoy from Wallops uses lightweight launchers fitted on both sides of the turret or hull of an AFV. These carry two or three 57mm rockets able to launch IR or chaff decoy payloads to an altitude of around 180ft (55m). The IR candle is ejected from the rocket then ignited. While descending by parachute it acts as a source of infra-red energy in the 3 to 5 and 8 to 14 micron bands.

Decoys have long been used in EW. Having retired the Quail, and allowed the proposed SCAD decoy to evolve into the ALCM air-launched cruise missile, the USAF no longer uses such equipment. But decoys still feature in USN plans. Given the Soviet use of long-range maritime patrol aircraft and radar-equipped satellites to track US task forces, the USN is keen to deploy suitable countermeasures. A combination of decoys and jamming can do much to confuse hostile surveillance radars, but more subtle techniques for masking or modifying the characteristics of US shipborne radars have also been studied. A rapidly inflatable corner reflector for shipboard launch has been demonstrated by the US Navy, but this is probably an anti-missile system.

Apart from radar-detection devices, the threat is from anti-radiation missiles, which, in addition to being used against land-based radars, are also highly effective against warships. When operating in a high-threat area, naval units may decide to operate with their radars switched off, to avoid announcing their position to hostile elint systems, but the ship's radar-guided weapons must be free to operate during an engagement. The enemy's ARMs must therefore be confused or decoyed. In the early 1980s, the USN developed prototype adaptive-control hardware and software for ARM countermeasure decoy systems, but subsequent developments are presumably classified.

Radio links are another prime target for EW, and one which has been exploited in several recent conflicts. Noise rather than deception techniques are the usual countermeasure in this case, with the output of the jammer transmitter either being

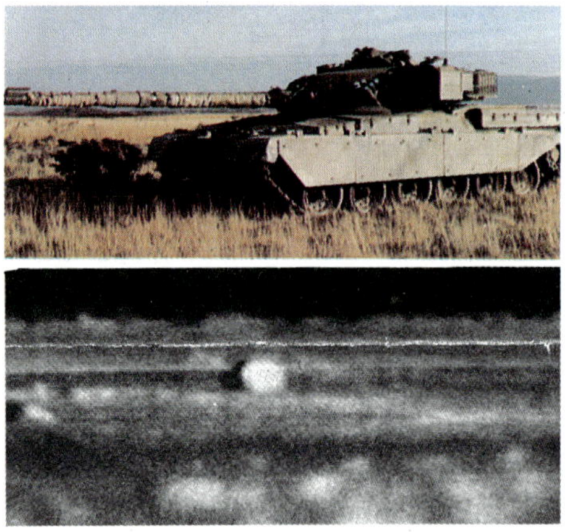

**Above: Armoured fighting vehicles are a prime target for infra-red viewing systems. Vehicles can easily be seen in IR images, although the example seen here is likely to be a test target rather than the actual tank shown above.**

**Above: Schermuly's Multi-Band Screen consists of a series of grenades launched from the vehicle's standard smoke launchers. After landing on the ground, these will emit a visual and IR screening material.**

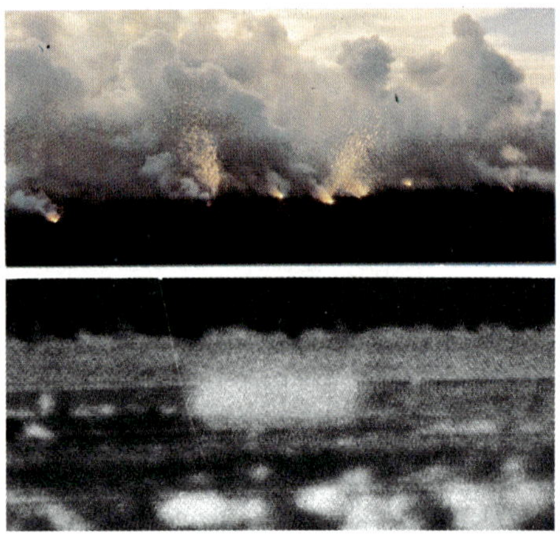

**Above: Within two seconds, the visual smoke is well dispersed, screening the vehicle from view. The IR image shows the vertical fountains of hot particles from each grenade starting to obscure the target.**

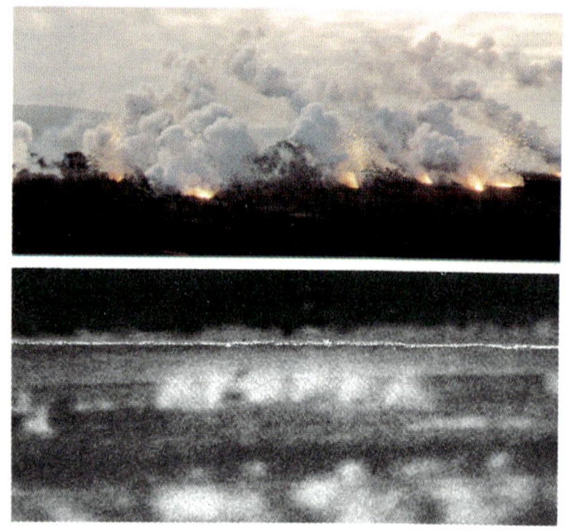

**Above: At +3 seconds, the screen is complete. A thick pall of smoke hides the vehicle from sight, while the infra-red image is completely obscured. Most anti-tank weapons would be useless in such circumstances.**

**Anti-radar ALARM**
Low-level, high-speed flight, evasive routing and ECM may not be sufficient to enable aircraft armed with short-range stand-off weapons or conventional ordnance to attack high-value targets well defended by layered SAM systems supplemented by radar-directed AA guns. So BAe have developed the ALARM weapon to destroy air defence radars and force radar-controlled SAMs and guns to revert to degraded, passive modes before the attack aircraft enters the defended zone. ALARM can be carried as part of a mixed weapon load and can be launched to loiter over the target area while it searches for targets, which may be forced to "shut down" briefly only to be destroyed when they "turn on" again.

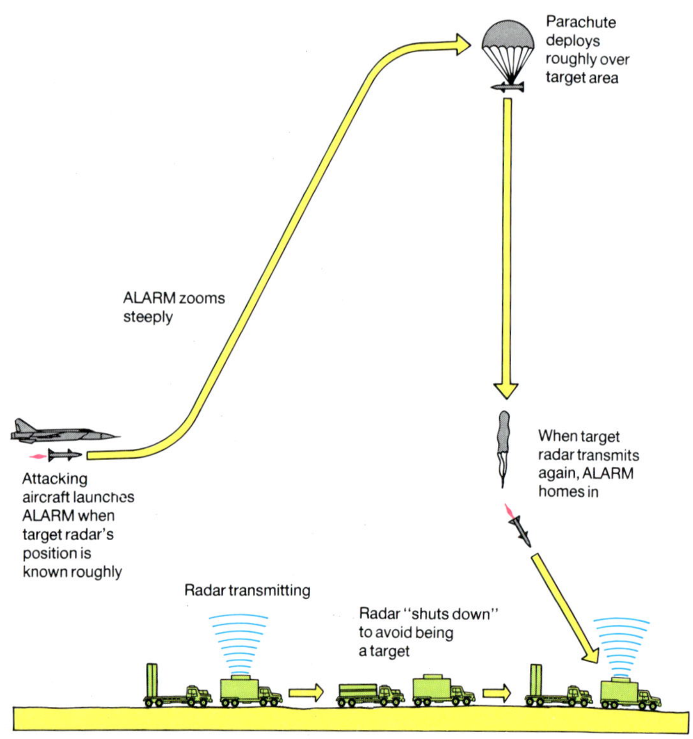

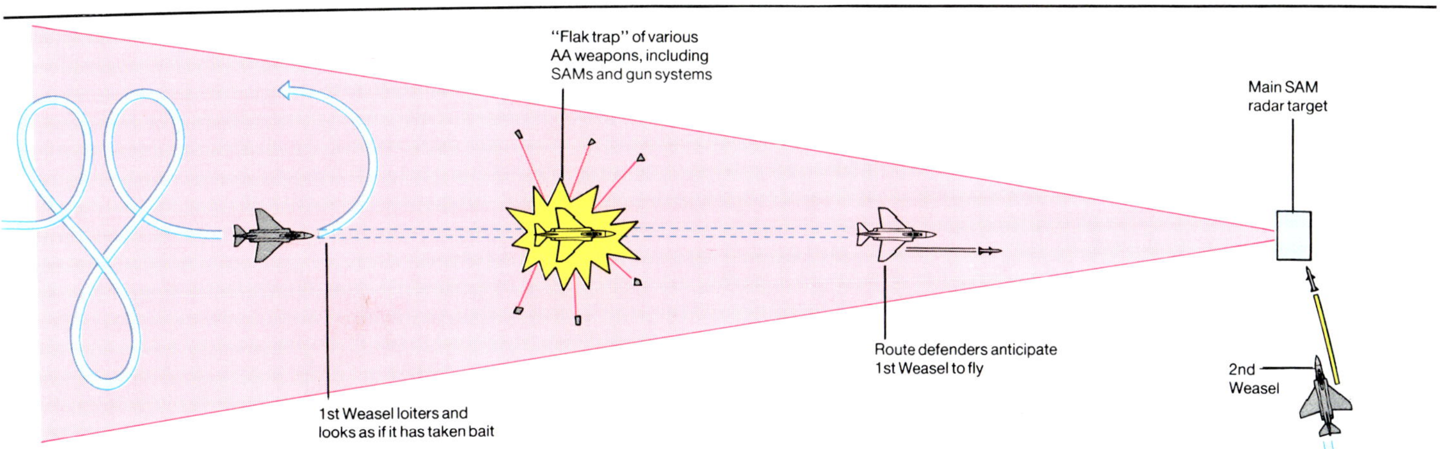


**"Weasel" versus SAM** One trick which North Vietnamese missile crews used against USAF Wild Weasel anti-radar strike aircraft was to create "flak traps" of anti-aircraft guns and light SAMs close to selected radar sites, then turn on the radar when a Weasel was in the vicinity in an attempt to lure the US aircraft into overflying the trap on its way in to a missile-launch position. The USAF response was to operate Weasels in pairs. As one lingered just out of range, pretending to have swallowed the electronic "bait", the seond would attack the radar from a different direction while the radar and its defences were concentrating on the first aircraft.

**Above: The BAe Alarm anti-radar missile is smaller than most rival weapons, so poses minimum drag penalty for the parent aircraft. It can supplement rather than supplant the normal ordnance load of a strike aircraft.**

spread over a range of frequencies used by the enemy, or selectively focussed on a few key transmissions. Once again, spot jamming is the most effective, since it concentrates the effect of the jamming while leaving most of the frequency band unjammed, and thus free for friendly communications.

Jamming of frequency-hopping radio is virtually impossible given the current state of technology. A hostile elint receiver may be fast enough to detect the short burst of signal on a given frequency which is transmitted before the radio it is trying to intercept hops to another frequency, but this achieves little. The elint receiver has no way of identifying to which of the many frequency-hopping systems active at any one time this tiny burst of signal belongs. The only electronic systems which know the frequencies on which subsequent pulses will be transmitted are the radios themselves. The elint system is faced with a large number of unidentifiable pulses. If several frequency hoppers are active, the signal-discrimination facilities of the elint system may be hopelessly overloaded.

Like radar, IR and electro-optical systems, sonar may be attacked by EW. Since World War II, surface ships have towed acoustic decoys designed to lure homing torpedoes. During the Falklands War, one Royal Navy ship is reported to have lost its towed decoy when the latter was attacked by a "friendly" ASW torpedo.

Submarine equivalents of chaff were used in the early 1940s. These were methods of creating disturbance of the water which would screen the submarine from sonar or at least confuse an inexperienced sonar operator. More recent systems involve sonobuoy-like canisters containing noisemaking equipment.

The US Navy's Project F34-371 is intended to develop improved countermeasures to protect surface ships from torpedo attack, and decoys which will allow submarines to evade hostile ASW vessels. Advanced technology is also being developed to counter perceived future sound-navigation and sonar systems, and also submarine-launched underwater weapons.

**The Sanctuary concept** One method of protecting radars from passive-homing anti-radiation weapons is the Sanctuary concept studied by the US DoD. The radar transmitter would not be located at the radar site but in an aircraft up to 270nm (500km) behind the FEBA, and reporting its position to ground by means of DME (Distance Measuring Equipment) stations. An alternative would be to mount the transmitter on orbiting spacecraft. Being passive, the radar "site" would then be difficult to locate.

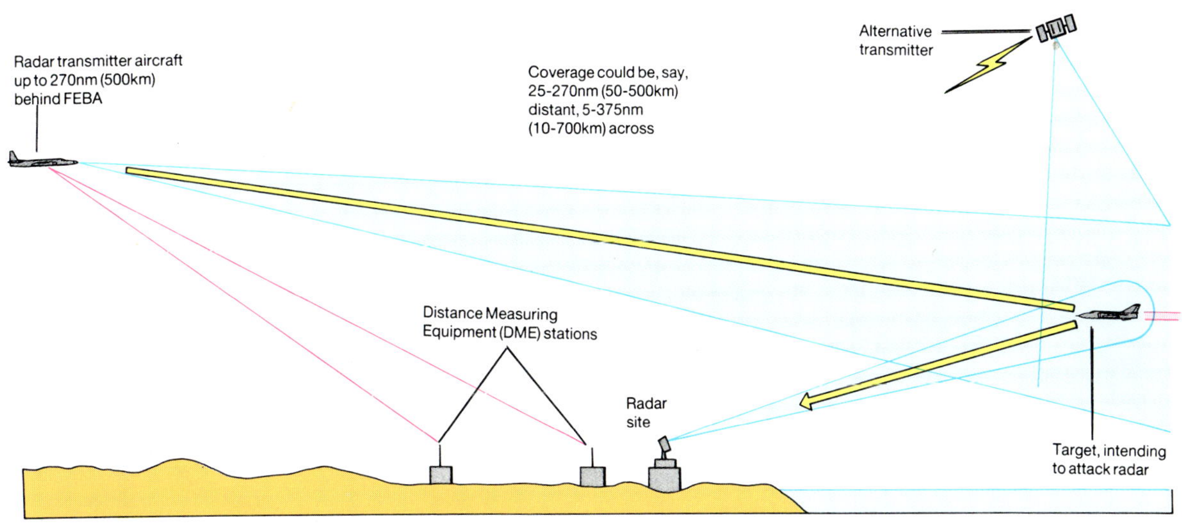

Anyone involved in a hot war and finding an F-4 Phantom diving steeply towards his position might be advised to take cover. A solitary SA-7 is hardly an effective air defence, as the writer discovered in early 1984. Two Phantoms had already flown passes against the sandbagged hilltop position, and a third was "loaded for bear" and on the way in for a dive-bombing attack. The dusty hilltop firing position dominated the surrounding terrain, so the SA-7 operator seemed to have everything on his side. He fired, only to have the gyros of the SA-7 guidance system topple – a common problem when attempting to engage high rate targets. As a result, the missile went rogue.

Pulling out from its dive, the F-4 rumbled off into the distance. As the sound and sight of the attacker vanished, the SA-7 operator laid down his weapon, walked a few yards out of the firing position, opened a door and stepped out into the sunlight of an English summer day.

The great advantage of simulation is that everyone gets to live for another day, hopefully to learn from experience the best way of using weaponry. The Phantom attack had taken place not on the Iraq/Iran border or any similar trouble spot, but in the prototype GQ Air Defence Tactical Training Theatre.

The interior of the dome-shaped trainer is large enough to permit large anti-aircraft weapons such as the four-barrelled ZSU-23-4 Shilka gun to be set up in a realistic firing position. The image of the surrounding terrain is reproduced by an array of slide projectors, while images of attacking aircraft are projected onto the dome by a computer-controlled film projector, the visual appearance, sound and infra-red signature of the attackers being varied according to aircraft aspect and range. The weapon in use is instrumented, and connected to a computerised system which allows instructors to monitor the trainee's performance, and to replay the engagement for assessment purposes.

As the cost of weaponry and ammunition escalates, so does the cost of training. Simulators can help cut the bill, while at the same time making training more realistic.

Simulators have been used for pilot-training since the days of the Link blind-flying trainer, but the latest designs offer full motion simulation, plus computer-generated external views which can reproduce any external scene, hostile aircraft and even air-to-air missiles. The writer recently watched the simulated downing of a MiG-23 by a Sidewinder heat-seeking missile.

More complex equipment in which two pilots may fight against one another in simulated dogfights has been pioneered in the USA. Northrop made use of such a system in the studies which resulted in the YF-17 (developed into the F-18 Hornet) and F-20 Tigershark. Similar equipment has now been developed in Europe by

**Above: British Aerospace's air combat simulator provides the pilot with the forward view and cockpit indications which he would experience in an operational fighter sortie.**

**Below: The French Air Force took delivery of this Mirage 2000 DA mission simulator in November 1984. Simplifying the terrain "features" eases the data-processing demands of view generation.**

British Aerospace as a spin-off from the company's fighter-design studies.

Navies can also make use of simulators to train personnel. Detailed simulations of ships' bridges, operations rooms, and engine-control positions can be used to give trainees initial familiarisation before attempting the real thing at sea, and to reinforce the experience gained during costly sea time.

## Laser-based training

For the training of tank crews, many nations use laser-based training aids. Probably the best-known is Simfire, now used by more than 35 armies throughout the world. Variants suitable for use by tanks, armoured personnel carriers and other AFVs, helicopters, direct fire anti-tank weapons, soft-skinned vehicles and even individual soldiers and their personal weapons have now been developed by Weston Simfire.

All versions may be used in conjunction with an umpire's gun. The latter can be used to check that trainees have not interfered with the system in order to make themselves invulnerable to Simfire "attacks". The umpire may also use his "gun" both to knock out Simfire-equipped targets, simulating the effect of minefields or indirect-fire weapons, and to reactivate casualties, which may return to "battle".

Simulation is even being applied to basic small-arms training. The newly-developed Ferranti Computer Systems "SMART" (SMall ARms Trainer) is intended to teach the basic skills of weapon aiming, holding and firing, and will bring full-bore live firing conditions into the classroom. The system uses a full-scale mock-up of the weapon, plus a large cathode-ray tube (CRT) screen used to display the target.

"SMART" uses the same principle as the "light pens" often used to input data to a computer CRT display. The latter are normally brought into contact with the screen but, in the case of SMART, the "stand-off" distance is several feet. As the electron beam scans the CRT face in order to build up the raster-scanned image, a weapon-mounted sensor is able to detect the moment at which the beam passes through its tiny field of view. By relating this with a timing waveform which controls the CRT scan pattern, the system is able to deduce the exact point on the CRT face to which the sensor and weapon are directed. Other sensors on the weapon monitor the first and second trigger pressures, butt/shoulder pressure and weapon verticality.

Targets can be scaled to simulate engagement ranges of just over 50 to almost 330 yards (50 to 300m), and the effects of day and night lighting conditions, mist, rain, smoke, and even hostile fire may be reproduced, along with the flag or smoke wind indication which the trainee would see on a genuine range. The sound of gunfire can be reproduced in stereo via a pair of headphones worn by the trainee, while the effect of recoil is reproduced by a mechanism using high-pressure compressed air fed to the gun via a flexible umbilical cable.

Ultimate "war games" are those practiced by NATO air arms – particularly the US services. To make US Air Force Red Flag air combat training more realistic, General Dynamics has devised replicas of Soviet anti-aircraft systems such as the SA-7 "Grail", SA-8 "Gecko", and SA-9 "Gaskin" anti-aircraft missiles and the ZSU-23-4 Shilka gun. These are not only visually correct, but also emit the appropriate radar signals.

Air combat ranges such as the NATO installation in Sardinia are the ultimate "computer game". The Air Combat Manoeuvring System involves a series of ground stations which receive data from special instrumentation pods fitted to aircraft participating in exercises, passing this to a central computer. After an exercise, the combat may be replayed on a 4ft×4ft ($1.5m^2$) video screen, allowing aircrew to analyse what went wrong – or right.

The replay may be stopped, speeded up, viewed from any angle, or any aircraft cockpit, with the electronic systems simulating the action, complete with missile firings, while replaying the R/T chatter. It may be an expensive way to train, but it ensures that NATO pilots have the realistic "combat" experience needed in order to maintain a qualitative edge over their numerically superior Warsaw Pact rivals.

**Left: Tank crew training can also be partly based on simulation. This M1 Abrams gunner is being shown a computer-generated "landscape" complete with buildings and manoeuvring targets.**

**Below: The USAF's Air Combat Manoeuvering System looks like the ultimate video arcade game, but serves a serious purpose – recording and replaying actual air-to-air sorties.**

# STRATEGIC WARFARE

For several decades, the nuclear forces of the East and West have co-existed in a mutually perceived nuclear balance. Strategic theorists in the United States even coined the acronym MAD (Mutual Assured Destruction) to describe US perception of the balance. A combination of Soviet force improvements and emerging technology now manages to upset this delicate situation, introducing new uncertainties into strategic planning, and perhaps increasing the risk of nuclear war.

The United States bases its strategic forces on the concept of a triad – a mix of land-based intercontinental ballistic missiles (ICBMs), submarine-launched ballistic missiles (SLBMs), and manned bombers. Although expensive, the triad concept offers flexibility, with each component possessing its own unique strengths and weaknesses.

The Soviet Union currently fields most of its strategic nuclear firepower in the form of ICBMs, with the SLBM force playing a secondary role. Having failed to develop effective long-range bombers, the Soviet Union has only a nominal strategic-bomber force, a situation which may change dramatically when the new Tupolev "Blackjack" enters service.

The only other nation to have developed and fielded its own "triad" is France. Britain has scrapped its bombers, and no longer has the capability to design or flight-test long-range ballistic missiles. The UK's "independent" deterrent now depends totally on the purchase of US missiles and equipment. China currently fields ICBMs and a token bomber force, but has SLBMs under development.

Manned bombers are flexible, recallable, but relatively slow. Highly-vulnerable to attack when on the ground, once committed to action they may have to fight their way through the enemy defences in order to reach their targets. Successful manned bomber operations in the face of the Soviet air defences requires a mix of penetrators and stand-off missile carriers. However, unlike ballistic missiles, they can be used to "show the flag" in support of national or international interests, and may be sent into action in the conventional role to provide massive fire-support.

**Below: The first B-1B swing-wing bomber of an eventual 100 which will bolster the aerial leg of the United States' strategic triad, the other two comprising land-based and sea-based missiles.**

In the late 1970s most defence experts would have predicted the imminent demise of the bomber, pointing to the cancellation of the B-70, B-1A, and several Soviet designs including a B-70 "look-alike" developed by the Sukhoi bureau, and a development of the Tupolev Tu-144 "Charger" supersonic airliner. Few would have dared to predict that in the less than a decade, the USAF Strategic Air Command would be planning to deploy not one but two new models of bomber.

Evolution of what is widely termed "stealth technology" has enabled designers to reduce the radar and infra-red signature of large strategic aircraft to the point where they become difficult targets for hostile radars and IR seekers. Stealth aircraft will thus be able to operate in the face of radar-based defence systems or the new-generation passive sensors now being developed for anti-aircraft use (such as the British Aerospace passive IR tracker being tested with the Rapier point-defence missile system).

Older bombers such as the US Air Force's B-52, whose ability to penetrate the Soviet Union's modern massed air defences and survive is now dropping off badly, will be used as stand-off cruise-

**A strategic balance has existed, rather uneasily, for some years, but modern developments – not confined to the "Star Wars Initiative" – threaten to create a high degree of instability, to everyone's disadvantage. Of especial danger would be a loss of invulnerability of the submarine-based second strike counter-value forces.**

missile carriers, with only the new B-1B and the "stealth bomber" being committed to the penetration role. A similar pattern is likely in the Soviet Union, with the new "Bear-H" acting as cruise-missile carrier back-up to the "Blackjack" swing-wing bomber.

US and Soviet land-based intercontinental ballistic missiles (ICBMs) are installed underground in vertical silos – concrete shafts capable of withstanding blast pressures of up to 2,000lb per square inch in the case of US designs, and greater pressures in the case of the latest Soviet types. These concrete structures are intended to protect the missile from all but direct hits by a nuclear weapon.

Given the accuracy of the first- and second-generation Soviet ICBMs, the silo concept was thought feasible when conceived, with both superpowers possessing ICBM forces which would survive in considerable numbers any attempt by the other side to launch a disarming first strike.

However, the exact location of missile silos can be determined by reconnaissance satellites, and attacking ICBMs targetted with great accuracy, starting their flight with virtually no built-in navigational error. This disturbing fact introduces a new and dangerously destabilising element in the design, deployment and control of strategic forces.

Above: The counter-force element of the triad is the land-based intercontinental ballistic missile (ICBM); seen here is a Minuteman III launch from Vandenburg.

Left: The sea-based deterrent is the submarine-launched ballistic missile (SLBM). The current US front-line system is this Trident 1 C-4 missile.

Like ICBMs, SLBMs are one-shot devices which cannot be recalled once launched, but have the unique advantage that their submarine launch platform is very difficult to locate given the current and projected levels of ASW technology. The highest accuracies possible with land-based missiles are not achievable with current SLBMs, since the latter start their flight from a mobile launch platform whose coordinates cannot be measured to the same degree of precision as those of land-based launch sites. New technology such as MARVs (Manoeuvrable Re-entry Vehicles) will eventually alter this situation, making SLBMs as accurate as land-based missiles.

Command and control is also a problem. Since the submarine is an autonomous weapon platform whose crew must be entrusted with the nuclear payload and the means of launching it, the complex command and control arrangements possible on land cannot apply to a freely roaming submarine. Silos may be linked by radio and land line to one another and to higher command authorities, but reliable long-range communication with a submerged submarine is still difficult, since most radio frequencies have virtually no ability to penetrate water.

## Soviet strategic build-up

During the 1970s, the US Government decided to restrict the future development of its nuclear forces, a move apparently carried out in the hope that the Soviet Union would reciprocate. In practice, the Soviets continued to develop and field new systems. The last new pattern of ICBM to be deployed by the USAF was the Minuteman III, fielded in 1970. In the years which followed, four new types of Soviet ICBM were fielded, and seven major modifications introduced to those systems. During the same time period, the sole USAF missile update programme saw only half the Minuteman III force modified.

Between 1967 and 1980, the Soviet Navy built and deployed some 70 new ballistic-missile submarines (SSBNs) carrying a total of around 900 missiles. During the same time period, the only new US Navy SLBM was the Trident, initially fielded aboard modified Poseidon submarines. By the time that the first new USN SSBN – the 24-round USS *Ohio* – was launched, the Soviets were already introducing the new Typhoon class, which eclipsed even the Ohios in size.

The long range of the SS-N-18 and -20 missiles carried by Delta III and Typhoon class SSBNs allows these vessels to target most of the continental USA from patrol areas within Soviet coastal waters or beneath the Arctic icecap. SSBNs cannot fire from beneath thick ice, but US Intelligence believes that the Soviet Navy has practised the technique of launching SLBMs through the smaller and less frequent patches of the thinner ice often found in the Arctic.

The decision early in 1984, to station Delta class boats in mid-ocean patrol areas was a move intended to offset the NATO deployment of USAF Pershing II and Ground-launched Cruise Missiles (GLCM) in Western Europe. Like the Pershing II, Soviet missiles fired from the new patrol areas will have a relatively short time of flight, giving the target much less warning time in which to react.

By the late 1970s, this massive re-equipment programme by the Soviet Strategic Rocket Forces had upset the strategic balance. During flight tests monitored by US Intelligence, Soviet third-generation ICBMs demonstrated a circular error of probability (CEP) of about 500 to 600 yards (450 to 550m) instead of the figure of more than 1,100 yards (1,000m) associated with earlier missiles. CEP is a term used by missile engineers to express the accuracy of a long range missile, and is expressed as the radius of a circle within which half of the rounds fired at a given target will fall. All rounds whose guidance and propulsion systems function correctly should deliver their warheads within a circle of around four times the CEP value. Considering the CEP figures demonstrated by recent Soviet IBCM flight tests, the theoretical vulnerability of the USAF Minuteman force has risen dramatically.

Soviet ICBM silos and other launch facilities are now hardened to what the US Department of Defense (DoD) terms "an unprecedented degree – far above the strength of our Minutemen silos". Such hardening is usually presented as evidence of Soviet nuclear warfighting capability – the prospect that it simply represents good silo design is rarely voiced.

An alternate – and perhaps equally valid – viewpoint is that the USAF was complacent in its threat assessment when the Minuteman silos were designed. Given the high degree of accuracy which the Minuteman would offer, and still higher accuracy promised by projected guidance improvements, the least that planners could have assumed was that equally accurate Soviet ICBMs would one day be fielded. Considering the Soviet tendency to use larger warheads than those on US strategic missiles, failure to predict the threat now being posed to US silos seems incompetent to say the least. The USAF's missile silos could have been built as "hard" as their Soviet counterparts but were not.

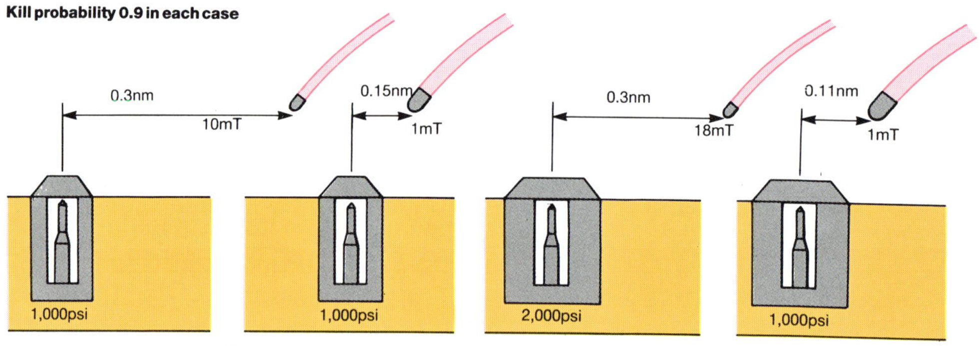


**Anti-silo missions**
The ICBM counter-force role requires them to destroy the enemy's ICBM force which, in all current cases, is located in hardened silos. The probability of a kill depends upon two variables: warhead yield and accuracy. As shown here accuracy is the most important; to achieve a 90 per cent kill probability on a 1,000psi silo, for example, requires a 10mT weapon at 0.3mm, but at half the distance (0.15nm) a 1mT weapon is required.

**Above: Work on a Mark 12 re-entry vehicle (RV) on a Minuteman III in its silo. The protective cone is ditched in space to expose individual nuclear warheads.**

**Below: US ICBM warhead is moved under close guard in the missile farm. All nuclear powers are very conscious about the security of their nuclear weapons.**

Protected by hardening measures, and facing a minimally-improved US strategic force, the Soviet Strategic Rocket Forces found themselves with a growing capability – at least on paper – to attack and largely wipe out the USAF ICBM force while using only a portion of their own missiles – the relatively heavy SS-18 and SS-19. A Soviet first strike could in theory destroy a large portion of the US ICBM force, while leaving the Soviet strategic force with sufficient firepower to retaliate against any counter-strike by the weakened US missile force.

The outcome of an attempted first strike is difficult to assess. Test flights involve the use of carefully surveyed missile ranges whose gravitational field is well known. Since the Earth is of slightly irregular shape, the gravitational effects on rounds fired on operational over-the-pole trajectories can only be theoretically assessed so, as far as the war planners of both sides are concerned, predicted combat effectiveness of ICBMs remains an unproven theoretical figure.

## "Smart" re-entry vehicles

This situation is unlikely to change until nuclear-armed "smart" re-entry vehicles (RVs) are deployed on ICBMs and SLBMs. Like that carried by the Pershing II, such payloads would have a terminal seeker able to correct the trajectory in the final stages of flight, compensating for errors in ballistic guidance, and giving ICBMs and SLBMs a CEP measured in just tens of metres. Given this class of accuracy, silo hardening against nuclear attack becomes a meaningless concept – the silo entrance would be within the fireball of the nuclear

**Above: One-eighth model of a US Peacekeeper silo which successfully survived a TNT simulation of an attack by the USSR's largest nuclear warhead.**

**Below: RVs being positioned on the warhead bus of a Peacekeeper ICBM. Each missile can carry up to ten such RVs each of which can be separately targetted.**

**Right: Time-exposure photograph of eight Peacekeeper RVs passing through clouds prior to impact in the south Pacific Ocean in a 1984 test.**

explosion of the RV's warhead. Hardening will still protect silos from the effects of nearby explosions or from warheads whose terminal homing system had failed, but would be useless against any enemy who had aimed two or more RVs against each target to insure against such failures.

Neither superpower has announced plans to deploy MARVs. The US company Martin Marietta once proposed a MARV payload which could be fitted to Titan but, as the Pershing II programme shows, such systems can now be packed into small RVs. The Mk500 Evader MARV once proposed as a Trident payload was designed to evade defensive attacks by anti-ballistic missiles (ABMs) and had no terminal-homing capability, but the creation of a "smart" strategic RV is well within US technological capability. The USAF and USN are already collaborating on tests of the Terminal Fix Guidance System, and will probably fly experimental hardware on obsolete Minuteman I boosters. The development of next-generation VHSIC (Very High-Speed Integrated Circuits) electronic technology will make MARVs even more feasible by reducing the amount of missile throw weight which must be diverted from nuclear explosives to missile guidance.

Flight testing of "operational" MARV payloads by the Soviet Union is likely within the next five years, says US Intelligence, but it is not clear whether these would have Mk500-style anti-interception capability or true terminal homing. The greater throw weight of Soviet missile designs should in theory make the installation of an RV guidance and control system more easy than on smaller US types, but this advantage may be partially eroded if the Soviets are unable to match US advances in micro-miniaturisation.

The wisdom of the 1975 decision by the US Congress to de-activate the Safeguard ABM system must now seem questionable. Seldom in the history of weapon development has a major weapon system been cancelled literally one day after being declared operational.

The problems posed by likely Soviet decoys and other penaids (penetration aids) were severe, but the fact that the Spartan and Sprint missile systems were accepted for production and deployment suggests that the Soviet decoy devices were not considered insuperable. Given the improvements in performance which digital signal processing has yielded in retrofit programmes for other types of radar, it is hard to believe that the retrofitting of new signal processing systems based on modern micro-miniaturised electronics would not have improved system discrimination.

Nuclear weapons are probably the most efficient weapons of mass destruction ever invented. Only biological weapons threaten greater loss of life from an equivalent payload. The most commonly discussed effects of nuclear explosions appear to be "fall-out" and radiation sickness, but in practice some 65 per cent of the energy output of most patterns of nuclear weapon is released as blast and shock (50 per cent of the total) and heat energy (15 per cent of the total).

Only around 15 per cent of the total output of the explosion takes the form of nuclear radiation, a mixture of radioactive byproducts of the explosion and a "flash" of nuclear radiation – alpha and beta particles, gamma rays, protons and neutrons. Like the heat and light from the explosion, this radiation lasts for only a short time, and is known as "immediate radiation".

Some time after the explosion, the radioactive byproducts – some two thirds of the weapon's radioactive output – will fall to Earth, contaminating the terrain around and downwind of the target area – the phenomenon known as "nuclear fall-out". If the weapon is exploded near ground level, particles of earth and other material will be swept up into the nuclear fireball and become radioactive, supplementing the fall-out created directly by the nuclear explosion.

Recent reports have suggested that the vast clouds of dust particles thrown up into the upper atmosphere as a result of any large-scale nuclear war would remain there for many decades, reducing the amount of heat and light from the sun which reaches the Earth's surface. The intensity of the resulting "nuclear winter" – as this situation has been called – has been the subject of speculation and debate, but must in the long term be difficult to predict – one more factor to weaken the case for proponents of the case that nuclear wars may be "winnable".

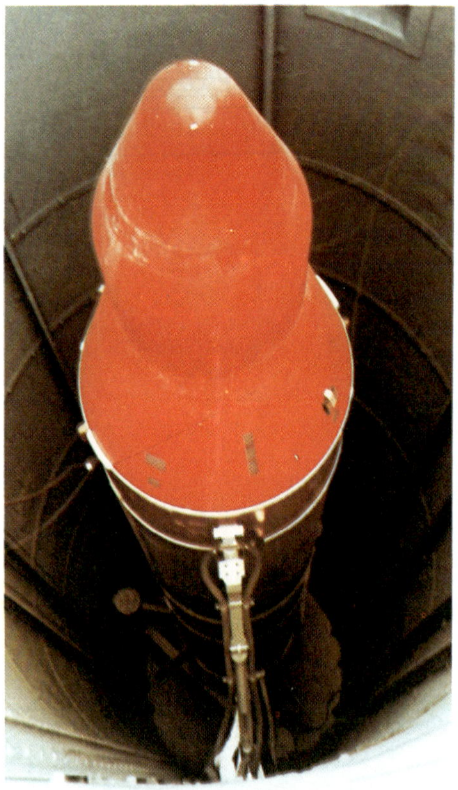

**Above: French SSBS S-3 in its silo; France possesses 18 such MRBM launchers as part of the *force d'dissuasion*. This system has been developed entirely within national resources with great skill and determination, but at great expense.**

**Below: A unique US test in which 600 tons of ammonium nitrate/fuel oil were exploded to simulate a 1 kiloton (1kT) nuclear airburst. The four plumes on the right were triggered prior to the main test in supporting experiments.**

Blast and heat produce most of a nuclear weapon's destructive power. Far from being militarily useful, radiation from fall-out may pose problems to the user of the weapon, particularly if he wishes to send his own forces forward to pass through or even occupy the target area. In order to minimise this, military commanders could attempt to use the smallest size of nuclear weapon capable of producing the desired destruction, perhaps arranging for the weapon to detonate high above the target – a technique known as "air-burst" – to ensure that the fireball does not touch the ground and thus generate additional fall-out.

## Fission and fusion

Most nuclear weapons designed for tactical applications are *fission* devices, deriving their energy from the fission or splitting of atoms of a heavy element such as uranium-235 or plutonium-239. Maximum yield attainable is several hundred kilotons, the level of radioactive fall-out produced – mostly xenon-140 and strontium-94 – being proportional to the power of the weapon. Above this size it is near-impossible to keep the mass of fissionable material intact for long enough to ensure that a sufficient amount takes part in the nuclear chain reaction. The energy liberated as the fission process gets under way in more and more of the material tends to disrupt the latter before nuclear reactions are fully active throughout its mass.

To create weapons of larger yield, designers use a different reaction – the *fusion* of hydrogen atoms to create helium. This gives rise to the popular designation "hydrogen bomb" for such weapons. Practical weapon designs use two isotopes of

hydrogen known as deuterium and tritium. Fusion is a more efficient process than fission. The complete fusion to helium of only 12 grammes of deuterium or tritium would result in a yield of around one kiloton. To achieve the same effect using fission would require around 56 grammes of plutonium-239.

Fusion reactions can take place only at very high temperatures which are normally above those which can be created in the laboratory. The only practical way of creating such temperatures in an operational weapon is by using a fission device. The use of fission promptly makes the resulting explosion "dirty" due to the radioactive byproducts mentioned earlier. Most fission weapons also use uranium to reflect and multiply the neutrons created by the nuclear reactions, and this too adds to the resulting fall-out.

Nuclear scientists consider that one alternative to a fission "trigger" might be a high-powered laser. Faced with the prospect of smaller nations bypassing the conventional route to nuclear status via elaborate and costly fission-weapon programmes, and creating a laser-triggered "poor man's H-bomb", the US Government has placed heavy restrictions upon high-powered laser technology.

Only about five per cent of the output from a fission weapon takes the form of immediate radiation, but during the 1970s work carried out at the Livermore Laboratory in California, one of the two US centres for nuclear weapon design, showed how more than 30 per cent of the yield of a weapon could be delivered in the form of immediate radiation. By releasing most of their energy in the form of immediate radiation rather than as heat or blast, such weapons would allow localised strikes against armoured formations while minimising civilian casualties in the surrounding area. The resulting Enhanced-Radiation Weapon (ERW) – a highly specialised type of fusion weapon – is commonly known as the "neutron bomb".

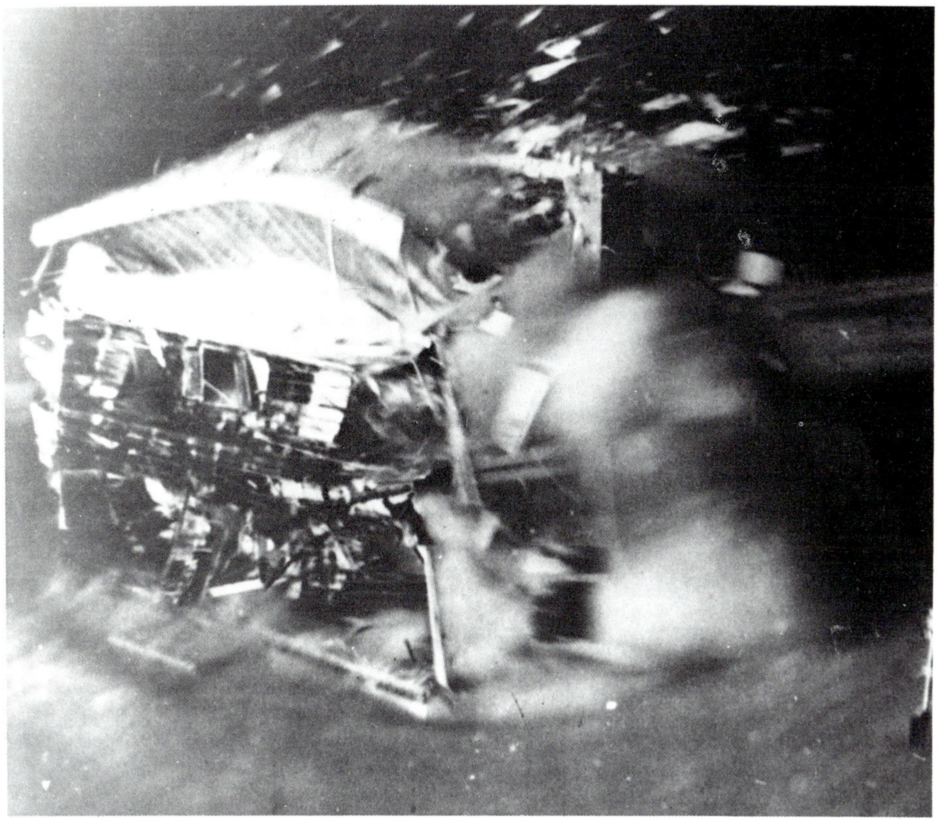

**Above: A typical American timber house is torn apart by the blast from a 15kT nuclear weapon in a test; a concrete basement remained intact. The effects of nuclear weapons have been very carefully assessed in a series of such tests.**

**Below: For many years heavy ICBMs, poised in their silos, have been the primary counter-force weapons systems. But as these are not now so invulnerable as they once seemed there is a definite trend towards smaller, lighter, mobile ICBMs.**

Neutrons produced by fusion reactions are much more energetic than those resulting from fission. In theory, a fusion weapon need be only about five per cent of the power of a fission weapon in order to produce the same neutron energy. In practice, the need for a fission trigger to initiate the fusion reaction, coupled with the absorption of neutrons by the bomb material, does introduce some penalty but, even so, a fusion weapon need be only some 20 per cent of the yield of a fission weapon to create the same neutron output.

Designers of ERWs must use the smallest-possible fission device to trigger the fusion reaction, and substitute another material for uranium in the reflector. The goal must be to develop as much of the energy as possible by fusion rather than by fission. At worst, some 50 per cent of the energy of an ERW explosion would be obtained from fusion, a figure which could be as high as 75 per cent in many practical weapon designs.

Fission may produce the biggest nuclear explosion, but nuclear fusion causes the long-lasting nuclear fall-out. The use of neutron weapons reduces the amount of thermal energy, blast and fall-out from a tactical nuclear explosion, but the issue of whether such weapons make nuclear war more likely is still under debate. If the prospect of lower casualties in areas closer to the explosion makes neutron weapons easier to use, argue proponents, such weapons represent a greater threat to a potential enemy – and thus are a better deterrent. Opponents have a simpler argument: if neutron bombs are easier to use, the chances that they *will* one day be used must be greater.

Early concepts of nuclear war planning were crude – the goal was simply to deliver as many warheads as possible in the shortest time possible. Such "nuclear spasm" plans have now been replaced - in the West at least – by a series of alternative actions of varying intensity. For the US strategic forces, these are specified in the SIOP (Single Integrated Operational Plan). The US President now has a series of options from which he may choose, but the reliable implementation of these by a command and control system which is vulnerable to attack remains debatable at present. Modifications are now being carried out to ensure that the US command, control and communications system remains capable of performing its basic functions throughout any potential sequence of Soviet attacks.

In the mid-to-late 1980s, the US DoD will upgrade the ground-based National Military Command Center (NMCC) in the Pentagon (Washington D.C.) and the Alternate NMCC located at Fort Richie, Maryland, using improved information-processing equipment, and enhanced protection against EMP (electromagnetic pulse). (These can devastate modern solid-state electronics by damaging or even totally destroying the microelectronic structure of semiconductor devices such as integrated circuits – Electronics section.) Equipment located at the North American Air Defense (NORAD) Cheyenne Mountain Complex is being modernised to handle the

**Above right: Command, control and communications ($C^3$) systems are essential for effective deterrence; this is NORAD HQ inside Cheyenne Mountain, Colorado.**

**Right: One sign of superpower mutual understanding is the "hot-line" between the leaders of the US and the USSR; this is the Washington terminal in operation.**

**Below: A Boeing E-4B National Emergency Airborne Command Post (NEACP) is refuelled from a Boeing KC-135 tanker during a routine deployment.**

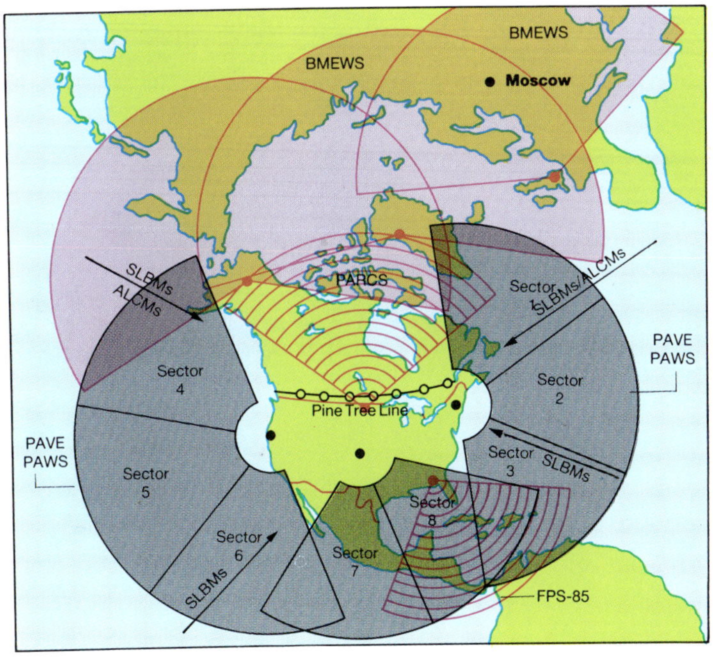


**US ballistic missile warning system**
Satellites over the Indian Ocean sense Soviet ICBM launches within 90 seconds and instantly warn ground stations at Guam and near Alice Springs, Australia. BMEWS then takes over to track and identify missiles Coming from the north. PARCS identifies the number of RVs and predicts impact sites. Satellites and the PAVE PAWS phased array system warn of SLBM and ALBM attack and the FPS-85 site in Florida gives further coverage of the threat from the Caribbean and FOBS attack from the south. The USN's Space Surveillance System (SPASUR) stretches across the southern USA. Despite these elaborate systems there are still serious and dangerous blind spots.

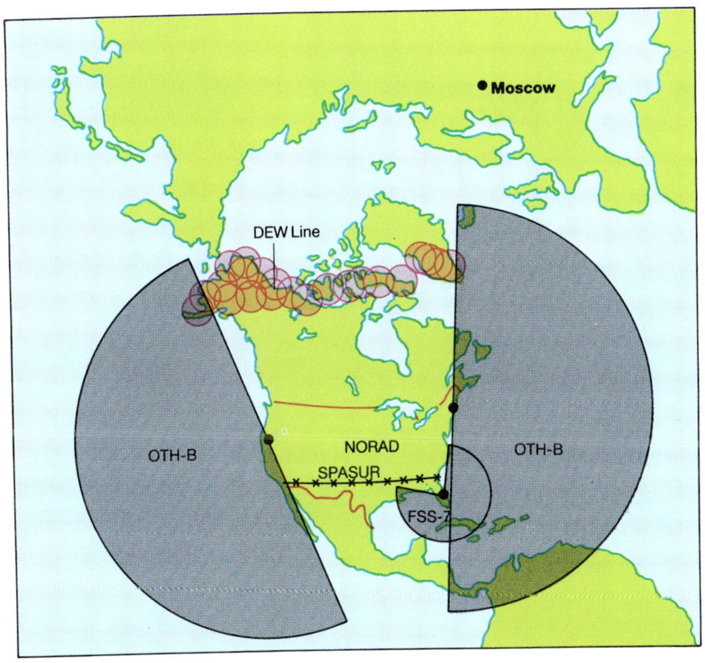


**US detection of manned bombers and ALCMs**
Because of the increasing threat from low-flying bombers and ALCMs the US has become concerned about the inadequate air defences of continental United States (CONUS). Experiments with Over-the-Horizon Backscatter radar (OTH-B) have proved successful and it is now planned to install eight stations each giving 60 degree coverage, which will give all-altitude coverage of the eastern, western and southern approaches to CONUS. Northern coverage is currently given by the DEW Line, but OTH-B is not suitable to replace in this area. AWACS surveillance aircraft and satellites supplement the ground-based coverage, but overall inadequacy is self-evident.

command, control and communications requirements unique to space operations. Although useful during normal crisis-management, and of obvious service during the early stages of a nuclear attack, these land-based facilities are likely to be high-priority targets for the Soviet SS-18 ICBM force.

Designed to provide strategic communications plus data processing and display facilities to National Command Authorities and subordinate commands, the US WorldWide Military Command and Control System (WWMCCS – usually pronounced "Wimex") gives information on the location and status of forces. Over the last few years, WWMCCS has acquired a reputation for unreliability, but these problems should be corrected by the WWMCCS Information Systems Program, an extensive and evolutionary upgrading of the total system.

Battle management of a nuclear exchange would largely be handled by airborne command posts. All four Boeing E-4B National Emergency Airborne Command Post aircraft are being fitted with improved data-processing equipment, and SHF satellite communication (satcom) terminals; by the end of FY85 the aircraft will also have been hardened against the effect of the powerful EMP generated by high-altitude nuclear explosions. EC-135 aircraft used as airborne command posts by the Commanders-in-Chief of Strategic Air Command and other nuclear forces are also being improved with EMP protection and better communications equipment.

## Satellite communications

Low-rate data communications to the US strategic forces, and two-way teletype communications for strategic-force management are provided by the Air Force Satellite Communications (AFSATCOM) system, which has been operational since 1983. Various military comsats such as the USN Fleet Satellite Communications (FLTSATCOM) system and Defense Satellite Communications System (DSCS) Phase III satellites carry AFSATCOM transponders, while ground terminals for the system are deployed throughout the US forces and command and control networks.

The DSCS-III spacecraft replace the earlier Phase II satellites. The first was launched in 1983, and the entire system is due to become operational in the mid-1980s. Because they use SHF as well as UHF communications, DSCS-III satellites are more jam-resistant than the Phase II spacecraft or the AFSATCOM system. They are used to pass missile attack warning data from the satellite early-warning system to command centres, and to provide a redundant link from the E-4B fleet to the strategic forces.

Even less vulnerable to jamming and the effects of nuclear detonations is the planned Military Strategic and Tactical

**Left: Distant Early Warning radar site in Greenland. Such sites have been essential to the security of the USA and Canada for many years, and the Soviet Union has similar systems.**

Relay (MILSTAR) satellite system. These spacecraft, currently in full-scale development, use EHF links, and incorporate survivability features intended to ensure their availability in any nuclear war. They will be used to provide two-way communications between commanders and their forces.

In the early stages of a missile attack, US commanders would be able to communicate with bomber and ICBM forces using the Ground Wave Emergency Network (GWEN) currently being deployed and tested. This consists of a network of EMP-hardened LF relay stations, each taking the form of a radio tower housing transmitters, receivers and a self-contained power supply. The basic system was due to be completed in FY85, while the entire system – with many more stations – should be available for service in the late 1980s.

Once the USAF's bombers are airborne, a reliable communications link will be provided from 1987 onwards by VLF-band Miniature Receiver Terminals mounted in each aircraft. Although these can handle only a slow data rate, they have a much longer range, plus a lower susceptibility to nuclear effects, than conventional UHF links.

Links to the US Navy's SSBN fleet are currently handled by land-based and airborne communications systems. Land-based ELF systems require very large antenna arrays and transmit massive levels of RF power – features which attract much hostile criticism from environmental pressure groups. After a long struggle, the USN now plans to deploy a two-site system with ground stations at an existing VLF test installation in Wisconsin, and at another site in Northern Michigan. Installation of ELF receivers in US Navy SSBNs should be under way by 1986.

This land-based system is backed up by airborne VLF facilities installed on the EC-130 TACAMO version of the Hercules. The designation TACAMO stands for "TAke Charge And Move Out" – a somewhat obscure means of designating a strategic communications system. Plans to replace these aircraft with E-6As based on the airframe of the E-3 Sentry were briefly postponed in the mid-1980s after a study showed that costs of such a scheme would be high, but despite studies of alternatives such as a modified EC-130, the programme was soon re-instated. The first E-6A will fly in FY86, and a fleet of 15 is planned by the early 1990s. The first squadron will be deployed in the Pacific around the end of the decade, followed by an Atlantic squadron some two years later. The aircraft will start its operational career carrying electronic systems removed from the EC-130s, but will receive new systems in the early 1990s.

All US nuclear weapons are fitted with the Permissive Action Link (PAL) – an electronic lock which prevents the weapon being detonated unless a pre-arranged numerical code is inserted. Used in conjunction with control and launch systems which require the co-operation of several individuals, this precaution is designed to prevent the unauthorised use of US nuclear weapons.

**Above: Aerial view of the above-ground antennas of the US Navy's Extremely Low Frequency (ELF) test facility at Clam Lake, Wisconsin, used for low data-rate signals to submerged submarines.**

**Below: USS *Michigan*, an Ohio-class SSBN, running on the surface. SSBNs must spend their entire patrol submerged, giving good security, but leading to severe communications difficulties.**

**Below: A retouched photograph showing the appearance of the proposed E-6 TACAMO a modified E-3 airframe, which has been selected to replace the EC-130. Fifteen are on order from Boeing.**

Above: Lockheed EC-130 TACAMOs are constantly on patrol and are essential elements in the strategic communications system to submerged submarines.

Below: Operators inside a patrolling EC-130 TACAMO. Signals to submerged SSBNs are by Very Low Frequency using an extremely long trailing-wire antenna.

## Soviet command and control

Less is known concerning the Soviet command, control and communications system. Ultimate control over Soviet nuclear weapons lies with the Politburo, with the General Secretary of the Communist Party probably acting as Commander-in-Chief. Ground-based command centres are located within an approximate 80-mile (130km) radius of Moscow, and are hardened to withstand overpressures of up to 1,000psi (70kg/cm$^2$).

Flying command posts based on the Ilyushin Il-62 airliner and Il-76 "Candid" military transports are known to exist, but these almost certainly lack the comprehensive data-processing and display functions of their USAF equivalents.

Since its northern regions are vulnerable to HF radio interference, the Soviet Union makes large-scale use of VLF radio and communications satellites. Communications with the submarine force, surface ships and long-range aircraft are handled by VLF radio. More than 25 Soviet stations have power outputs of more than half a megawatt.

The Soviet military have their own system of several dozen small comsats, and may also use the Molniya (multi-purpose), Statsionar (TV/communications) or Volna (maritime) communications satellites. The existence of military comsats in geosynchronous orbit seems likely, but has not been confirmed.

Operational control of nuclear warheads is shared between the military and the KGB. As a result, the Soviet strategic nuclear forces are generally at a lower state of readiness than their US counterparts. Precautions against unauthorised use of nuclear weapons are probably similar to those of the USA. If any Soviet equivalent of the US PAL system has been deployed, this is probably under the control of the KGB.

In a world where the term "guided missile" is commonplace, it may come as a surprise to many people that, unlike shorter-ranged homing missiles, the payload of a long-range ballistic missile is normally unguided for most of its flight path. Once the missile booster has burned out, and the warhead-armed re-entry vehicles have been released, these follow a pre-planned trajectory from release point to target. With the earlier weapons, the task of the booster was to bring one or more warheads onto the correct course and speed, making final adjustments either with low-power vernier motors or small solid-propellant thrusters. In the case of a missile fitted with multiple re-entry vehicles (MRVs), a simple release mechanism gave the individual RVs the slight velocity increment needed to produce the planned spread of impact points.

With the introduction of MIRVs (Multiple Independently-targetable Re-entry Vehicles), a more complex system was required. The individual re-entry vehicles (RVs) are mounted on a powered and guided "bus" - virtually a final stage of the launch vehicle. Once the booster has brought the bus onto the correct velocity for impact in the general target area, the bus uses its thrusters to refine the aim on to the target for the first RV. Once this is correct, one RV is released, and the bus begins the task of trimming its trajectory to match the next target. A second RV may then be released, and the process continued until the full number have been released on their correct trajectories.

The bus can also release decoys intended to confuse enemy ABMs, plus chaff, jammers or other EW systems. These are known as penetration aids – often abbreviated to "penaids". Decoys intended to confuse radars may be relatively simple devices which simply need to match the radar cross-section of the real RV. Latest advances in ABM technology now allow infra-red systems to attempt the task of discriminating decoys from RVs, so the decoys must also attempt to match the thermal signature of real RVs, to continue the deception as long as possible.

In the past, Soviet booster technology has severely lagged behind that of the USA. First-generation missiles on both sides used liquid propellants, initially of the non-storable variety. Cryogenic propellants such as liquid oxygen were difficult to store, and to pump into the missile during the rapid pre-flight countdown, but offered the high performance needed for intercontinental-range missions. While the Soviet Union opted for storable liquid propellant for its second-generation weapons and for its SLBM force, the USAF adopted this combination only for the Titan II ICBM, preferring to use solid-propellants for the Minuteman ICBMs and the US Navy's SLBMs.

Solid-propellant rocket motors are simpler than liquid-propellant motors, and easier to store and maintain at launch readiness. Problems of thrust-vectoring for control purposes, and of thrust-termination at the end of powered flight, were more difficult than those associated with liquid propellant, but US industry soon solved these. With the exception of early

**Ballistic missile attack**
A ballistic missile flight has four phases. In the boost phase first and second stage engines burn, followed by the post-boost phase in which the bus deploys the multiple warheads and "penetration aids". Third is the mid-course phase in which these various objects cruise through space, prior to re-entry and final descent to the target in the terminal phase.

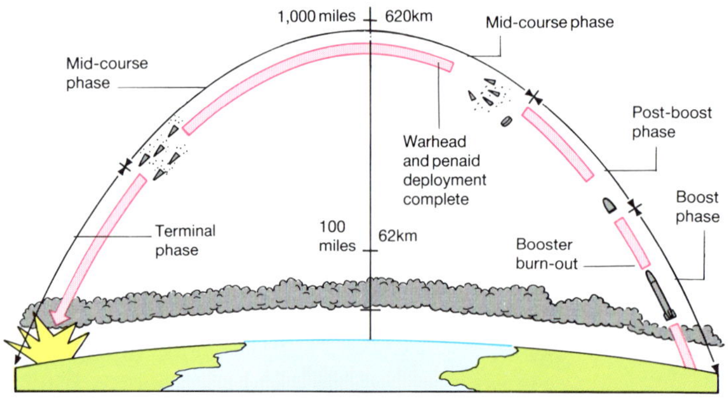

**Below: The post-boost phase of a Soviet ICBM. As the front-end covers and skirt fall away, the bus then deploys MIRVs and penaids according to a programmed plan.**

**Right: A Soviet MRBM blasts off on a "hot" launch, showing the great amount of efflux that has to be dissipated. "Cold" launches require less clearance in silos.**

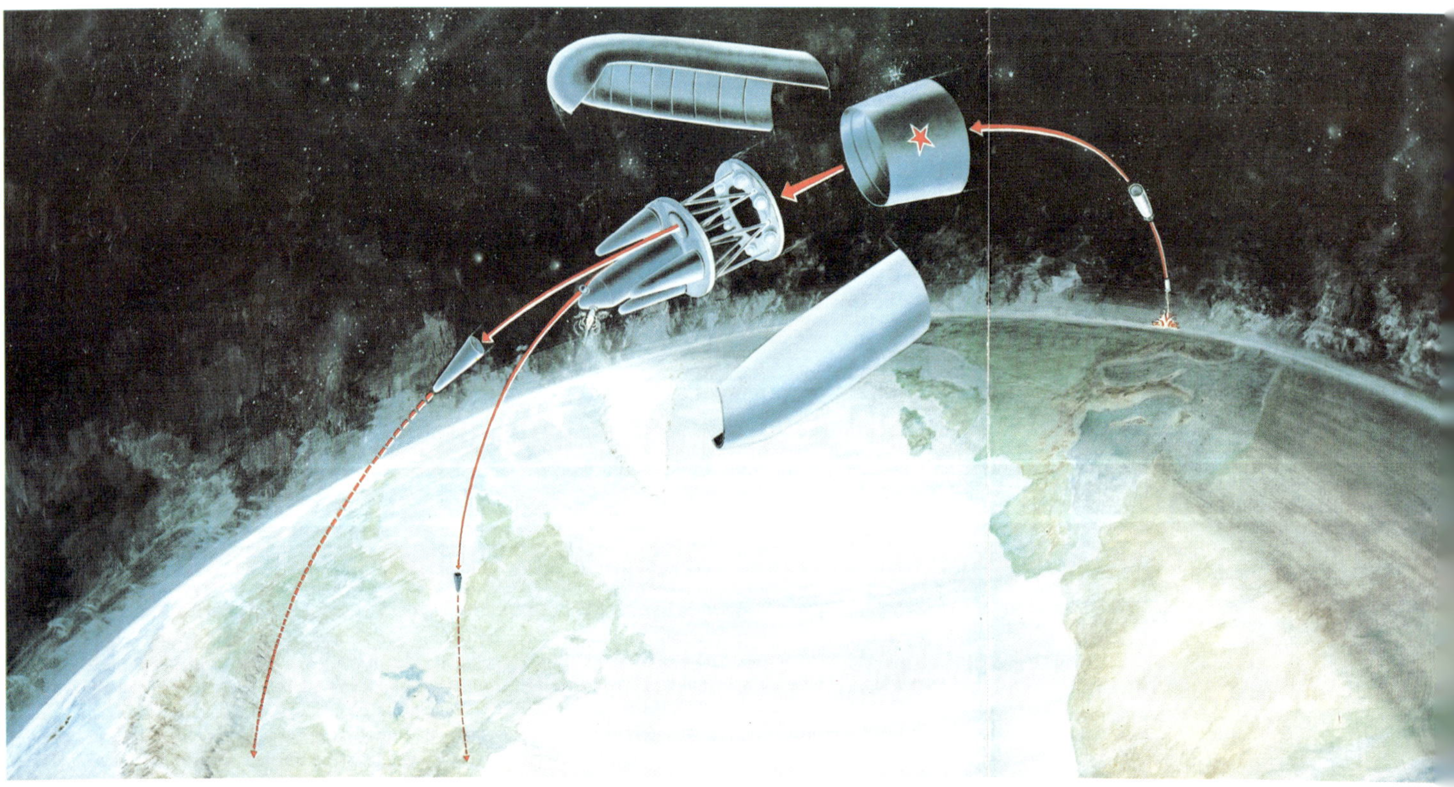

**Above: Extendable rocket motor nozzle attached to a Minuteman III third-stage rocket motor is fired at a simulated height of 100,000ft (30,770m) in a Peacekeeper technology demonstration.**

attempts to create high-energy rocket motors for the US Navy's Trident missile, solid-propellants have been trouble-free in US service.

Soviet solid-propellant technology suffered from technical problems, however. Their first solid-propellant ICBM was the SS-13 "Savage". Deployed in only limited numbers, it entered service in 1968, some six years after Minuteman I. Only with the latest generation of strategic missiles has the Soviet Union used solid-propellant successfully. Even so, early trials of these weapons did meet with limited success. Several early test flights of the SS-X-24 failed, while the first four flights of the SS-N-20 SLBM were unsuccessful.

One technique which the Soviet Union pioneered with great success, however, was that of cold launching. In a conventional "hot-launch" vertical silo, the missile is ignited within the silo, with the hot efflux either being ducted away or allowed to flow past the missile and out of the silo mouth. Conventional land pads used for test flights or space shots often use large amounts of cooling water in an attempt to minimise efflux damage to the launcher, but operational ICBM silos are denied such luxury, and they therefore need to be refurbished before another round can be loaded.

Cold-launch was originally devised for SLBM use, and involves firing the round clear of the submarine launch tube or land-based silo by means of high-pressure gas, then igniting the rocket motor at a safe distance. By adopting this technique for ICBMs such as the SS-17 and -18, Soviet designers were able to maximise the size of the round which could be fitted into existing silos originally designed for hot-launch.

The SALT 1 agreement attempted to restrict future missile developments by forbidding increases in silo dimensions, but Soviet designers were able to use this rule to good effect when designing new missiles intended for installation in existing silos. By switching to cold-launch on the SS-17 and -18, missile size could be increased without altering silo dimensions. Larger missile size resulted in greater missile weight and, as a result, bigger rocket motors and a longer range or heavier nuclear payload. While observing the limits of the SALT 1 treaty, the Soviet Union was able to violate its intention by deploying the more powerful replacement missiles which the treaty had attempted to ban.

A second advantage was that since silo damage was minimal, the possibility of reloading silos in wartime could circumvent the effects of arms-control treaties. Some Western sources allege that Soviet silos now have reload rounds, but the fragmentary evidence published to date is not yet convincing. Reloading of silos has certainly been tested by the Soviet Union, but the US Government has never officially claimed that reload rounds have been built or deployed.

**Below: A partially open cover on a Soviet ICBM silo. An attempt to camouflage the cover has been made, using a pile of logs, but this would not mislead current highly sophisticated satellite sensors.**

**ICBM "hot" launch – 1**
Developed in the 1960s this system allows the rocket efflux to flow up past the missile body as it lifts off. This system confines the missile to one self-contained silo, and is used by the US Minuteman and Soviet SS-13 and (possibly) SS-11 and SS-19.

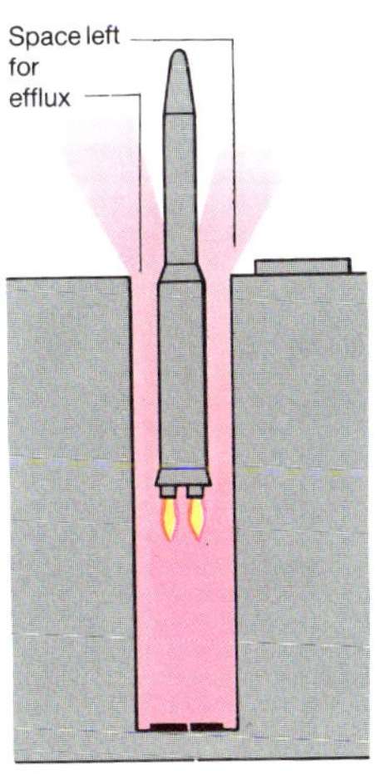


**ICBM "hot" launch – 2**
The initial system for "hot" silo launches featured ducts which led the rocket efflux away from the missile silo. This system was used for the US Titan II, now being phased out of service, and possibly some of the earlier Soviet liquid-fuelled ICBMs, as well.

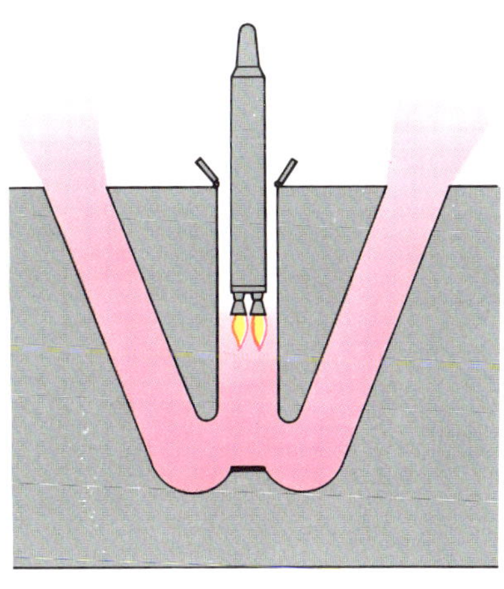

**ICBM "cold" launch**
Several modern ICBMs (including the SS-17 and SS-18) are sited in hardened silos from which they are "cold-launched". The missile is ejected at high acceleration by a powerful gas generator, its first-stage motor only firing when clear of the silo, which can be re-used within a short period. The new US ICBM (Peacekeeper) will use a similar system.

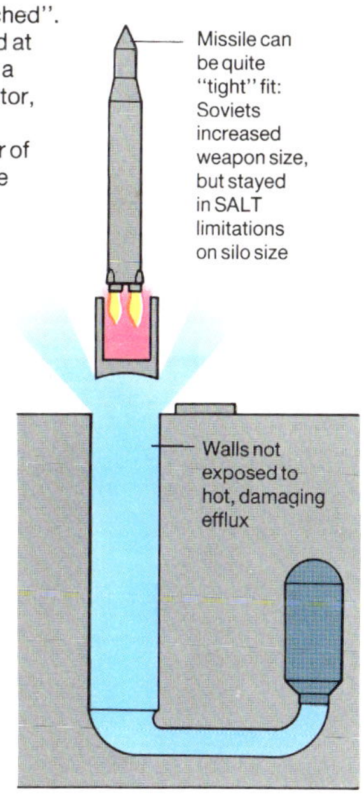

Since the mid-1960s, the Boeing LGM-30 Minuteman II and III solid-propellant missiles have been the main component of the US ICBM force. Despite a programme of modernisation completed in 1981 to improve the guidance system and safety of the Martin Marietta LGM-25C Titan II liquid-propellant missile – the last survivor of the US first-generation ICBMs – de-activation of the six squadrons still deploying them started in November 1982, and should be completed by 1987.

Keeping Minuteman operational is beginning to pose problems. Spare parts for the missile and ground electronic systems will become more difficult to obtain, given the fast pace of electronic component development, while the life of solid-propellant rocket motors is not indefinite. The USAF has embarked on programmes of limited modernisation. The problem is more acute with the older Minuteman II, but the provisions of the unratified SALT 2 treaty forced the USAF to abandon plans to upgrade 50 silos to the Minuteman III standard. Installation of the triple-MIRV Minuteman III in place of the 50 single-warhead IIs would have added another 100 warheads to the US strategic force.

A warhead-upgrading programme of the early 1980s retro-fitted 300 Minuteman III missiles with higher-yield Mk12A re-entry vehicles carrying the 330kT W-78 warhead, but the remaining 250 rounds still have the older Mk12 RV with the 200kT W-62 warhead. This is lighter than the Mk12A, so the unmodified weapons have a slightly longer range and larger MIRV "footprint".

Upgrading of the silos to improve the level of hardening was completed in early 1980, but plans to install the Airborne Launch Control Systems (ALCS) Phase III were abandoned.

Development of the Martin Marietta MGM-118 Peacekeeper (formerly known as MX) has been delayed by the persistent failure of previous US administrations to devise an acceptable basing system. The original scheme to transport missile transporter/erector vehicles around a vast network of underground tunnels was never really practical, while the later plan to base the weapon in a system of dispersed multiple shelters fell victim to environmental pressure groups.

A "Dense Pack" basing system proposed would have required the installation of the missiles in superhardened silos spaced only about 600 to 700 yards (550 to 650m) apart. This scheme relied on the assumption that the blast and material thrown up by attacks upon one silo would interfere with warheads arriving on its neighbours – what planners dubbed the "fratricide effect" – but Congress was unwilling to trust the future of the ICBM force to such an unproven theory.

In an attempt to reach a definitive solution, President Reagan set up the Scowcroft Commission to investigate ICBM basing. Previous assumptions that a single weapon system or basing mode could solve the long-term ICBM vulnerability problem were rejected by the Commission, which recommended two separate lines of development.

As a short-term solution, 100 Peacekeeper rounds will be fielded in converted Minuteman silos in the northwest part of the current Minuteman wing at F.E. Warren AFB, Wyoming. Political opposition to the weapon continues, however, so this plan could follow earlier schemes into oblivion.

Test flights began in 1983 from aboveground canister launchers at Vandenberg AFB, California. Following the final series of test-shots from modified Minuteman silos, deliveries of operational Peacekeeper rounds to SAC are due to begin in May 1986. By December of that year, the first ten rounds should be operational, the entire 100 being in place by late 1989.

Peacekeeper is cold-launched from its launch canister by pressure from a gas generator. Main-stage ignition takes place some 50 to 100ft (15 to 30m) in the air. Under current plans, the missile will carry 10 Mk21 re-entry vehicles – an improved and slightly heavier version of the Mk12A – although it has sufficient throw weight to carry 12. CEP is expected to be about 390ft (120m), ending the current Soviet monopoly in hard-target kill capability.

## Developing a small ICBM

MIRVing had exacerbated the vulnerability problem, not only by allowing a single missile to knock out several enemy silos but also by concentrating several warheads on each US missile, making these even more attractive targets for a disarming "first strike". The Scowcroft Commission proposes that the second part of this trend may be countered by the development of a single-warhead small ICBM, a concept

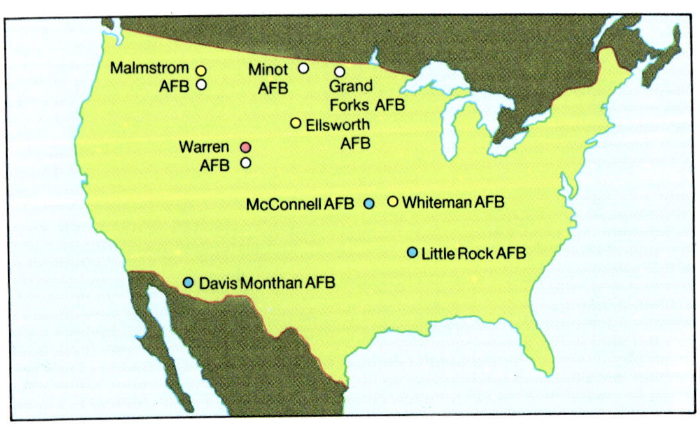


**Strategic bomber and ICBM sites in CONUS**
US ICBM missile fields are sited in the centre of the country. Titan IIs will be finally withdrawn from service shortly, but the US Congress has authorised only 50 Peacekeeper missiles (to be located in current Minuteman III silos). Bomber bases are more widely scattered.

- MX
- Titan II
- Minuteman II
- Minuteman III

**Left: Canister test for the Peacekeeper (MX) missile on 25 February, 1982. Such tests are essential to the success of the "cold-launch" system.**

**Above: An early engineering mock-up of the "Midgetman", the proposed "small" ICBM which would have one re-entry vehicle and be launched from a mobile launcher.**

**Current Soviet (red) and US (blue) SLBMs and ICBMs**
**1.** Poseidon C-3 SLBM.
**2.** SS-11 Mod 3 ICBM.
**3.** SS-N-18 SLBM.
**4.** SS-N-8 SLBM.
**5.** Titan 2 ICBM.
**6.** Minuteman III ICBM with Mark 12A RV.
**10.** Trident I C-4 SLBM.
**11.** SS-127 Mod 3 ICBM.
**12.** Trident 2 D-5 SLBM.
**13.** Titan 2 Mod ICBM.
**14.** Peacekeeper ICBM.
**15.** SS-19 Mod 3.
(Missiles are drawn to scale in both length and diameter. Soviet missile outlines are provisional. In the performance diagram below the missiles are not to scale.)

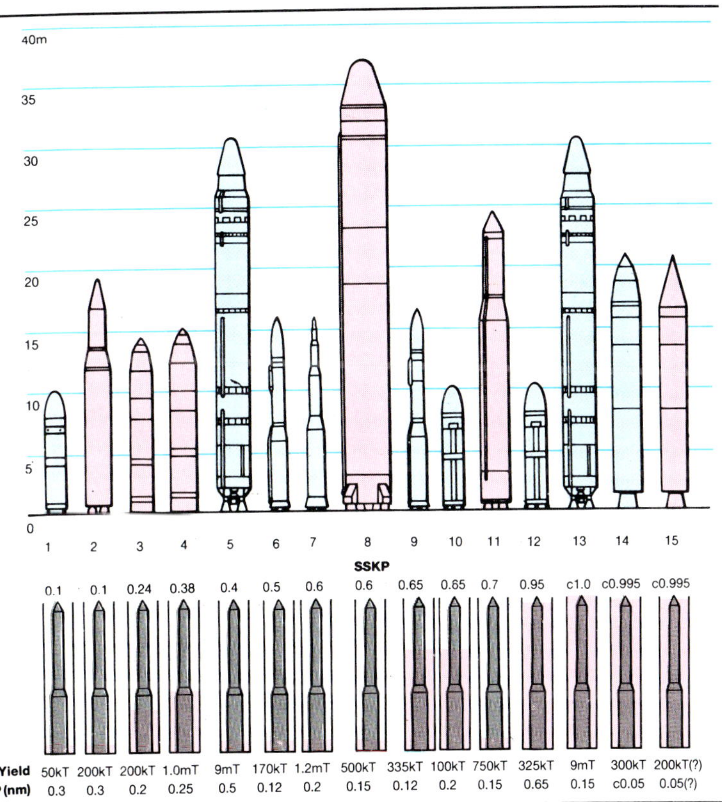


**Performance of current ICBMs and SLBMs**
Red shading shows comparative Single-shot Kill Probability (SSKP) for one warhead of relevant missile in above line-up against a 2,000psi hardened target.

promptly nicknamed "Midgetman". By distributing the total number of land-based warheads over a larger number of potential targets, this programme could greatly increase the Soviet targetting problem.

The resulting weapon will be suitable for mobile basing, probably on custom-designed blast-resistant wheeled vehicles. New technology will be used to keep weight and size to a minimum. Lightweight structural materials of high strength are being investigated for the design of the motor casings, while possible high-accuracy guidance systems include a lightweight version of the Peacekeeper system, and high-technology INS designs featuring ring-laser gyroscopes and stellar inertial updates.

Studies suggest that the Small ICBM may be around 36ft (11m) long and 42in (106cm) in diameter, and weigh about 30,000lb (13,500kg). Such a design would be only marginally larger than the Martin Marietta MGM-31 Pershing II, but more than twice the weight. Several designs of mobile launcher have been evaluated, and blast tests carried out on sub-scale prototypes have given good results. Full-scale development will start in FY87.

## Soviet missiles

The 1970s saw a massive upgrading programme by the Soviet Strategic Rocket Forces, with four new types entering service. Although a small number of solid propellant SS-13 missiles remain in service, most of this generation of Soviet ICBMs stayed with liquid-propellant.

Known to the Soviet Union as the RS-16, 150 SS-17 cold-launched ICBMs are currently deployed in ex-SS-11 silos. Most are Mod 3 versions with a payload of four 750kT MIRVs

Deployment of the hot-launch SS-19 (Soviet designation RS-18) ended in 1982 with some 300 in service, but about 60 are probably assigned to Eurostrategic targets. Earlier models are currently being upgraded to the Mod 3 standard with six 500kT MIRVs.

Heavyweight of the SRA force is the massive SS-18 (Soviet designation RS-20)

**Below: At one stage in its development the Peacekeeper (then known as MX) was to have been based in a "racetrack" system, but this proved politically unacceptable. This shows a test of the hardening of the racetrack.**

cold-launch ICBM. Most of the 308 in service are Mod 4, carrying 10 500kT MIRVs. According to US Intelligence, silos have been modified to withstand over-pressures of up to 6,000psi (420kg/cm$^2$).

In parallel with these liquid-propellant weapons, the Soviets also developed the solid-propellant SS-16 (Soviet designation RS-14) intended for silo or mobile deployment. Under the unratified SALT 2 agreement, the Soviet Union undertook not to deploy this missile, but US Government testimony in 1984 to the Senate Armed Services Committee claimed that limited deployment may have taken place.

SALT 2 allows both superpowers to deploy only one new type of ICBM but, according to the US Government, two new Soviet designs have been developed, both using solid-propellant. SS-X-24 is similar in size to Peacekeeper, and carries up to 10 MIRVs of unknown yield, with a CEP of 850ft (260m) or less. Initial deployment will probably be in super-hardened silos, although a mobile version is expected.

According to the Soviet Union, the second solid-propellant weapon is not a "new" ICBM as defined by SALT, but an improved version of the SS-13 "Savage". Designated SS-X-25 by the US DoD, this system – like the SS-X-24 – is likely to be deployed both in silos and on mobile launchers.

This weapon is likely to prove a problem during arms-control talks, since the US Government regards it a violation of SALT 2. According to the US DoD, the SS-X-25 is "probably a second new ICBM type, prohibited by the SALT 2 agreement". Even if it should prove to be simply a modified and improved SS-13, the provisions of SALT 2 are still violated by flight tests, since the round has apparently been flown with payload which "violates the SALT 2 provisions regarding the permitted ratio between the weight of an ICBM re-entry vehicle and the missile's total throw weight".

According to some reports, the SS-X-25 could be deployed in what could be a multiple-shelter basing scheme similar to that originally proposed for Peacekeeper. US reconnaissance satellites have located a network of railway lines running from a central track and terminating in what could be semi-hardened protective shelters. Located at Plesetsk, the test range which also serves as the SS-13 base, this could be intended for the SS-25.

Although two further Soviet ICBM projects were reported in the late 1970s, these liquid-propellant weapons are now thought to be new variants of the existing SS-18 and SS-17 or -19. The currently deployed generation of Soviet ICBMs "probably marks the end of significant Soviet investment in silo launchers and in the development of wholly-new liquid-propellant ICBMs", says the Pentagon.

The mobile SS-20 is based on the first and second stages of the SS-16 ICBM. Fired at pre-surveyed sites from a vehicle-mounted launcher, it can have a CEP as small as 2,460ft (750m), giving it a useful hard-target capability. Normal payload is three 150kT MIRVS, although single RV versions have been reported.

**Soviet strategic bomber and ICBM sites**
The map of the Soviet strategic bases shows the main weight of these forces lying in the Western USSR, with a second focus in the Far East. The sites across the centre of the country are perforce near to the Trans-Siberian Railway, which imposes major logistical constraints. The SS-20 does not count, in SALT II terms, as a strategic weapon, but is shown here since it must be considered "strategic" by the Western European powers and Far Eastern powers such as China and Japan, upon whom it is quite clearly targetted. Although mobile, it normally operates from 45 fixed sites.

- Bomber bases
- Interceptor bases
- SS-11
- SS-13
- SS-17
- SS-18
- SS-20

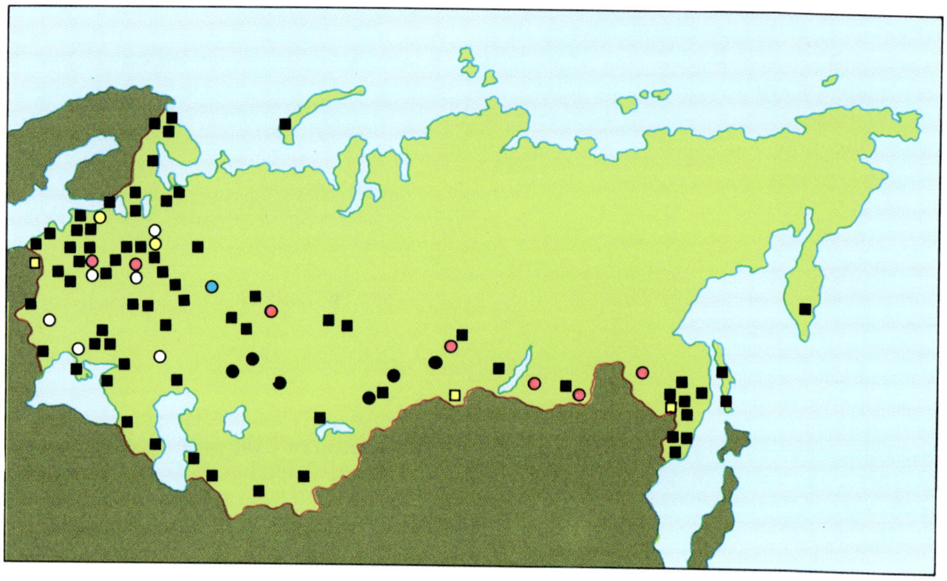

**Below: The mobility, performance and range of the SS-20 are of great concern to the West. Its high mobility makes it difficult to keep under surveillance.**

**Below: The new Soviet SS-X-24 will probably be silo-deployed when first fielded. It may become rail-mobile later, much enhancing its survivability.**

**LAND BASED BALLISTIC MISSILES**

| Country | Designation | Type | Payload | Range: miles (km) |
|---|---|---|---|---|
| USA | Minuteman II | hot-launch solid | 1 1-2MT RV | 6,835+ (11,000+) |
| | Minuteman III | hot-launch solid | 3 MIRV (see text) | 8,080+ (13,000+) |
| | Pershing II | mobile solid | 1 low-yield RV | 2,070 (3,330) |
| Soviet Union | SS-17 Mod 1 | cold-launch liquid | 4-6 200kT? MIRVs | 6,215 (10,000) |
| | SS-17 Mod 2 | cold-launch liquid | 1 high-yield RV | 6,835 (11,000) |
| | SS-17 Mod 3 | cold-launch liquid | 4 750kT MIRVs | 6,215 (10,000) |
| | SS-18 Mod 1 | cold-launch liquid | 1 RV | 7,455 (12,000) |
| | SS-18 Mod 2 | cold-launch liquid | 8-10 550-900kT MIRVs | 6,835 (11,000) |
| | SS-18 Mod 3 | cold-launch liquid | 1 20MT RV | 9,940 (16,000) |
| | SS-18 Mod 4 | cold-launch liquid | 10 500kT MIRVs | 6,835 (11,000) |
| | SS-19 Mod 1 | hot-launch liquid | 4-6 200kT? MIRVs | |
| | SS-19 Mod 2 | hot-launch liquid | 1 RV | |
| | SS-19 Mod 3 | hot-launch liquid | 6 500kT MIRVs | |
| | SS-20 | mobile solid | 3 150kT MIRVs | 2,485 (4,000) |
| | SS-20 | mobile solid | 1 RV | 560 (900) |
| | SS-X-24 | cold-launch solid | up to 10 MIRVs | 7,455? (12,000?) |
| | SS-X-25 | cold-launch solid | MIRVs | 5,590+? (9,000+?) |
| France | SSBS S3 | hot-launch solid | 1MT RV | 1,865 (3,000) |
| China | CSS-1 | pad-launched liquid | 20kT RV | 1,090 (1,750) |
| | CSS-2 | pad-launched liquid | ????? | 1,555 (2,500) |
| | CSS-3 | hot-launch liquid | 1 2MT RV | 4,350 (7,000) |
| | CSS-4 | hot-launch liquid | 1 5MT RV | 6,835 (11,000) |

**Above: Soviet SS-X-25. The US claims that the testing of this missile violates the provisions of SALT II, which banned more than one new ICBM system.**

**Below: French S3 IRBM. There are 18 launchers on the Plateau d'Albion and these missiles are scheduled to remain in service until the mid-to-late-1990s.**

**Above: Chinese CSS-4 ICBM. The effect of China's growing strategic capability upon the overall strategic balance has yet to be fully taken into account.**

Although mobile, the SS-20 does not normally spend its time "on the road". Like the USAF's GLCM cruise missiles, it is normally operated from fixed bases, and would be sent off-base only in an emergency. As this was written, 45 launch sites had been identified – 28 covering NATO targets plus 17 covering China. Each is equipped with nine launchers, which may have reload rounds. The eventual total number of sites could exceed 50.

The essentially-tactical SS-22 with a range of 560 miles (900km) is a major component of the Soviet response to the USAF's GLCM and Pershing II deployments of 1984/85. It originally entered service in the late 1970s, but was forward-deployed to locations in East Germany and Czechoslovakia in early 1984 in retaliation for the US deployments. This forward-based Soviet force is reported to consist of 100 rounds carried on 50 twin-round launch vehicles. The US designation SS-22 has been the subject of comment from the Soviet Government, which does not regard the weapon as a new missile, but simply an improved version of the earlier SS-12 "Scaleboard".

## France's land-based nuclear missiles

The current French IRBM is the Aerospatiale SSBS S3, two nine-round squadrons of which are based in silos in the Plateau d'Albion. Each missile carries a single 1MT RV, and has a reaction time of about three and a half minutes. Its eventual replacement in the mid-to-late 1990s will be the Aerospatiale SX. Although a solid-propellant two-stage weapon like the S3, this could be mobile-based on trailer-mounted erector/launchers, or deployed in silos. Range will be 2,485 miles (4,000km), and the weapon could use the guidance system of the MSBS M4 submarine-launched missile. Alternative single RV or triple MIRV payloads of around 150kT yield have been studied, and the force will be sized to give the desired attack force of around 100 warheads which France requires. SX deployment will be either 100 single-warhead missiles or about 30 triple-MIRVed rounds.

## China's ICBM force

China's modest ICBM force consists of less than 10 CSS-3 missiles plus a small number of CSS-4 weapons; all are liquid-propellant silo-based weapons. CSS-3 is a two-stage limited-range ICBM which uses storable liquid propellants, and carries a single 2MT warhead over a range of 4,350 miles (7,000km). The larger CSS-4 (Chinese designation thought to be DF-5) is a full-range design able to carry a 5MT warhead 6,835 miles (11,000km) or more. These ICBMs are supplemented by two older designs, the 1,090 mile (1,750km) range CSS-1 (based on the Soviet SS-3) and the 1,555 mile (2,500km) range CSS-2 IRBM. Both are probably deployed at fixed sites.

The future of this small force is unclear. The large number of types currently in service – all in modest numbers – must be a logistical nightmare for Chinese planners, and it is likely that all will eventually be replaced by smaller silo-based or mobile weapons using solid propellant.

*"Move deterrence out to sea,*
*Where the real estate is free,*
*. . . And where it's far away from me!"*

Originally used in the late 1950s as a slogan by the supporters of Polaris and the US Navy's nuclear-powered ballistic missile submarine (SSBN) programme, this simple jingle sums up the advantages of submarine-based strategic nuclear systems, the cornerstone of the West's nuclear forces. More than three decades after the USN sent its first Polaris boat to sea on an operational patrol, the SSBN is still the most survivable strategic missile platform. Despite an unparalleled expansion of the Soviet surface fleet, and a huge investment in anti-submarine systems, the Soviet Navy is still unable to track and follow patrolling SSBNs.

Until 1982, most of the US SLBM force was deployed aboard the 31 Benjamin Franklin and Lafayette class submarines, 19 carrying the Lockheed UGM-73A Poseidon C3 missile. Many of these have now been fitted with extra MIRVs, raising the total payload from 10 to up to 14. This upgrading was done to compensate for delays in the Trident programme.

Heavier and longer ranged than Poseidon, the Lockheed UGM-93A Trident I (C4) carries a payload of seven Mk4 MIRVs, each with a 100kT W-76 warhead. To speed service deployment, Trident first became operational aboard the converted Poseidon-class submarine *Francis Scott Key*. Initial Operational Capability (IOC) was achieved in October 1979.

Twelve Benjamin Franklin and Lafayette class SSNs were reworked for Trident. The 16 launch tubes on each vessel were modified to handle the new missile, fire-control and missile-checkout equipment, and the modified submarine was re-ballasted to cope with the greater weight.

*Ohio*, the lead-vessel of a planned class of at least 15, was commissioned on 11 November 1981, two and a half years behind schedule. Despite the delay, the USN is pleased with these new submarines, whose performance during sea trials has met or exceeded design specifications. Despite carrying 24 rounds (eight more than on earlier classes), these boats require only 13 more crewmen, and can spend a greater portion of their time at sea. Ohios are faster, quieter, and harder to detect than the earlier US SSBNs, and can launch their missiles faster than one a minute.

By the time that the ninth vessel is delivered by General Dynamics in 1989, the USN will have switched to the definitive Trident II. Longer and wider in diameter than Trident I, it will be the largest missile compatible with Ohio-class launch tubes. Its maximum range of more than 4,000nm (7,412km) will give the USN vast areas of ocean in which to deploy its SSBN fleet, a potential headache for Soviet ASW forces. Payload will be 10 to 15 Mk5 MIRVs carrying the W-87 nuclear warhead. This design of RV is based on the Mk21 RV carried by Peacekeeper – a rare example of USAF/USN strategic co-operation.

The Mk6 stellar-inertial guidance system should give a CEP of only 390ft (120m), rivalling that of land-based ICBMs, and allowing SLBMs to be targeted against hard targets such as missile silos and command bunkers. Once Trident II is operational, the case for land-based missiles will become more difficult to argue, since ICBMs will no longer be the only weapons – or even the optimum weapons – able to destroy hard time-critical targets. The shorter flight times of SLBMs (a result of their shorter range) will make such weapons the fastest method of knocking out time-urgent targets.

Production of Trident II will probably begin in 1987. It will be retrofitted into earlier vessels – a process which could cost around $125 million, not counting the cost of the missiles. By 1998, a total of 312 rounds will be deployed on 13 submarines.

**Above: Poseidon C-3 SLBM, launched from the submerged USS *Bancroft*, exits the ocean on a live-firing test. Poseidon can carry up to 14 MIRVed warheads.**

**Below: USS *Woodrow Wilson*, capable of launching strategic missiles. They are the survivable counter-value deterrent, if the SSBN cannot be detected.**

**Below: Lifted hatch covers reveal the deadly SLBMs on an SSBN. With MIRVs a US SSBN can attack up to 160 counter-value targets (Poseidon) or 192 (Ohio).**

**Right: Trident I C4 missile in the boost phase. The increasing accuracy of US SLBM warheads may soon give them a counter-force capability.**

## SSBN IN THE COUNTER-VALUE ROLE

| Submarine type | Type | Missiles No. | RVs | Yield | Targets per submarine | EMT per submarine |
|---|---|---|---|---|---|---|
| Washington (US) | Polaris A1 | 16 | 1 | 500kT | 16 | 11.3 |
| Lafayette (US) | Poseidon C3 | 16 | 10 | 50kT | 160 | 35.8 |
| Lafayette (US) | Trident C4 | 16 | 8 | 100kT | 128 | 40.5 |
| Ohio (US) | Trident C4 | 24 | 8 | 100kT | 192 | 60.7 |
| Delta III (USSR) | SS-N-18 | 16 | 7 | 500kT | 112 | 79.2 |
| Taifun (USSR) | SS-N-20 | 20 | 7 | 500kT | 140 | 99 |

**SSBNs in the counter-value role**
The primary role for the SSBN/SLBM combination is that of counter-value strike against the enemy homeland. The table shows the increasing firepower available, especially in number of targets resulting from MIRVing.

A quarter of a century of Western ASW research has failed to develop weapons capable of reliably locating patrolling SSBNs, and there are no indications that the Soviet Navy has made any greater progress. Despite this fact, the US Navy is not complacent when contemplating the long-term future of its SSBN fleet. Under the SSBN Security programme, the service is reviewing all current ASW techniques, and considering possible technological breakthroughs for the future. If this evaluation shows potential threats to the Trident and Poseidon force, suitable countermeasures will be developed, says the Pentagon.

### Britain's SSBNs

The pioneering Lockheed UGM-27 Polaris ended its career with the US Navy in 1981, but in its UGM-27C Polaris A3TK form, remains in service aboard four Resolution class submarines of the Royal Navy until the planned UK Trident fleet becomes operational in the early 1990s. Development of the British-designed Chevaline payload system has been completed and the system has entered service. Exact details of the system have never been announced, but it is widely believed to consist of six 40kT warheads plus advanced penetration aids. Although the payload can be manoeuvred in space, with a maximum spacing of 43 miles (70km) being possible between individual impact points, the Chevaline system does not constitute a true MIRV system. This strongly suggests that the RVs may only be manoeuvred in formation and not individually.

Having decided in the summer of 1980 to replace the RN Polaris force with Trident, the UK Government eventually opted in 1982 to choose the Trident II. Missiles and MIRVs will be purchased from the United States, along with the launch systems and checkout equipment. The nuclear warheads fitted within the RVs will be of UK

**Current SSBN types**
Present types of SSBN, including those in the British, French and Chinese navies. The grand total is 113, but a very much smaller number will be on active patrol at any one time. Numbers at sea are usually of the order of: USA, 23; USSR, 13; UK, 1 (possibly 2 on occasions); and France, 2 (the Chinese SSBNs are not thought to be yet in service). Any significant increase in numbers at sea would obviously suggest that some form of action was imminent.

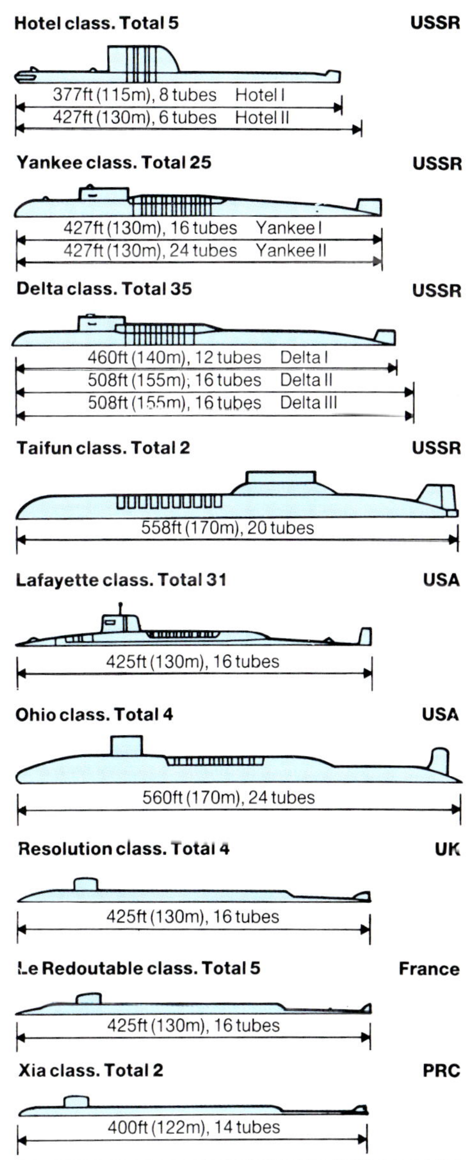

**Typhoon under-ice operation**
One known method of operation for the Soviet Navy's latest SSBN – the Typhoon class – is to penetrate far under the Arctic ice-cap, which offers substantial ASW protection, particularly from maritime patrol aircraft and satellites. If required to launch its missiles the submarine can surface, either in open water or by smashing thinner areas of ice. The SS-N-20, 20 of which are carried, has a range of 5,157 miles (8,300km) bringing the entire CONUS within range from anywhere under the Arctic ice-cap.

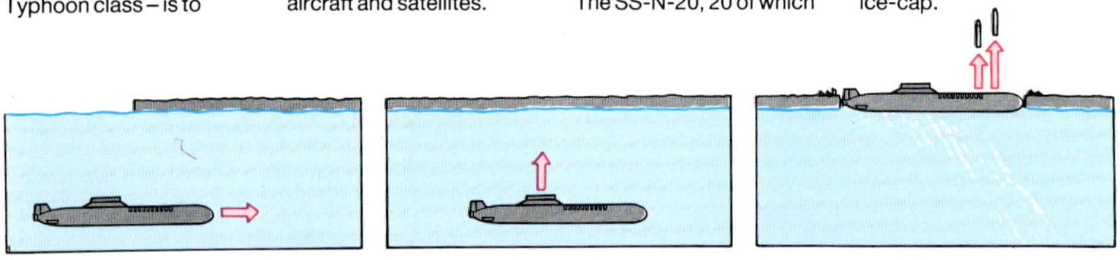

Above: A new version of the Delta class – the Delta IV – is about to join the Soviet fleet, able to launch its 16 SS-N-23 SLBMs from the safety of "home" seas, Barents and Okhotsk.

Below: The huge Typhoon SSBN, first seen publicly in 1985. This picture fails to emphasise its size; it is 558ft (170m) long, with a beam of 81ft (25m), the largest submarine ever built.

design, as will be four submarines which will carry the weapon. Designed and built by Vickers, these will carry only 16 launch tubes. Filling all tubes in the planned fleet will require a total of 64 rounds. Under the agreement between the two Governments, the UK will purchase 72 rounds at an estimated cost of $3,000 million.

Trident is scheduled to enter RN service in the mid-1990s, but like the US Peacekeeper ICBM, it faces a stormy political future. Given the present unilateralist views of the British Labour Party, entry into service would seem to depend on the current Conservative Government winning the next two general elections. A more significant problem is that of funding. Although the UK Government is reluctant to admit the fact, the cost of Trident could result in significant cutbacks in other UK defence programmes, particularly those of the Royal Navy.

## Soviet SLBMs

The most important Soviet SLBM is the 4,320nm (8,000km) range SS-N-8 which arms 18 Delta I class submarines each with 12 launch tubes, and four Delta II class submarines each with 16. This missile comfortably outranges all current American SLBMs, including Trident I. Ranges of up to 4,965nm (9,200km) have been reported on test flights, although these could have been trials of more modern missiles. Three versions have been reported, having single RV, MRV and MIRV payloads respectively. CEP is reported to be about 1,310ft (400m).

The SS-N-18 (Soviet designation RSM-50) first entered service in 1979 on the 16-round Delta III class, 14 of which are currently in service. An adaption of the earlier Delta designs, this class may be recognised by the height of the massive raised missile section needed to accommodate the 46ft (14m) length of these bulky liquid-propellant missiles. Earlier Deltas had similar raised aft sections, but on the Delta IIIs the height of the missile compartment gives the class a visual appearance which borders on the grotesque. Missile variants have been reported with a single 450kT RV, three 200kT MIRVs, and seven MIRVs of unknown yield.

To maintain this force into the 1990s, the Soviet Navy has embarked on the deployment of two new types of SLBM, and two new patterns of SSBN. All Soviet records for size, range and throw weight were broken by the new SS-N-20 missile. Arming the huge Typhoon class ballistic missile submarines, this 50ft (15m) long weapon has a range of up to 4,480nm (8,300km) and a payload of from six to nine MIRVs, and is the first successful Soviet solid-propellant SLBM.

Larger than even the US Navy's Ohios, Typhoon class submarines are around 590ft (180m) long, and 25,000 tons in displacement. The unusually wide beam of 75ft (23m) suggests that the outer casing may contain two pressure hulls mounted side by side. Twenty launch tubes are mounted in an unconventional location forward of the sail. At least four vessels have been sighted, of a class of at least 10.

Above: *Le Redoutable*, SSBN of the French Navy, which has an ambitious and entirely national nuclear ship programme. The missiles – M4 and M20 – are also French.

Above: French M4 submarine-launched ballistic missile has six 150kT warheads, all with EMP protection, and penetration aids to confuse enemy defences.

Above: The Chinese SLBM is test-fired. This missile arms the Xia class SSBNs, of which two have been commissioned; whether they are operational is unclear.

**The modern capital ship**
The submarine was, until the 1960s, confined in its capabilities to the use of torpedoes against surface ships or other submarines. Today, however, it has a very considerable potential against land targets. Submarine-launched cruise missiles (SLCM), with either nuclear or conventional warheads, can be fired against targets deep inland, and they have a very high probability of penetrating the defences. SSBNs can also launch ballistic missiles, either with single or MIRVed warheads. So long as the SSBN remains virtually undetectable it will be one of the major elements in the strategic balance. Should there be a major breakthrough in ASW technology, however, the current balance will be seriously upset.

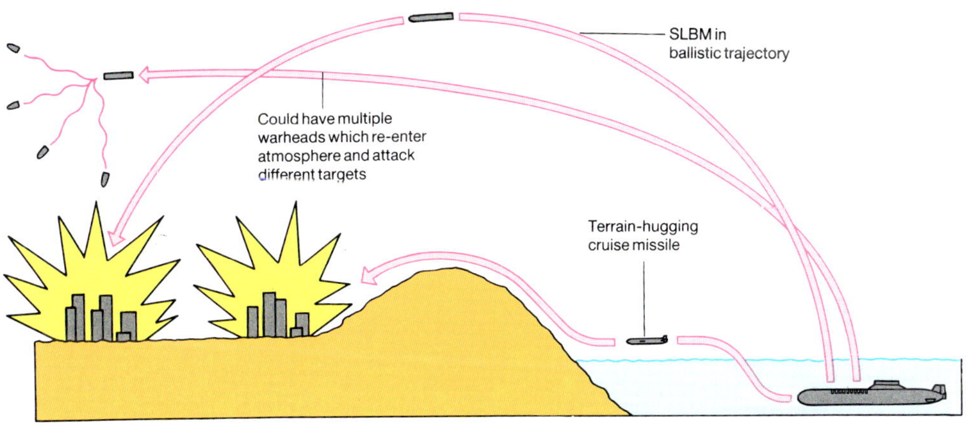


In 1985 the US DoD revealed the existence of another new Soviet SSBN. This is designated Delta IV, scheduled for deployment in 1985/6. This will probably be the first launch platform for the new SS-NX-23, which started flight trials in 1983. Similar in size to the SS-N-18, this retains well-proven liquid propellants, but offers increased throw weight and higher accuracy. According to the US DoD, it may be retrofitted to the Delta III fleet.

## The French SSBN/SLBM fleet

Most of France's fleet of six missile-carrying submarines are equipped with the Aerospatiale MSBS M20, but 1985 saw the service debut of the improved M4 missile aboard the newly-commissioned sixth vessel, *L'Inflexible*. Twice the weight of the M20, the M4 carries a payload consists of six radiation-hardened 150kT re-entry vehicles, plus penetration aids. Some sources credit these with MIRV capability, but MRVs are more likely. M4 gives greater operational flexibility than the M20; rounds may be fired more rapidly and from a greater operating depth. All of the earlier vessels except *Le Redoubtable* (the eldest) will be rearmed with the M4 in a retrofit programme due to end by 1989.

Two French developments are planned for the future. A seventh submarine is projected, but funding problems could result in delay or even cancellation. New technology being developed under the M4C programme could be retrofitted into M4 rounds or applied to an improved missile likely to be designated M5. This could include a MIRV payload.

## Chinese SLBMs

First successful test flight of China's CSS-NX-3 SLBM took place on 12 October 1982, when a missile flew a range of about 995 miles (1,600km) following an underwater launch. This trial followed the reported loss of a Golf class submarine in August 1981 as the result of an explosion during an underwater test firing, and a CSS-NX-3 launch from an underwater launch pad on 30 April in the same year.

A submarine-launched missile was displayed during a Beijing parade in 1984 celebrating the 35th anniversary of the Chinese revolution. The missile seen on trucks was some 23ft (7m) long and 39 inches (99cm) in diameter, well below the 27.8ft×53 inch (8.5m×135cm) dimensions of the Polaris A1. Its final identification as a strategic SLBM cannot be regarded as confirmed. This may not be the CSS-NX-3, but an experimental test vehicle.

Operational SLBMs are likely to be deployed aboard indigenously-designed SSBNs based on the Chinese Han-class nuclear attack submarines. The first Xia-class vessel was laid down in 1978 and probably launched in the spring of 1981. It was probably the launch vessel used for the October 1982 CSS-NX-3 test. Early reports credit these vessels with six launch tubes, but more recent reports suggest they may have 12 or even 14.

Given an operational range of 1,000 to 1,300nm (1,853 to 2,410km), the Xia class will be able to cover targets in the eastern USSR from patrol areas in the Western Pacific and Philippine Sea. Considering the number of Soviet missiles targeted on Chinese territory, the Chinese Navy may be planning to increase its strike depth into the Soviet Union by establishing patrol areas in the Yellow Sea, but these relatively shallow waters are not good locations for SSBN operations.

Four decades of superpower rivalry have failed to teach American defence planners the simple fact that the Soviet Union is capable – often with surprising speed – of developing a counter to most of the high-technology weapons deployed by the West. Nothing illustrates this fact better than the Soviet counter to the US cruise missile programmes. Promoted in the late 1960s as an almost unstoppable wonder weapon, these prompted the development of Soviet defenses sufficiently effective to result in one early design being taken out of production.

Cruise missiles are winged weapons of aircraft-like configuration. Both superpowers developed and deployed cruise missiles in the 1950s, but these early weapons were physically large – pilotless equivalents of contemporary fighter and bomber aircraft. Some Soviet designs were literally pilotless variants of manned fighters.

Development of most current patterns of US cruise missile started in the early 1970s with the goal of using advances in micro-miniaturised electronics and small turbofan engines to create relatively small weapons for USAF and USN use.

Subsonic cruise missiles have flight times of several hours when flying strategic-ranged missions. Since inertial navigation systems (the favoured self-contained guidance system for strategic missile applications) exhibit a growing error with passing time, design of these 1970-vintage strategic cruise missiles required the creation of Tercom (terrain contour matching), a supplementary guidance system able to update and correct the inertial system at regular intervals. This uses a radar altimeter within the missile to sense the profile of small pre-mapped patches of terrain which lie along the flight path. By comparing this with geographic data held in its electronic memory, Tercom can deduce the missile's current position.

En route to the target, cruise missiles can take advantage of the local terrain contours to screen themselves from hostile radars – exactly the same technique as is used by manned aircraft at low level. Even without such terrain-hugging flight, US cruise missiles are still elusive targets for Soviet SAMs and interceptors. Being much smaller than manned aircraft, they reflect very little radar energy, while the small turbofan powerplant will be a difficult target for infra-red homing weapons.

**Above: US BGM-109B Ground-Launched Cruise Missile (GLCM) is being deployed in those NATO countries willing to accept it, causing controversy in some of them.**

The course flown need not be a direct one, but will probably have built-in changes of direction to confuse the defences. The US intends to map a large enough number of patches of terrain to allow individual missiles to approach their target from any direction.

The theory may appear perfect, but practical implementation of Tercom has not been trouble-free. The detailed profile of terrain exhibits seasonal variation due to the growth of crops and vegetation, and can be altered by heavy snowfalls. Tercom is reported to have experienced problems in this respect, and a 1983 agreement between the US and Canadian Governments allows ALCM to be tested against terrain and climate similar to that of some areas of the Soviet Union by flying rounds over a weapons range near Cold Lake, Alberta.

Not all terrain is rugged enough to yield good Tercom data, so cruise missile planners must choose their landfalls accordingly. Soviet air-defence experts have apparently taken advantage of this fact by concentrating surface-to-air defences in coastal areas thought likely to be overflown by cruise missiles. Tactics of this type, plus the introduction into Soviet service of new air-defence weapons including interceptors armed with "look-down" radars and "snap-down" missiles, are already affecting USAF cruise missile plans.

Boeing's AGM-86B ALCM-B (Air-Launched Cruise Missile B) entered operational service in December 1982 on USAF B-52G bombers at Griffiss AFB. Ninety aircraft were fitted with a six-round launcher under each wing, while the Common Strategic Rotary Launcher currently under development will allow a further eight ALCM-B or SRAM (Short-Range Attack Missiles) defense-suppression missiles to be carried internally. SAC is now modifying its B-52H fleet to carry the missile.

ALCM-B has a maximum range of 1,555 miles (2,500km) – sufficient to cover 85 per cent of potential Soviet targets from launch points outside Soviet territory. In any conflict with the Soviet Union, cruise missile-equipped B-52s would try to avoid flying

**Below: AGM-86B ALCM-B (Air-Launched Cruise Missile-B) is launched from a low-flying Boeing B-52G. This role has given the old bomber a new lease of life.**

**Below: T-ASM version of the submarine-launched Tomahawk SLCM has an anti-ship role. The visually similar TLAM-C is a nuclear-armed land-attack version.**

**Right: An unarmed ALCM is examined by technicians at the completion of its first free-flight test over Canada; the test was a complete success.**

Above: Conventionally-armed Tomahawk SLCM approaches its target, having been launched from a submerged US Navy submarine off the California coast.

Above: The Tomahawk SLCM (seen on the left) impacts on and demolishes the target, a reinforced concrete structure the size of a warehouse (or a static HQ).

into Soviet airspace. When attacking most targets, they would release their missiles from launch points 220 miles (350km) or more from the Soviet coast.

Vigorous development of US Navy's BGM-109 Tomahawk Sea-Launched Cruise Missile (SLCM) has resulted in a family of variants, both nuclear and conventionally armed. SLCM will be deployed on over 140 surface ships and submarines, including the reactivated New Jersey class battleships, Long Beach, California, Virginia and Ticonderoga class cruisers, Spruance and new DDG-51 class destroyers, and Los Angeles class submarines.

The simplest installation was the armoured box launchers planned for most surface ships, but DD-963 destroyers and CG-47 missile cruisers will be fitted with multi-purpose vertical launchers which may also be used for weapons such as Standard and Vertical-Launch Asroc, while late-production Los Angeles class boats will be built with 12 vertical-launch tubes located between the forward end of the vessel's internal pressure hull and the rear of the bow-mounted sonar array. Earlier submarines of the same class will receive this installation as a retrofit. Tomahawk was originally designed for launch in encapsulated form from standard USN torpedo tubes, but the service now realises the degree to which this plan would reduce the already limited torpedo capacity of submarines.

Nuclear strike version of Tomahawk is the BGM-109A Tactical Land Attack Missile Nuclear (TLAM-N). This carries a W-80 200kT nuclear warhead, and follows a flight profile similar to that of the air-launched ALCM.

Although operational evaluation was completed in October 1983, the planned deployment was briefly delayed by Congress, with the US House of Representatives voting in May 1984 to limit the deployment until the end of FY85, unless the Soviet Union deployed a similar weapon.

The remaining sea-based Tomahawk variants are non-nuclear systems. The 280 mile (450km) range BGM-109B Anti-ship Tomahawk (TASM) has a 990lb (450kg) high-explosive warhead and an active-radar terminal guidance seeker, while the BGM-109C Land Attack Tomahawk (TLAM-C) uses Tercom and a digital scene-matching area correlation system (DSMAC) electro-optical terminal homing seeker to deliver a 990lb high-explosive warhead or a submunitions dispenser for the BLU-97/B multi-purpose bomblet on small land targets. Development of air-launched BGM-109H and L non-nuclear versions of Tomahawk was cancelled in FY84.

Perhaps the most controversial version of Tomahawk is the US Air Force's BGM-109G Ground-Launched Cruise Missile (GLCM) version of Tomahawk. Deployment started in December 1983 with the arrival at Greenham Common in the UK of the first of the planned batch of 96, while the first of 112 for Comiso in Sicily followed in March 1984.

Other bases currently planned are Florennes Air Base, Belgium (48 rounds), Woensdrecht, in the Netherlands (48 rounds), a second UK base at Molesworth (64 rounds), and 96 rounds in West Germany (possibly at Wuscheim). The USAF will procure 560 missiles, 137 TELs (transporter erector launch vehicles) and 79 Launch Control Centres (LCCs). Most of the missiles will be deployed in Europe, 96 being kept in USA as a reserve.

Each GLCM flight has four four-round TEL vehicles, plus two LCCs. These are wheeled vehicles weighing about 35,000lb (15,875kg), and hauled by M1013 tractors. Designed to protect their missiles and launch crews from small-arms fire or NBC attack, they have a limited degree of resistance to nuclear blast. Normally located within hardened shelters at the bases, these vehicles will be dispersed to pre-surveyed launch sites in an emergency. Tractors and vehicles can operate off-road, and do so during practice alerts.

At the time that the US Services committed themselves to this generation of cruise missiles, there was no sign of an effective Soviet defence against such

weapons. Soviet interceptors deployed during the 1970s lacked radar systems able to look downwards and detect low-flying targets, while many of the SAM systems could not engage targets flying at altitudes of less than 100m.

Prototype cruise missiles took part in a highly-classified series of survivability tests in which they were test flown against US air-defence systems, captured Soviet missiles and equipment designed to simulate next-generation Soviet equipment. The results of these flights did much to convince US defence planners that the new missile posed a severe threat to the Soviet Union, and that the creation of realistic anti-cruise-missile defences would be a large financial burden to the Soviet defence budget.

The first Soviet weapon to offer a degree of anti-cruise missile capability is the new SA-10 surface-to-air missile, which entered service in 1980. This vertically-launched weapon has a top speed of Mach 6 and a range of 31 miles (50km), and may have been extensively redesigned in the course of development to cope with the cruise missile threat. Two versions are thought to exist. The original variant, although transportable, is deployed at fixed sites positioned to defend high-value targets in the Soviet Union. Deployment of a second version based on eight-wheeled vehicles started in the mid-1980s. Transporter/erector vehicles each carry a cluster of four launcher/containers, while the associated planar-array radar is carried on another chassis of the same type.

US Department of Defense analysts believe that at least 500 SA-10 sites would be needed in order to create an effective defence of the Soviet homeland against cruise missiles. By early 1984, around 40 fixed sites had already been deployed.

Other potential threats are the recently-deployed Mikoyan MiG-31 "Foxhound" interceptor with AA-9 shoot-down missiles, and the even newer Sukhoi Su-27 "Flanker". "Foxhound" has had some success in intercepting simulated cruise missile targets using AA-9 during trials at the Vladimirokva test range on the Caspian Sea, while Flanker's "fire-and-forget" AA-X-10 may also prove effective in this role.

Rapid development of these defensive systems has forced a re-think in US cruise missile plans. SAC had originally planned to buy 3,000 Boeing ALCMs to arm its bomber fleet. Early deliveries would have been ALCM-B, with the longer-ranged ALCM-C coming off the line from 1984 onwards. This improved model would have had sufficient range to reach all targets from "safe" offshore launch areas, and to allow USAF planners more flexibility in routing ALCM rounds along indirect flight paths.

Improved Soviet defences caused this plan to be scrubbed. Production of ALCM-B is being phased out with only 1,499 having been built, while the ALCM-C project was abandoned. The remaining 1,300 cruise missiles for Strategic Air Command will be of an all-new design – the General Dynamics Advanced Cruise Missile (ACM). Production of this weapon will probably begin in FY85, with operational deployment starting in 1989.

**Above: ALCM over the Arctic wastes of Northern Canada, an area selected for test flights because of its similarity to large parts of the Soviet Union.**

**Anti-ship crusie missile launchers**

The recent adverse publicity about US cruise missiles of various types has totally ignored the fact that for many years the Soviet forces have deployed large numbers of these weapons. This table shows that only ten years ago the Soviet Navy had an absolute monopoly on anti-ship missiles and that they still have a virtual monopoly in such missiles with a range in excess of 100nm. The table also illustrates just how long it takes to redress an imbalance in weapons systems, even in today's circumstances and with defence priorities being applied. Virtually all these Soviet missiles are nuclear-capable.

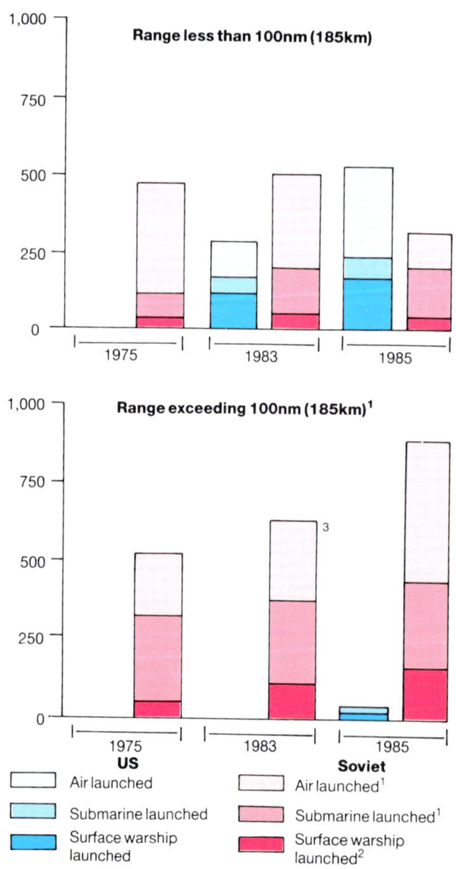


1. All Soviet weapons in these categories are nuclear-capable.
2. Includes carriers, cruisers, destroyers and frigates.
3. These ASM launchers could carry longer-range (100nm plus) ASMs.

The ACM will be subsonic, but will have have a longer-range than ALCM-B – around 1,990 (3,200) rather than 1,550 miles (2,500km). This should allow virtually all Soviet targets to be covered from offshore launch points, and permit planners to route missiles along missile flight paths which circumnavigate or confuse Soviet defences. The ACM will be powered by the new Williams International F112 turbofan, a follow-on to the powerplant of ALCM and Tomahawk.

Few details of the design are available – a result of the application of "stealth" technology. The guidance system could use technology developed under the USAF's Cruise Missile Advanced Guidance (CMAG) programme, but will be too early in timescale to use the carbon dioxide laser-based all-weather guidance system being investigated by this project. Designed to operate at low altitudes, providing terrain-following, terrain-avoidance and obstacle avoidance, terminal homing on high-value fixed targets, and the ability to attack mobile targets, this laser-guidance system will probably be applied to future cruise-missile programmes.

For the future, the US Navy is also believed to be planning "stealth-technology" cruise missiles. Designs were prepared by Boeing and GD under the classified Teal Dawn programme, while another missile developed by Lockheed and rejected by the USAF in favour of the ACM is reported to be continuing for the Navy.

Back in the late 1950s, the USAF briefly deployed the Northrop Snark, a first-generation cruise missile of intercontinental range. Virtually a pilotless bomber, this had a wingspan of 42ft (13m), and weighed 60,000lb (27,215kg) when ready for launch. It may have seemed like a good idea at the time, but was withdrawn from service after a two-year deployment. But the concept of intercontinental-range cruise missiles is not dead. Before the end of the decade the US Defense Advanced Research Projects Agency (DARPA) hopes to develop the technology for 6,835 mile (11,000km) range air-launched cruise missiles, Vought is already evaluating suitable technology under a a $12.1 million USAF contract.

Above: The Pentagon fears that the new Soviet SS-NX-21 cruise missile became operational in 1985 and could be deployed near US coasts. It is said to have a range of 1,864 miles (3,000km) and to be launched from standard 21in (533mm) torpedo tubes of new and older attack submarines.

Above: US DoD impression of the new nuclear-armed AS-15 ALCM launched from the "Bear-H" bomber. The AS-15 is similar to the American Tomahawk.

Below: Dassault Mirage IVP strategic bomber of the French Air Force carrying an ASMP nuclear weapon on centre-line hardpoint. ASMP range is 62 miles (100km).

## Soviet cruise missiles

If the Pentagon thought that the Soviet Union would be unable to develop its own ALCM/Tomahawk-class cruise missiles, it was greatly mistaken. Soviet weapons were available for deployment by 1984/85. By designing a single air-breathing multi-purpose type, the Soviet Union has been able to near-simultaneously deploy land, sea and air-launched weapons. All have a range of around 1,865 miles (3,000km), although it remains to be seen how effective Soviet engineers have been at devising Tercom-style guidance.

The Soviet equivalent of ALCM is the AS-15 carried by the new "Bear-H" version of the Tu-95. Much to the surprise of defence analysts, this elderly design was returned to production in the early 1980s in order to provide low-life airframes for use as cruise-missile carriers. The AS-15 may also be fitted to the Tupolev "Blackjack" when this large VG bomber enters service in the late 1980s.

Like the AS-15, the naval SS-N-21 was deployed for the first time in the autumn of 1984 as a response to US cruise missiles. Fired from standard torpedo tubes, it will probably be fitted to the Victor III class, the new Mike and Sierra classes, and what the US DoD has described as "a new Yankee class SSN". The Yankee class is armed with the SS-N-6 SLBM, but the DoD reference may refer to the nine vessels of this class which have been converted into cruise-missile launchers, losing their SS-N-6 missile tubes.

Like the SS-20 IRBM, the SSC-X-4 land-based cruise missile will probably be based only on Soviet territory, while operating procedures will probably draw on experience gained with the ballistic missile.

Two larger cruise missile are also under development, and could enter service in 1986. Winged weapons some 39ft (12m) in length and 20ft (6m) in wingspan, these use a common airframe for sea and land-based applications.

## French cruise missile

The only other air-breathing strategic weapon currently under development is the wingless Aerospatiale ASMP (Air-Sol Moyenne Portee) which entered service in 1985 on 18 Mirage IVP bombers – reworked Mirage IV aircraft which originally carried the AN52 free-falling nuclear bomb. Each aircraft now carries a single round on the fuselage centreline, and has been fitted with new avionics including an inertial navigation system used to pass pre-launch navigational data to the weapon.

About 16ft (5m) long, and with a launch weight of around 1,985lb (900kg), ASMP is powered by an integral rocket/ramjet and has a range of around 62 miles (100km) while carrying a 100 to 150kT warhead. After release from the parent aircraft, it flies at Mach 3 at high altitude, dropping to lower altitudes and speed for the final terrain-following approach to the target. ASMP will also be carried by 50 carrier-based Aéronavale Super Etendards and the planned fleet of 85 Mirage 2000N due to enter service from 1988 onwards.

The Boeing B-52 Stratofortress has been the principal manned bomber of USAF's Strategic Air Command since the mid-1950s. Less than a third of the 744 aircraft originally built remain operational, and these have an average age of more than 20 years. In 1985 the oldest version in the inventory was the B-52D, which was heavily modified for the conventional bombing role – a legacy of Vietnam – and could carry up to 84 500lb (225kg) bombs in the weapons bay, plus up to 24 750lb (340kg) bombs on underwing pylons. Only a handful of the 170 produced remained in service by late 1984, and all will probably have been retired by the time this book is published.

The SAC fleet is now made up of B-52G and -52H models. The G introduced a redesigned wing with integral fuel tanks, a revised tail fin of lower profile, a remotely controlled tail-turret armed with 50-calibre machine guns, and the ability to carry air-to-surface missiles such as the AGM-69 SRAMS. A total of 193 were built, and 168 were still operational in late 1984. Up to 20 ALCM cruise missiles will also be carried on external pylons and internal rotary launchers.

Following the cancellation of the B-70 bomber, a further version of the B-52 was introduced in 1961. The B-52H was the first to use turbofan engines – Pratt & Whitney TF33s – and had an improved structure better suited to low-level flight. A 20mm rotary cannon replaced the 50-calibre tail guns used in earlier models. The last of 102 B-52Hs was produced in June 1962, and 96 remain in service.

Maintaining this aging fleet of veterans requires a considerable engineering effort. Modification programmes currently under way or planned are intended both to prolong the aircraft service life and to expand the mission profile. This work is expensive: in fiscal year 1984 the USAF spent $460.7 million on B-52 modifications, and more than $76 million on developing future improvements.

Cruise missiles will be fitted to 99 B-52G and 96 B-52H, externally in the case of the -52G, externally and internally in the case of the H model, which will be fitted with the Common Strategic Rotary Launcher for the internal carriage of ALCM, SRAM, and other weapons. By the end of FY84, 90 B-52G aircraft were equipped to carry ALCM, and work on converting the B-52H force began in 1985. By the end of the decade, the "new" Rockwell B-1B will be the primary penetrating bomber, and most B-52s will either be serving as cruise missile carriers or assigned to general-purpose missions.

**Above: AGM-69 SRAMs on a rotary launcher in the bomb-bay of a Boeing B-52 strategic bomber. Maintaining this aging fleet is becoming increasingly costly.**

The B-52's ECM systems are regularly upgraded, and the latest projects include Pave Mint (updating the ALQ-117 ECM set of the B-52G to counter new surface and airborne fire-control and missile radars), and the installation of the ALQ-172 in the B-52H to cope with current and projected airborne interceptors.

Other B-52 modernisation programmes will replace outdated radar equipment with modern solid-state components, install a new environmental-control system, and improve communications by fitting VLF/LF receiver terminals, and upgrading the SATCOM terminals. The existing electro-optical viewing system on the B-52G is to be updated, and a digital unit will replace the existing FLIR signal processor. The 69 aircraft not scheduled to receive ALCM will be fitted with a new integrated stores management system.

The General Dynamics FB-111A was built in only modest numbers, and 55 still fly with two SAC wings. As the B-1B enters service, surviving FB-111As will be reassigned to Tactical Air Command.

Contracts for development of the B-1 were signed in 1970, and Rockwell International flew the first prototype on December 23, 1974. Approval for a 200+ production run was granted two years later, but then President Carter's decision

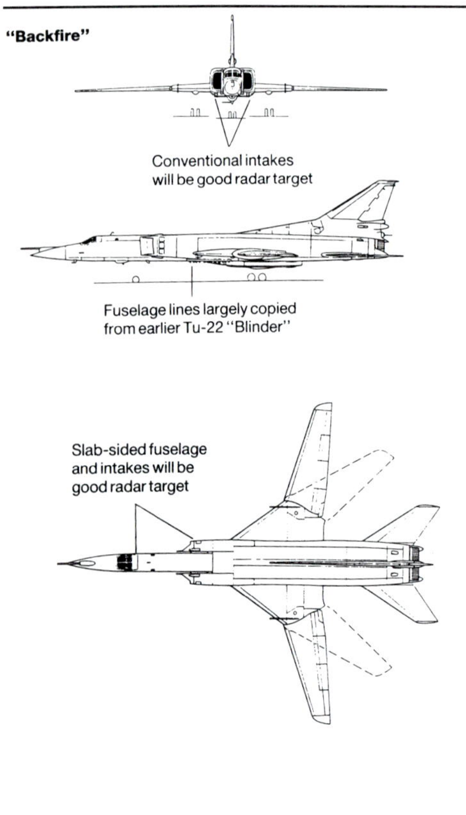


**Tupolev Tu-22M "Backfire"**
The "Backfire" achieved notoriety in the 1970s breakdown of SALT II, due to arguments as to whether it was "strategic" or not. The fuselage lines are largely copied from the earlier Tu-22 "Blinder" and make an easy target for modern radars.

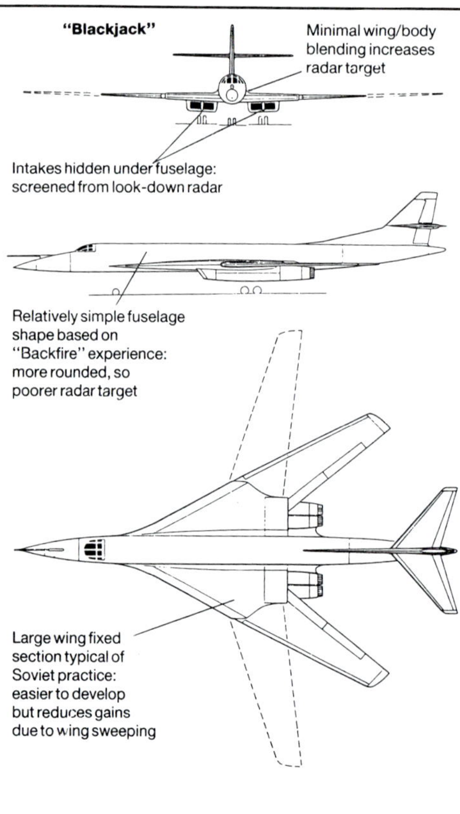


**Tupolev "Blackjack"**
"Blackjack" is likely to prove to be the Soviets' first genuine strategic bomber. It is not such a great advance over the "Backfire" as is the B-1B over the B-52 and follows Soviet practice with a large fixed wing section, tending to negate the gains from wing sweeping.

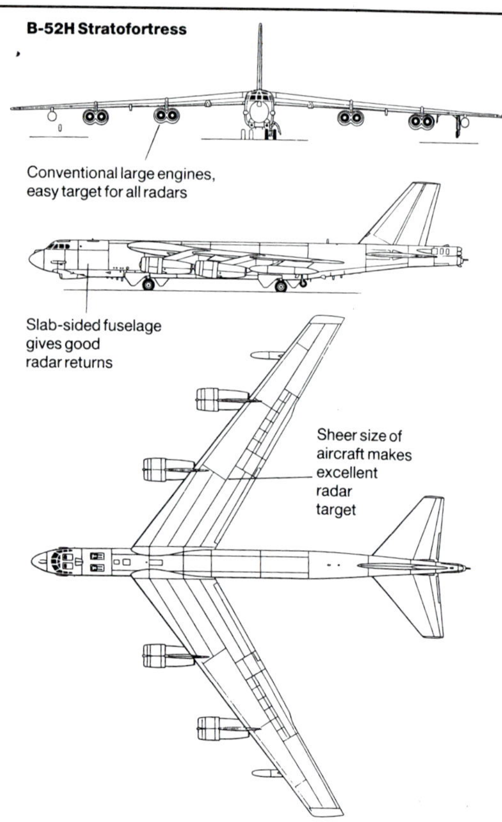


**Boeing B-52**
Despite its age the B-52 will remain in service in large numbers for many years to come. Modern technology – especially in electronics and avionics – is being used to enhance survivability and increase its capability in modern conditions.

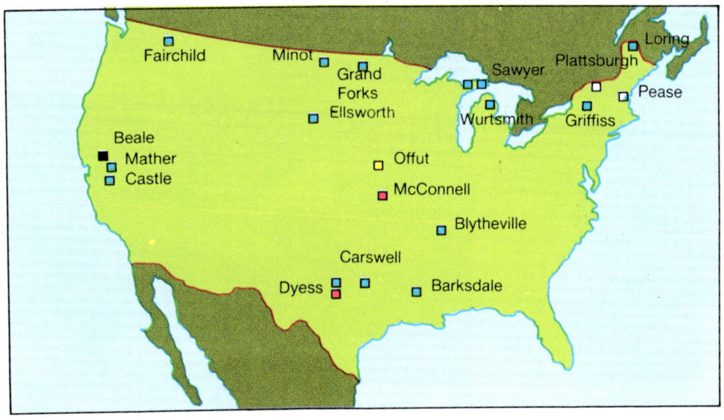


**USAF strategic aircraft bases in CONUS**
The currently planned disposition of SAC's major assets once the B-1B has entered service. These aircraft, together with the essential airborne tanker fleet of KC-135s and KC-10s, comprise the "air-breathing leg" of the USA's strategic triad.

- B-1
- B-52
- EC/RC-135
- SR-71
- FB-111

in June 1977 to cancel the programme came as a bombshell. Four years later, the US House and Senate Committees for Defense Appropriations gave the aircraft a second chance when they ordered the USAF to develop a new bomber. The alternatives were soon narrowed down to a derivative of the B-1, a stretched derivative of the FB-111, and an all-new design based on "stealth" technology.

The chosen solution was to order 100 B-1B bombers and to begin development of a new Advanced Technology Bomber (ATB) for service in the 1990s. Compared with the B-1A, the B-1B has a lower top speed, a lower radar cross-section, and will carry a heavier payload including cruise missiles.

In January 1982, Rockwell received an $886 million contract for the new B-1B. On March 23 of the following year, the second B-1A prototype was restored to flying status, allowing tests to be made on the B-1B fixed-geometry engine inlets, flight control system, and composite bomb bay doors.

The first B-1B production aircraft was rolled on 4 September 1984 (a week after the crash of the No. 2 prototype at Edwards AFB), and flew on 18 October. Delivery of the first operational B-1B was scheduled for the summer of 1985, with the first squadron being declared operational in August 1986. By early 1987, Rockwell hopes to hit a production rate of four aircraft per month, completing deliveries of the planned fleet of 100 in June of the following year. At a programme unit cost of $283 million, the B-1B is expensive, and

**Above: Like the Soviet "Bear" bomber, the B-52 keeps on finding new roles. This aircraft is laying sea mines, a potentially very valuable mission.**

**Below: The SAC strategic inventory still shows 55 FB-111As, armed with four SRAMs. As B-1B enters service FB-111s will transfer to Tactical Air Command.**

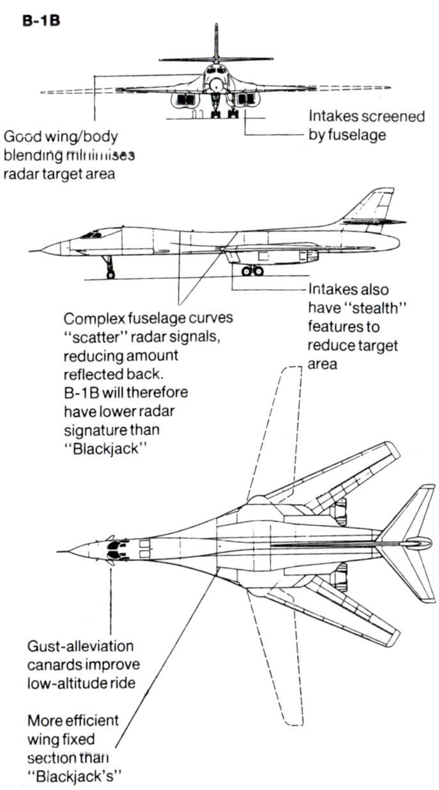


**Rockwell B-1B**
The B-1B is a major advance in design and is not just a revamp of the earlier B-1A. It possesses extremely efficient aerodynamics giving a good ride at low level. It also has a number of "stealth" features to reduce its radar signature.

there are no plans for a follow-on order.

Although planned as an interim bomber, the B-1B should have a long operating life. It should be able to penetrate deep into heavily defended airspace well into the 1990s, and to survive in the face of less sophisticated defences beyond the year 2000. Planned alternate roles include stand-off missile carrier, and naval-support missions such as maritime surveillance and minelaying.

The development contract for the highly-classified Advanced Technology Bomber was awarded to Northrop, a company whose prior bomber experience was confined to the unsuccessful YB-49 Flying Wing of the late 1940s, but an organisation which has significant experience in carbon-fibre technology and ECM.

Production of this new "stealth" bomber is expected to follow close behind that of the B-1B. ATB is expected to be smaller than the B-1, and will probably have a metal load-bearing structure. First flight is planned for 1987, and the first batch from a planned run of up 150 could enter service in the early 1990s, replacing the B-1B in the penetration role, and allowing the latter to replace the B-52 as a stand-off missile carrier. According to the US DoD, the ATB should maintain SAC's ability to penetrate Soviet air defences into the 21st century.

**Above: USAF's B-1 was designed to meet an urgent need to replace the ageing B-52 and to provide a stategic low-level penetration capability.**

**Above: This satellite picture shows the Tupolev "Blackjack" strategic bomber prototype beside two Tu-144s, apparently being used for experimental purposes.**

**Above: AS-6 cruise missile beneath a Tupolev Tu-22M "Backfire" bomber. These nuclear-armed missiles would be used to attack targets in Europe and Asia.**

## The Soviet bomber force

The Soviet Union does not have – and never has had – a viable bomber force capable of flying deep-penetration strategic missions into US airspace. The turboprop-powered Tu-95 "Bear" was always vulnerable to interception. Early-model "Bison" bombers used Mikulin AM-3 turbojets and were short of range; by the time that the definitive model powered by D-15 two-spool engines was available, the aircraft was incapable of avoiding USAFs F-102 interceptors, let alone the later F-106. The only other long-range bomber project – the massive Mach 1.8 Mya M-52 "Bounder" - was as bold a design for the late 1950s as the "Blackjack" is by today's standards. Several prototypes were built, but the type never entered production – presumably due to performance deficiencies.

Surviving "Bears" and "Bisons" still serve in modest numbers, along with the shorter-ranged Tu-16 "Badger" and Tu-22 "Blinder", but many are now used as tankers, ECM aircraft or as air-to-surface missile carriers. The Tu-95 "Bear" has recently been returned to production as a cruise missile carrier.

At least 130 Tupolev Tu-26 "Backfires" - approximately half of the production run to date – are in service with Long Range Aviation, 40 in the eastern part of the Soviet Union, and the remainder in the northern and southern flank of the western USSR. Although once considered by the USAF as a threat to the continental USA, Backfire lacks the range for such missions, except as a "one-way trip" with a final landing attempt in Latin America. Production of this aircraft continues at around 30 per year.

Operational capability of the Soviet Air Force's Long Range Aviation units should receive a boost from 1987 onwards when the Tupolev "Blackjack" variable-geometry bomber enters service. Larger and faster than the B-1, it will be the first effective intercontinental bomber fielded by Long Range Aviation, arriving in service at a time when the USA is protected by less than a dozen USAF fighter squadrons plus about half that number of Air National Guard units. US-based Nike Hercules SAMs have all been withdrawn from service. By the early 1990s, it will have replaced the "Bear" and "Bison" fleet.

The massive airframe of "Blackjack" has sufficient internal volume for a large fuel load. Powered by four large turbofans – probably a new-technology turbofan in the thrust class of the Tu-144 supersonic airliner's Kuznetsov NK-144s – the aircraft has the range to fly combat missions against the continental USA, and enough performance to give it a fair chance of surviving the mission, returning to base, and being able to repeat the process. It will also give the Soviet Government the heavy high-explosive firepower needed to intervene in Third World conflicts in support of Soviet allies, and a useful long-range maritime-strike capability.

According to the US DoD, "Blackjack's" tactical radius operating from bases in the northern USSR but not refuelling in flight would allow coverage of Canada, the north-western USA, Morocco, Algeria, Egypt, most of Saudi Arabia, northern India, Laos, and the northern part of Vietnam. Given air refuelling, the bomber could range as far as Mexico, Nigeria, Ethiopia, and Sri Lanka.

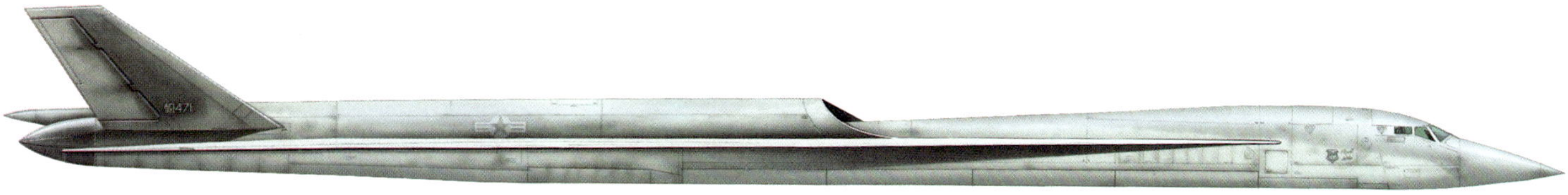

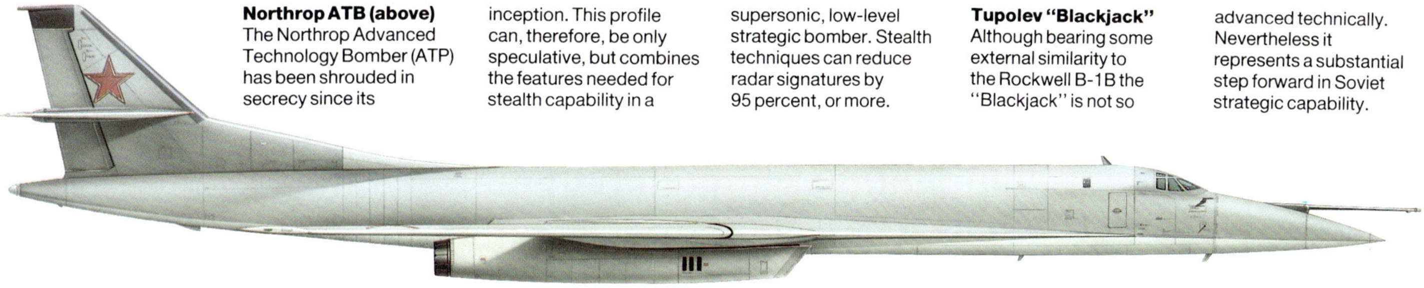

**Northrop ATB (above)** The Northrop Advanced Technology Bomber (ATP) has been shrouded in secrecy since its inception. This profile can, therefore, be only speculative, but combines the features needed for stealth capability in a supersonic, low-level strategic bomber. Stealth techniques can reduce radar signatures by 95 percent, or more.

**Tupolev "Blackjack"** Although bearing some external similarity to the Rockwell B-1B the "Blackjack" is not so advanced technically. Nevertheless it represents a substantial step forward in Soviet strategic capability.

**Below: Tupolev Tu-22M "Backfire" flying high above the Baltic, seen by a Draken of the Royal Swedish Air Force. The "Backfire" caused very considerable controversy when it entered service, as it was seen to pose a major threat to the USA, but it is now appreciated that it lacks the range for anything more than a one-way flight, recovering to a Latin American country. "Backfire" continues in production at 30 per year, and serves with the Soviet Navy and Air Force.**

How effective it would be at avoiding the depleted US defences is hard to assess, since this would depend on the performance of the aircraft's ECM systems. With the deployment of the AS-15 cruise missile, "Blackjack" will be able to avoid most defensive systems, launching its weapons from stand-off range.

## Europe's meagre bomber force

For Britain's Royal Air Force, the 1982 Falklands War was the finale of some 65 years of manned bomber operations. The last surviving Vulcan bombers – a handful of K.2 tankers – were retired in May 1984. France continues to operate around 35 Dassault Mirage IVA bombers, but this large delta-winged bomber is beginning to show its age. The original armament was a single first-generation nuclear weapon carried in a semi-buried ventral location, but 18 are being retrofitted to the Mirage IVP standard as carriers for the ASMP cruise missile. The modification will add an inertial navigation system (the IVA uses a Doppler radar as its primary navaid), computers, new cockpit displays, and a digital datalink used to pass data to the missile prior to launch. An earlier rebuild programme modified 12 aircraft as Mirage IVR photo-reconnaissance aircraft.

## China's aging bombers

The only other nation to operate strategic nuclear bombers is China, which has a fleet of more than 100 Xian B-6 (Tu-16 "Badgers"). Despite its age, the piston-engined B-29 remains in Chinese service in the form of around 80 Tu-4 copies. The military usefulness of such an old design must be minimal, but like the Soviets, the Chinese seldom throw military hardware away. Data sheets on Chinese high-explosive bombs issued in 1984 even included Soviet-built World War II combat types such as the Tu-2 as being suitable carriers for such stores.

Small numbers of Tu-16 bombers serve with the Egyptian Air Force, while Libya and Iraq operate the Tu-22 Blinder. Although able to carry free-falling high-explosive bombs, these constitute only token long-range bombing forces. Today, only the superpowers can afford to deploy long-range bomber arms.

# SPACE WARFARE

Space has been used for military purposes for over thirty years. The United States launched its first successful reconnaissance satellite in 1960 and the Soviet Union followed suit in 1962. Since then, satellites have been developed to perform a wide variety of military tasks – ranging from electronic eavesdropping to communications – and space systems have played an ever-increasing role in terrestrial military operations.

The military exploitation of space is dominated by the Soviet Union and the United States. France, the United Kingdom and Italy have military communications satellite programmes, and China has launched some reconnaissance satellites, but these efforts are dwarfed by those of the two superpowers.

The Soviet Union launches about 100 satellites a year, of which about 70 per cent are for exclusively military purposes and 15 per cent perform dual military and civilian functions. The United States generally launches less than 20 military satellites per year. This contrast in Soviet and American military satellite launches is misleading, however. American satellites have much longer lifetimes and are more versatile; the number of active satellites maintained by each nation is about 120.

In the future, both nations will become even more active in space as the United States takes advantage of the Space Shuttle and the Soviet Union brings into service a new, heavy-lift rocket or re-usable shuttle similar to the American one.

At present, the overwhelming majority of both superpowers' space effort is devoted to "non-weapon" purposes, such as intelligence gathering, communications and navigation. Space-based assets are "force multipliers", allowing many traditional military missions to be conducted more efficiently, although as technology advances space systems are evolving into "force enablers", opening up new mission possibilities. To cite just a few examples, satellite systems now existing or in development will enable global control of forces, improved tactical communications, all-weather navigation, precise weapons delivery independent of range, and long-range naval target acquisition.

As reliance on military satellites has grown, so too has the awareness that the loss of satellites in wartime could be a crippling handicap. This has led to the development of techniques for disabling an opponent's satellites – either by destroying or jamming them – and has also led to the development of methods of protecting

**Space systems are becoming increasingly crucial for the conduct of terrestrial military operations, so the loss of key satellite systems could be a critical handicap in a conflict. The outcome of war on Earth could depend on the outcome of war in space.**

**Above: Launch of the Soviet A2. More than 700 launches have taken place. The A2 and a variant, the A2e, are used for manned and unmanned launches.**

**Left: The Space Shuttle Challenger lifts off on its first night launch. The Shuttle plays a major part in US military space plans as both a launcher and a test bed.**

satellites from enemy action. Thus, whereas space systems were formerly viewed merely as a means of aiding terrestrial military operations, space itself is now also seen as a new theatre of actual warfare.

War in space, however, may not be simply confined to a battle between satellites. Strategic ballistic missiles travel through space on their way to their targets, and technological progress in many areas has now opened up the possibility of destroying these missiles in space. Both superpowers have large research programmes to assess the feasibility of a host of possible space weapons which might be suitable for this purpose. Before such weapons can be produced many enormous technical obstacles must be overcome, but progress is being made on many fronts and some spectacular results have already been achieved.

The decision to deploy space-based ballistic missile defences will depend on the outcome of research projects now in hand as well as on the political and strategic climate prevailing in the future. But, as mentioned earlier, the decision to use space for military purposes was made over thirty years ago with the result that space is already highly militarised. What follows is a description of the military use of space now, plus an account of where research is leading.

**Below: Testing of the NATO IIIB military communications satelite. The US DoD uses this in return for equivalent capacity on a DSCS-2 satellite.**

There are essentially six tasks performed by military satellites: early warning and attack assessment; surveillance and reconnaissance; communications; navigation; meteorology; geodesy.

## Early warning and attack assessment

Intercontinental ballistic missiles (ICBMs) take only about 30 minutes to reach their targets and submarine-launched ballistic missiles (SLBMs) might take only 10 minutes, depending on the launch and target locations. These short flight times make it essential to detect and assess a nuclear attack as rapidly as possible. To achieve this, the superpowers employ early-warning satellites which use infra-red sensors to detect the tremendous heat from a ballistic missile's exhaust, seconds after launch. These satellites also monitor missile tests and satellite launches.

The United States maintains three early-warning satellites in geostationary orbit. Launches from the Soviet Union and China are monitored by a satellite over the Indian Ocean, and satellites over the Pacific Ocean and South America monitor SLBM launches from the Pacific and Atlantic Oceans.

These satellites – known as Defense Support Program (DSP) satellites – are equipped with a Schmidt telescope 12ft (3.63m) long, with a 3ft (0.91m) aperture. At the telescope's focus is an array of 2000 lead sulphide infra-red detectors, each of which scans an area 3.7 miles (6km) across. Orientation is maintained by spinning the satellite at about 6rpm around the Earth-pointing axis. The telescope is offset from this axis by 7.5°, producing a conical scanning pattern. By plotting an infra-red source over several scans, it is possible to determine whether the source is stationary – a forest fire, for instance – or moving. A missile can be identified within a minute of initial detection, which usually occurs as it breaks through cloud cover.

In 1975, an early-warning satellite was temporarily blinded and there was concern that a Soviet ground-based laser was responsible, though the US Defense Department later stated that blinding was caused by an intense fire at a broken gas pipeline.

**Below: The Defense Support Program Satellite (DSPS). This spin-stabilized, early-warning satellite monitors missile tests and would provide the first information about a nuclear attack.**

However, the Defense Department does believe that the Soviet Union has two ground-based lasers that could have anti-satellite capabilities. Fears about laser blinding – or the prospect of more powerful laser weapons – are taken so seriously that the United States has embarked upon several programmes to increase the survivability of early-warning satellites and plans are also in hand to improve their sensing capability. The High Altitude Low Observable (HALO) programme is intended to produce laser-resistant, high-resolution sensors which stare constantly at their target areas instead of scanning them periodically. Ablative materials are also to be used to make early-warning satellites less vulnerable to laser weapons. These features should be integrated into the next-generation Satellite Early-Warning System (SEWS) to be deployed in the early 1990s.

Early-warning satellites are also equipped with devices to detect nuclear explosions. In peacetime, this helps verify compliance with the treaty banning atmospheric nuclear tests. In a nuclear war, these sensors would enable rapid damage assessment to be carried out. By 1988, however, the US DoD will perform this function using the Integrated Operational Nudet (nuclear detonation) Detection System (IONDS). IONDS sensors will be fitted to Navstar navigation satellites and will be able to provide precise data on the location, altitude and yield of nuclear explosions.

Virtually all Soviet satellites are known simply by a number in the Cosmos series and early-warning satellites are no exception. Cosmos 520, launched in September 1972, was the first early-warning satellite to be identified and gradually over the next ten years a 9-satellite early-warning system was assembled. The satellites are spaced at regular intervals around a highly elliptical, semisynchronous orbit inclined at 62°. Geostationary orbits are difficult to achieve from the Soviet Union's northern launch sites, but the orbit used by Soviet early-warning satellites provides good coverage of the Northern hemisphere, and each satellite can view American ICBM sites for up to six hours per 12-hour orbit.

## Surveillance and reconnaissance

Surveillance and reconnaissance satellites – often known as spy satellites – are used to obtain information about a multitude of military activities. Surveillance is a relatively regular monitoring activity, whereas reconnaissance is a search for specific intelligence, possibly of a more urgent nature. Although the tasks are different, they are increasingly combined on a single satellite platform equipped with several sensors.

Photographic surveillance and reconnaissance satellites use optical, infra-red and – possibly – radar techniques to obtain highly detailed pictures of areas of interest. The information produced can be of incalculable value. The strengths and locations of forces can be determined with considerable precision, and weapons under construction or testing can be observed in remarkable detail. The resolution of satellite photographs is so fine that, for instance,

**Orbital characteristics**
Low-earth orbits are used mainly by weather, surveillance and reconnaissance satellites. These orbits lie between 90 and 1,240 miles (150 and 2,000km) with inclinations between 65 and 115 degrees. Highly elliptical Molniya orbits are used by the USSR for early-warning and communications satellites. Inclination is typically about 63 degrees and satellites descend to altitudes of a few hundred miles over the Southern Hemisphere and rise to c.24,855 miles (40,000km) over the Northern. Semi-synchronous orbits are used mainly by navigation satellites. Altitude is c.12,425 miles (20,000km): inclination is 63-65 degrees. Geostationary orbits (altitude 22,370 miles/36,000km) are used mainly by the US for early-warning, communications and some elint systems. These satellites remain above a fixed point on the Equator.

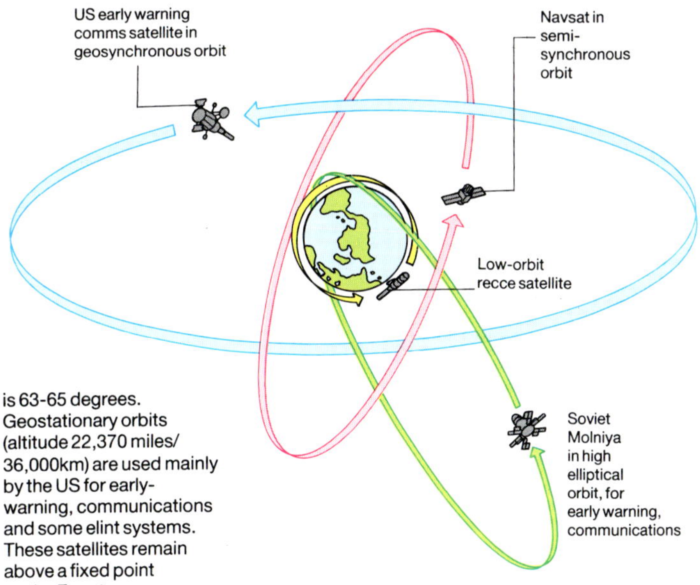


**Below: Impression of a satellite incorporating technology based on the High Altitude Low Observable programme. Sensors would enable monitoring of terrestrial activity from high orbits.**

**Bottom: In 1986, a Teal Ruby satellite is due for launch. This will be equipped with a mosaic of hundreds of infra-red sensors and will test the feasibility of tracking aircraft from space.**

the size of an aircraft engine's air intake can be measured. These satellites enable a weapon's performance to be estimated while still in the prototype stage, perhaps years before it is fielded. Military exercises and actual conflicts can be monitored and such satellites are also the primary means of verifying compliance with treaties. Both superpowers observed events during the Falklands and Iran-Iraq wars using satellites.

Probably the best-known spy satellite in the American inventory is Big Bird. This can perform both wide-area surveillance and "close-look", high-resolution reconnaissance with multi-spectral scanners. Its cameras can identify objects as small as 12in (30cm) across, its film being processed on board, scanned by an optical system and transmitted to receiving stations on Earth. These transmitted images are of poorer quality than photographs so, if exceptionally fine detail is required, Big Bird can jettison film capsules which are recovered in mid-air by specially equipped HC-130 aircraft based in Hawaii.

Big Bird satellites are placed into low altitude, sun-synchronous orbits so that they pass over their targets at the same time each day. The orbit can take the satellite as low as 100 miles (160km) where atmospheric drag would normally cause re-entry in about a week. Big Bird, however, has rocket motors which are used periodically to nudge the satellite back into position, extending its life to about 200 days.

The United States also uses two other photographic reconnaissance satellites designated Key-Hole (KH) -8 and KH-9. Their existence only became public knowledge in 1983 and details are sparse, though they are known to be film-return type satellites whose use is confined to photographing intelligence targets of the highest priority. Their orbits are low and lifetimes are limited: KH-9 can probably remain aloft for about four months.

Production of new Big Bird, KH-8 and KH-9 satellites has ceased and increasing use is being made of KH-11 which, unlike its predecessors, does not use a film-return system. Instead, it uses a digital imaging device which reportedly can achieve resolutions comparable to Big Bird's despite the fact that it operates at a higher altitude. The higher altitude extends orbital life to perhaps more than two years, and the imaging system makes information available almost instantaneously. KH-11 has multi-spectral and infra-red sensors and – according to some reports – a radar sensor capable of penetrating cloud cover.

KH-12 is reportedly under development, almost certainly with radar sensors, with improvements over KH-11 likely to include higher resolution imaging.

The Soviet Union still relies heavily upon photography rather than data-imaging and two types of photo-reconnaissance satellite are in current use. The third-generation type was introduced in 1968 (Cosmos 208) and can remain in orbit for up to 14 days. A fourth-generation, thought to incorporate some Soyuz hardware, was introduced in 1975 with the launch of Cosmos 758. This type is highly manoeuvrable and has a mission life of up

**Left: A Titan IIID carrying a Big Bird reconnaissance satellite is launched from Vandenberg Air Force Base, Calif. KH-11 is also launched by the Titan IIID.**

**Above: An HC-130 aircraft catches a simulated film-return capsule of the sort jettisoned by reconnaissance satellites such as Big Bird.**

to 60 days. These short life-times explain why the Soviet Union launches about 35 photo-reconnaissance missions per year, compared with the American rate of around three per year.

The Soviet Union also uses its Salyut space stations for photo-reconnaissance. Although all missions perform a mixture of civilian and military functions, Salyuts 3 and 5 were of a predominantly military character. The 33ft (10m) focal length camera, ostensibly for solar observations, was better suited for reconnaissance and the lower orbit of Salyuts 3 and 5 also implied this function. Furthermore, the crew were military officers, code words were used to conceal the meaning of conversations between the crew and ground-control, and telementry was transmitted in forms previously associated with reconnaissance satellites.

In April 1982, Salyut 7 was launched on a primarily civilian mission, but the station exhibited a capability which will prove extremely useful in sustaining military tasks. In March 1983, Cosmos 1443 docked automatically with Salyut 7, which at that time was unmanned. The Cosmos vehicle carried a large propellant load which was transferred to Salyut. This ability to refuel Salyut automatically, thereby increasing its orbital life and manoeuvrability, will be particularly useful for lengthening low-orbit reconnaissance missions which require frequent nudges to stay aloft.

Photographic reconnaissance is by no means the only method of intelligence gathering from space. Electronic intelligence (ELINT) satellites – also known as ferrets – have many applications. Ferrets can locate a radio transmitter, eavesdrop on communications and monitor the telemetry from missile tests. A radar's purpose can be established by examining

**Big Bird and KH-11**
Big Bird's very high resolution cameras provide broad views from which specific targets are selected for discriminatory analysis. Some pictures are returned to Earth in recoverable capsules as shown. Infra-red pictures can reveal the presence of heat-generating equipment and camouflage systems. Big Bird is backed up by the KH-11 digital reconsat which can take TV images of high clarity. These images are available in near real-time for detailed analysis.

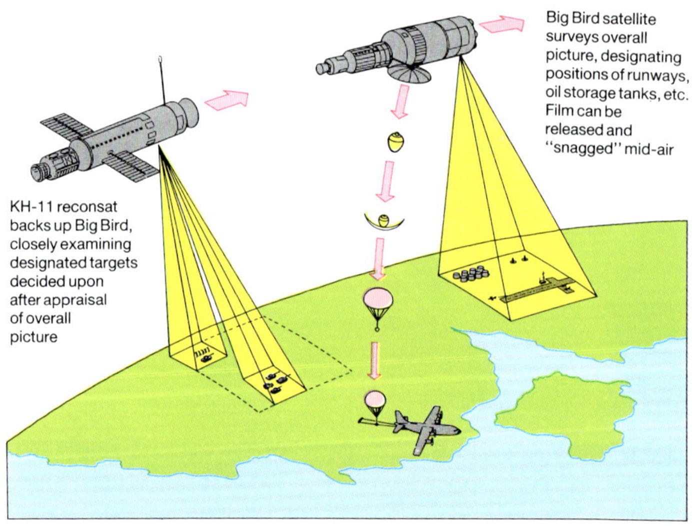


pulsewidth, pulse-repetition frequency, transmitter frequency, and modulation. This type of information is useful for studying peacetime military activities and it greatly facilitates the planning of wartime operations. Radar sites which have been located and identified by ferret satellites can be attacked – or avoided – by penetrating aircraft and effective electronic countermeasures can be developed to jam radio communications and deceive radars.

At least two types of American ELINT satellites are known to be in service. One is known as a subsatellite ferret usually launched along with Big Bird but subsequently boosted into a circular orbit at an altitude of about 300 miles (500km), or in some cases over 900 miles (1,500km). The principal purpose of subsatellite ferrets is to conduct general surveys, detecting new radar sites and monitoring changes at known ones.

The other type of US ferret is known as Rhyolite and two of these are maintained in geostationary orbits, one above the Horn of Africa and the other over the Indian Ocean. Rhyolites bristle with antennae, the largest being over 65ft (20m) in diameter. On-board electronics filter out background noise and jamming signals. It is believed that one of Rhyolite's primary missions is to monitor missile launches from the Tyuratam and Plesetsk launch sites. The Rhyolites now in operation were launched in the early 1970s and may already have been supplemented by their eventual successor, known as Aquacade. Virtually no information about Aquacade has been published, but in view of the remarkable progress made in electronics during the last decade there can be no doubt that Aquacade will be more sensitive and powerful than Rhyolite.

The Soviet Union employs various sorts

**Above: The Soviet Union's Salyut 7 space station. Salyut space stations are used for both military and civilian purposes. Surveillance and reconnaissance tasks form a major part of military missions. Docking mechanisms at each end of the space station can be used by manned capsules or unmanned re-supply vehicles.**

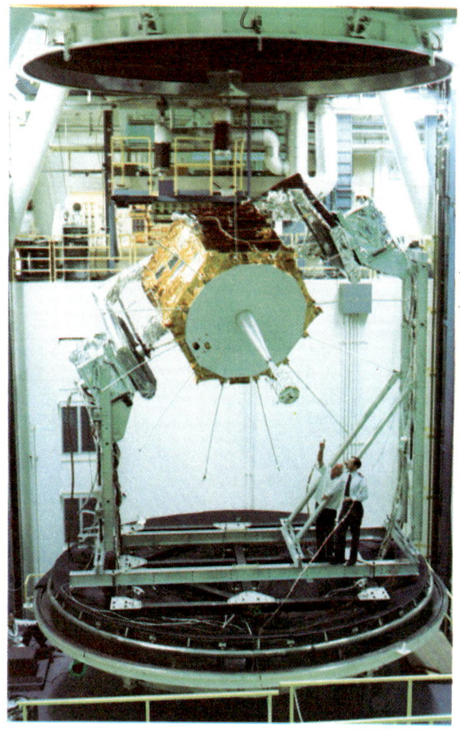

**Above: A prototype of the US Navy's Fleet Satellite Communications (FLTSATCOM) satellite undergoing tests. FLTSATCOM became operational early in 1981.**

of ELINT satellite but, as is generally the case, little is known about them. A series of eight ferrets operates in a 400 mile (650km) orbit with an inclination of 81.2°. Their task is to identify and pinpoint military radars and radio transmitters. Other ferrets are launched into a variety of orbits.

Satellites fitted with ELINT equipment and various other sensors are also used for ocean surveillance. The main American satellite system for this purpose is White Cloud, which consists of three clusters of satellites spaced at 120° intervals around a 680 mile (1,100km) orbit inclined at 63.3°. Each cluster consists of three small satellites, each weighing only a few tens of kilograms, dispersed from one "parent" satellite. The clusters orbit in a formation which allows the data from infra-red and millimetre wave sensors on each satellite to be collated, providing information on surface vessels more than 1,865 miles (3,000km) distant.

Work is taking place with a view to fitting satellites operating along the lines of White Cloud with radar, colour scanners, scatterometers and sensitive infra-red detectors. These will provide detailed information about surface vessels and may even enable the detection of warm water trails from nuclear-powered submarines. Equipment for this is being developed as part of the National Oceanic Satellite System which will also provide ocean data for civilian applications. Another programme known as the Integrated Tactical Surveillance System (ITSS) will provide the techniques for monitoring naval vessels, aircraft and missiles.

The Soviet Union operates ELINT ocean reconnaissance satellites (EORSATs) at altitudes of about 280 miles (450km) with inclinations depending upon mission. They frequently operate in pairs and have low-thrust engines to maintain correct height and spacing. EORSATs are used to monitor communications and radar emissions from naval missions. Radar-equipped ocean-surveillance satellites (RORSATs) are used to detect and track surface vessels.

RORSATs generally operate at altitudes of around 155 miles (250km) and, even using their thrusters to maintain altitude, orbital lifetimes are limited to about 70 days. However, these satellites are powered by nuclear reactors containing around 50kg of slightly enriched uranium so at the end of a RORSAT's life the reactor section separates and is boosted into a 560 mile (900km) orbit, while the rest of the satellite falls to Earth. On at least three occasions this process has failed and the reactor has re-entered. In one case, on 24 January 1978, reactor fragments from Cosmos 954 fell over the Canadian Northwest Territories.

EORSATs and RORSATs are known to be used to track American carrier task forces in particular, but they are also used to monitor "targets of opportunity", such as the Falklands War.

## Communications satellites

The role of communications satellites (comsats) is obvious, though it is less well-known just how heavily military forces rely on them: over 70 per cent of all American overseas military communications are relayed by satellite, for instance. Furthermore, the importance of communication satellites is growing. As technology advances, sensors and weapons operate at greater ranges and the complexity of warfare increases dramatically. More than ever, real-time, long-range communications are essential.

The main American military satellite communications systems in use are the Defense Satellite Communications System (DSCS), the Air Force Satellite Communications System (AFSATCOM), the Fleet Satellite Communication System (FLTSATCOM) and the Satellite Data System (SDS).

DSCS is the Defense Department's primary network for long-haul, high-volume communications between major facilities. One network, DSCS-II, consists of four geostationary satellites plus two in-orbit spares. DSCS-II can handle 1,300 voice channels or 100 megabits per second. DSCS-III satellites are now being phased in and will provide a greater data-handling capacity. In addition, DSCS-III satellites are highly ECM resistant, hardened against EMP and have propulsion systems to manoeuvre in case of attack.

FLTSATCOM also consists of four geostationary satellites whose primary purpose is to provide naval communications. More than 900 relay links are in use on ground stations, surface vessels, submarines and naval aircraft.

AFSATCOM is not a satellite system in its own right, but comprises communications packages carried on other satellites such as FLTSATCOM, SDS, DSCS and others. AFSATCOM terminals are used by E-4B airborne command posts, RC-135 reconnaissance aircraft, Strategic Air Command bombers, TACAMO aircraft, and various ground stations such as ICBM command posts.

SDS is a three-satellite network in highly elliptical orbits to fill the polar "gaps" in communications which arise from using geostationary systems. This system may also be used to relay data from KH-11 reconnaissance satellites.

By 1990, the Military Strategic-Tactical

and Relay (Milstar) system should be operational, complementing and eventually superseding FLTSATCOM, AFSATCOM, DSCS and possibly SDS. Milstar will be the first system to employ the millimetre-wave region which will provide an enormous communications capacity. Four satellites will be in geostationary orbits – plus one in-orbit spare – and three will be in highly elliptical polar orbits. An unspecified number of spares will also be placed in supersynchronous orbits at altitudes of 110,000 miles (177,000km). Eventually, all Milstars may be placed in supersynchronous orbits to decrease vulnerability. The satellites will be hardened against EMP and lasers, and will also be cross-linked by laser and other systems.

Despite the sophistication of all these satellites, none can solve a specific communications problem which has plagued strategic planners for over 20 years: that of communicating with nuclear submarines. At present, submarines have to approach the surface to receive command instructions and this makes the submarines more liable to detection. Work is being done on extremely low frequency (ELF) transmissions which can penetrate the ocean depths but these would require huge antennae and have very poor data transfer rates. An alternative approach is to communicate using a satellite-borne blue-green laser. Blue-green laser light penetrates the atmosphere and sea water to a considerable depth. Tests using an airborne laser to communicate with a submarine have been very encouraging, so laser communication satellites may soon remove the chink in submarines' armour.

The Soviet Union currently employs three main types of military communications satellite. The Molniya I operates in an elliptical orbit inclined at 62.8°. A series of Molniya Is is maintained, stationed at 45° intervals. The orbital period is 12 hours and each satellite remains visible from the Soviet Union for about eight hours per orbit. Molniya I satellites are used for direct, point-to-point communications.

**Above: The US Navy's FLTSATCOM-A, first in a series of military communications satellites, undergoing weighting and balancing before launching into geosynchronous orbit in 1978.**

The second type of Soviet communications satellite is used for tactical communications. These satellites are launched eight at a time into 930 mile (1,500km) circular orbits inclined at 74°. To ensure global coverage 24 are needed, and 30 or more are generally operational, ensuring redundancy and resistance to jamming. These satellites are used for real-time command and control communications over the Soviet Union and Eastern Europe. They also have a store-dump facility whereby a transmission is recorded, then played back and re-transmitted. This is used for global communications.

The third type of Soviet comsat is a store-dump type which operates at 500 miles (800km) altitude. Only one is active at a time and it is believed that this type of satellite is used to collect data from remote sensors and spies.

Not yet operational but believed to be in development by the Soviets is a system known as GALS. This will be a four-satellite, geostationary system along the lines of the American DSCS.

## Navigation satellites

The first American navigation satellite (navsat) was known as Transit, originally developed to enable Polaris submarines to update their navigation systems accurately in all weathers. A series of Transit satellites operate in 680 mile (1,100km) circular orbits transmitting oscillating radio signals which can be picked up by small receivers. Position on Earth can be calculated to within 165 yards (150m) by measuring the Doppler shift in the transit signals and combining this with knowledge of the satellites' orbital motion. Since becoming operational in 1964, the system has been regularly improved. Most recently, Nova satellites incorporating more powerful transmitters, and with a better ability to compensate for orbital disturbances, have been used in the Transit network. Time lapses between position updates vary between 30 minutes (at high altitudes) and 100 minutes near the equator. The system is used by several thousand civilian and military operators world-wide.

Some years ago, military planners realised that technology had progressed to the extent that a navigation satellite system could be produced which would provide extremely precise position and velocity information to users equipped with a very

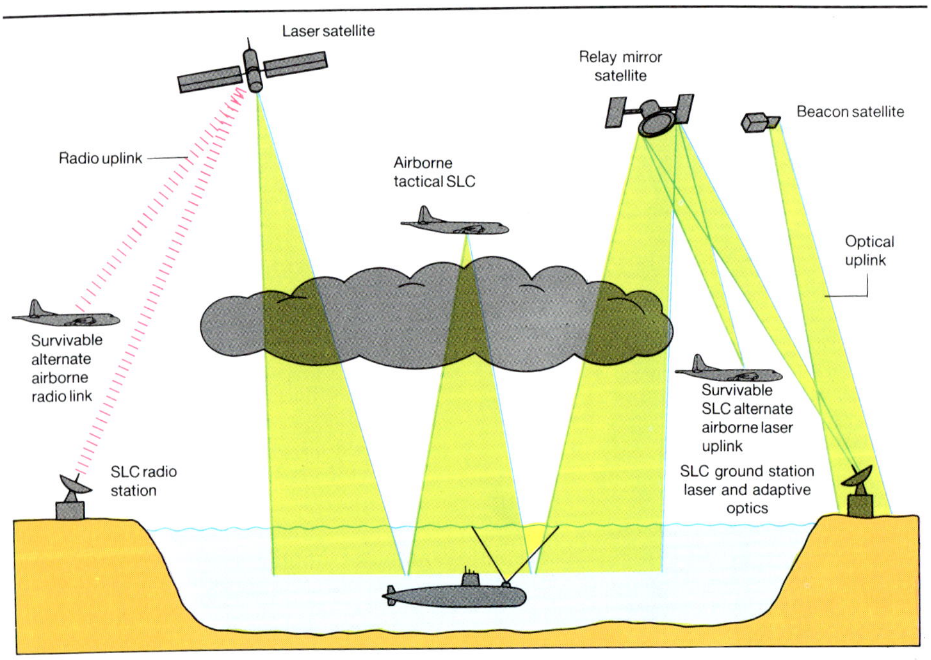


**Submarine laser communications (SLC)** Signals might be sent to submarines by laser techniques. Up-links could use radio to laser communications satellites which would transmit laser signals to submarines equipped with laser receivers. Or, laser signals could be relayed from a land-based laser by a space-based mirror using adaptive optics. Up-links could also be air based for survivability. Direct air-to-submarine signals may also be used.

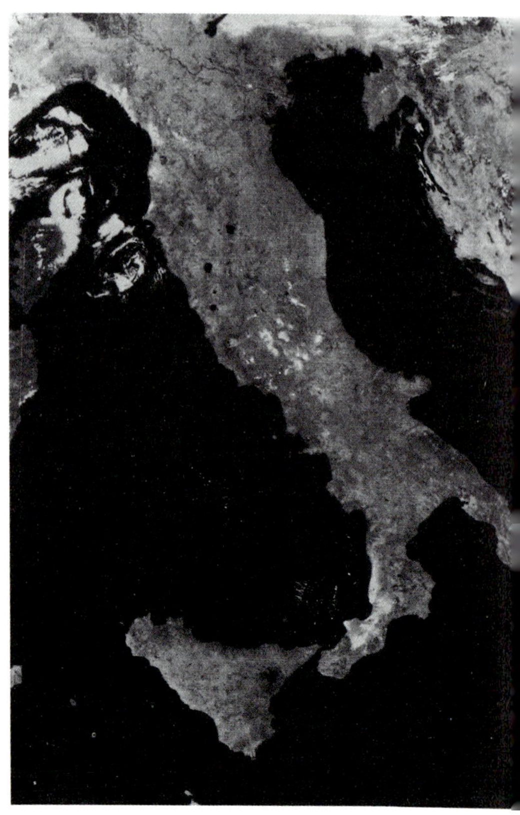

small receiver. Consequently, the Navstar Global Positioning System (GPS) programme was initiated. GPS will consist of 18 Navstar satellites placed at regular intervals around three orbital rings, each inclined at 63° to the Equator. The satellites' altitude will be about 12,425 miles (20,000km) with a period of 12 hours. Each satellite will transmit in two different codes, one for military use and the other for civilian use. The former will enable a user to establish position on Earth in three dimensions to within 16 yards (15m) and velocity to within a few centimetres per second. The military signal will be encrypted and highly resistant to jamming. User terminals will be man-portable and will remain passive. Navstar prototype satellites have been tested with considerable success and the full system should be operational by the end of 1988.

The applications for Navstar are almost limitless. It is anticipated that Navstar will service 20,000 users. Ships, aircraft, missiles and ground forces will be able to establish position with unprecedented accuracy using passive receivers. Weapons will be able to be delivered with remarkable precision over any range. SLBMs such as Trident will thus be able to achieve the same accuracy as ICBMs, and cruise missiles will be able to navigate without emitting any signals. Air support for ground forces will be improved, minesweeping will be more efficient since swept areas will be mapped more precisely. Reconnaissance satellites will be able to provide more accurate information about potential targets. Aircraft will be able to rendezvous for in-flight refuelling without radar or radio assistance.

Soviet navigation satellites follow the American pattern almost exactly. One system – comparable to Transit – is in 620 mile (1,000km) polar orbits and is maintained by about four launches per year. The Soviet Union's latest system will be operational shortly and is almost a carbon-copy of Navstar. The system – known as Glonass – will have both military and civilian users. The satellites will be in three orbital planes at an inclination of about 63° and an altitude of 12,425 miles (20,000km). Twelve satellites are planned, so coverage will be incomplete. This has led to speculation that Glonass military receivers might be designed to process Navstar's signals (possibly the "coarse", civilian frequency).

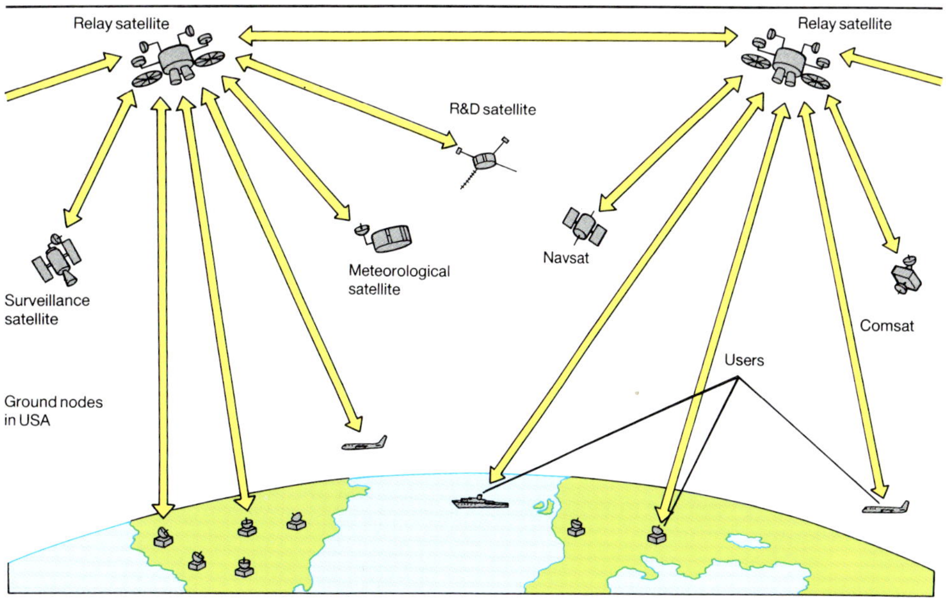


**Satellite Control System (SCS)**
In war, communications and satellite control facilities could be key targets. SCS has been put forward as a data relay system which would "network" huge amounts of command, control and communications systems. Surveillance, navigation, meteorological and communications satellites would be cross-linked with terrestrial armed forces and command centres to provide survivable data links.

## Meteorology by satellites

The United States uses civilian weather satellites for some purposes, but resolution is insufficient for many military requirements. Consequently, the Department of Defense's primary weather information system is the Defense Meteorological Satellite Program. Two satellites known as Advanced Meteorological Satellites (AMS) are maintained in 12-hour orbits and these provide optical and infra-red photographs with resolutions of either 2.3 or 0.37 mile (3.7 or 0.61km). Other sensors provide information about atmospheric water vapour, ozone content, and temperature. A gamma ray detector is also carried to provide data about nuclear explosions.

Meteorological satellites provide data about cloud cover so that reconnaissance satellite missions can be planned efficiently. At a tactical level, a mobile receiving station can provide real-time weather photographs covering a region 4,600 by 1,865 miles (7,400km by 3,000). Such information is useful for any military mission in peace or war, and is particularly valuable in planning air operations.

Soviet weather satellites – Meteor 2s – are in slightly lower orbits than their American counterparts and, presumably to compensate for this, three are kept operational. Resolution is believed to be poorer than AMS, but the sensor suite is otherwise very similar.

**Left: A picture of Italy produced by the US Defense Meteorological Satellite Program (DSMP). Optical and infra-red images can be obtained.**

**Below: DMSP satellite undergoing tests in an anechoic chamber. The sunshade is used to protect the sensor suite from direct sunlight.**

## Geodesy

Geodetic satellites are used to produce maps of the Earth using photographic and radar techniques. They also provide data about the Earth's gravitational and magnetic fields. Among other things, this information enables the trajectories of ballistic missiles to be predicted accurately and is essential for the guidance systems of cruise missiles.

The United States' Defense Mapping Agency has launched a variety of satellites since the mid-1960s in its Geodetic Satellite Program, and the Soviet Union has maintained a similar programme.

Satellites will play a key role in any terrestrial conflict. It is often said that satellites are "force multipliers", meaning that satellites allow forces to be used more effectively and efficiently. However, as systems such as Navstar and Milstar become operational towards the end of the decade, space systems will become essential rather than just useful.

This increasing dependence upon satellites has inevitably caused military planners to be concerned about the survivability of space systems while also considering the merits of destroying an adversary's space assets. Consequently, steps are being taken both to increase satellite survivability and to develop anti-satellite weaponry.

The first requirement for operating any anti-satellite (ASAT) weapon is a means to identify and locate an adversary's space systems. However, this requirement is less demanding than might be imagined since both superpowers have the necessary detection and tracking equipment in place or under construction as a result of other mission needs. Monitoring satellites is already essential to avoid collisions, to prevent re-entering satellites from causing false attack alarms, and simply to determine what the other side is doing. Also, many satellite-tracking facilities exist as a by-product of the infrastructure intended to detect and monitor a ballistic missile attack.

**Above: a USAF Pave Paws phased-array radar. Pave Paws is intended primarily to provide early warning of attacks by Soviet SLBMs and aircraft.**

**Below: A radar in the Ballistic Missile Early Warning System (BMEWS) is used to track Soviet missiles shortly after initial detection.**

But tracking objects in space is not an easy task. About 15,000 man-made items have been placed in orbit since the launch of Sputnik 1 in 1957. Many of these objects are rocket casings, dead satellites and various sorts of debris. Most have fallen back to Earth but the North American Defense Command (Norad) still tracks about 5,000 objects. In addition to these known space hazards, incidentally, there are more than 10,000 pieces of debris – mainly the products of over 70 explosions in space – which are too small to track.

The "front line" in the American space monitoring system is provided by early-warning satellites which produce initial tracking data for rocket launches from Soviet territory. Tracking is then handed over to the Ballistic Missile Early Warning System (BMEWS) which consists of three large radar installations at Fylingdales (England), Clear (Alaska) and Thule (Greenland). More sophisticated "Pave Paws", phased-array radars at Otis (Massachusetts) and Beale (California) also feed satellite information to Norad, though their prime purpose is to detect SLBMs launched from the Atlantic and Pacific Ocean. (Two more Pave Paws systems are planned in Georgia and Texas.) A global network of radars and cameras known as Spacetrack also makes 18,000 satellite observations per day and is able to assess the exact shape and size of space objects. (It was this system which in 1973 was able to

Above: The Ground-based Electro-optical Deep Space Surveillance (GEODSS) system. The low light level TV camera is clearly visible on the back of the telescope.

Below: The Twin 26ft (8m) domes of the GEODSS Experimental Test Site at White Sands, N.M. GEODSS is expected to be fully operational by 1987.

determine that a solar panel aboard Skylab had failed to open.)

By 1987 the Ground-based Electro-Optical, Deep Space Surveillance (GEODSS) system will also be fully operational. GEODSS will employ telescopes coupled with photo-imaging tubes such that objects "no bigger than a football" can be identified in orbits between 3,420 and 23,000 miles (5,500 and 37,000km). GEODSS stations will be sited at White Sands (New Mexico), Taegu (Korea), Maui (Hawaii), Diego Garcia (in the Indian Ocean) and one site in Portugal.

All these systems – and more – are being integrated into a Consolidated Space Operations Center (CCSOC) which will be manned by 1,800 personnel.

The Soviet Union has systems broadly comparable to BMEWS and Pave Paws. There are installations at Pechora, Kiev and Komsomolsk-na-Amure, all on the periphery of the Soviet Union, plus a large installation at Abalakova in central Siberia. The Soviet Union maintains that this is for satellite tracking but its real purpose may be to guide anti-ballistic missiles. The Soviet Union also operates a fleet of satellite communications vessels as well as ships equipped with spacecraft tracking radar. (Ships are used since the Soviet Union lacks suitable overseas bases.)

The United States and the Soviet Union are both developing ASAT weaponry, though the systems operate in completely

Below: A geodetic satellite fitted with reflectors to facilitate laser measuring techniques. Geodetic satellite data enhances ballistic missile accuracy.

Below: Due to its lack of overseas bases the Soviet Union operates a fleet of satellite communications vessels named after dead cosmonauts. The *Kosmonaut Yuri Gagarin* is the largest of these. Data is relayed to and from the appropriate flight control centre either direct by short-wave or via satellite.

different ways. The American system consists of a two-stage missile attached to a device known as a miniature homing vehicle (MHV). The missile is launched from a slightly modified F-15 which is directed to its launch area by Norad. After launch, the missile carries the MHV into space and hurls the MHV towards its target. The MHV employs an infra-red homing system to close in on the target and is intended to collide directly with it. Collision velocity should be about 8 miles per second (13km/s) so the impact is roughly comparable to a direct hit by a 16-inch shell.

This interception method – known as direct ascent – gives very little warning time for a satellite to conduct evasive manoeuvres. Furthermore, although initial deployment is planned only at Langley and McChord Air Force Bases, in the USA, many other sites could be used and with only minor alterations the system could be employed by aircraft carriers. Furthermore, although the system under development is intended for low-altitude interception (up to around 340 miles (550km) the technique could be applied to higher altitudes relatively easily. Barring a political decision cancelling development, the American ASAT system could be operational in 1987.

The Soviet Union already possesses an ASAT capability based on a technique known as co-orbital interception. The method is simple in principle but difficult in practice, as the high incidence of failures has shown. The intercept mechanism weighs about 4,400lb (2,000kg) and is equipped with either radar or infra-red sensors (both types have been tested). This too is a low-altitude system: all tests have taken place at altitudes of less than 1,550 miles (2,500km).

An SS-9 rocket launches the intercept mechanism into an orbit which "grazes" that of the target satellite and the launch is timed such that within one or two orbits the two objects are close to one another. The interceptor uses thrusters to approach to within about one kilometre of the target and then explodes sending a cloud of fragments, rather like a shotgun blast, towards the satellite. This technique can take up to 3½ hours from launch to interception, ample time for a satellite to attempt to maneouvre out of harm's way or take some other evasive action.

A slight variation of this interception method, however, speeds the process up dramatically. Here the interceptor is placed into a low parking orbit, beneath the target's orbit. The lower orbit means a shorter orbit time so the interceptor rapidly catches up its target. At the appropriate point, the interceptor fires its engines and "pops up" into the target's orbit and follows the normal homing-destruction sequence. Using this technique, the time between launch and interception can be as short as 30 minutes, leaving little time for evasive action.

The overall success rate so far achieved by this system is around 50 per cent. It is too early to tell what success rate the US system might achieve but, for various reasons, the US system will have comparative advantages. To cite just two, the US ASAT will enjoy considerable launch flexibility and, due to their orbital geometry, more varieties of Soviet satellites are vulnerable to low-altitude interception.

These systems are by no means the only ASAT possibilities. A threat which satel-

**Above: Impression of the Soviet co-orbital anti-satellite weapon. This system, incorporating radar or infra-red sensors, has achieved about a 50 per cent kill rate.**

**Right: Cutaway of US miniature homing vehicle (MHV). Diameter is 12in×13in (30×33cm). Steering is by 56 single-shot rocket tubes around the periphery.**

**Below: An F-15 Eagle performing captive flight tests with a mock-up of the US direct-ascent anti-satellite weapon. This system offers considerable flexibility.**

lites already face is electronic countermeasures. An adversary can attempt to interfere with a satellite's data and command links in an effort to throw a satellite off-course or to jam signals transmitted to and from a satellite. Another approach is to illuminate a satellite with a ground-based laser. Some American satellites may already have been temporarily blinded in this way and Soviet laser facilities at Saryshagan may be suited to this purpose. By increasing their power, however, ground-based lasers might destroy satellites by causing them to overheat or – eventually – by causing them physical damage. Another possibility is to place lasers on high-flying aircraft to reduce atmospheric interference, though the power of such lasers would be limited.

The "natural" environment for laser weapons, though, is space where the beam can propagate free from atmospheric absorption and distortion. According to some reports, the Soviet Union is well on the way to building satellites equipped with lasers for use against satellites, and feasibility studies have also been conducted in the United States. One estimate predicts that the Soviet Union may have six anti-satellite, high-energy lasers in orbit by 1990.

A less exotic ASAT weapon is the space mine. This concept calls for satellites armed with conventional explosives to be placed in orbit near their intended targets. At the outbreak of hostilities, these satellite "shadows" would be commanded to explode, destroying the target satellite.

Nuclear explosions in space have also been proposed as anti-satellite weapons. These would be effective in two ways. The electromagnetic pulse could upset the electronics on satellites thousands of miles from the detonation and the explosion would also interfere with radio transmission. This is an indiscriminate ASAT technique though, since the effects would be felt by both "hostile" and "friendly" satellites.

Another means of degrading satellite systems is to attack their ground facilities. Options available include sabotage, conventional weapons, nuclear attack and high-altitude electromagnetic pulse.

All these potential threats are being taken so seriously that current satellite designs incorporate countermeasures against them. Solar panels are being developed which are more resistant to nuclear effects. Data and command links are being encrypted to avoid electronic spoofing while frequencies and transmission modes are being selected to provide resistance to jamming and EMP interference. Hardening against lasers is increasingly becoming a standard feature and satellites are also being given additional manoeuvring capability to evade interceptions. Also being studied is the use of "stealth" technology to render satellites less visible to radar.

Further into the future, satellites may be fitted with more aggressive survivability aids such as lasers or homing projectiles, intended to destroy approaching interceptors. Despite these planned aids to survivability, losses are clearly deemed inevitable and spare satellites are being placed in orbits ready to plug any gaps which may appear.

The survivability of ground stations is being improved by producing mobile facilities, and efforts are also under way to increase satellite autonomy whereby a satellite can perform "housekeeping" functions – such as correcting orbital anomalies – independent of ground stations.

The incorporation of so many survivability features in satellite design clearly indicates that ASAT devices – such as interceptors and laser weapons – are deemed to be a genuine threat. This is not surprising since both the United States and the Soviet Union have major research programmes in progress to explore the possibility of using new technologies in space weapons applications. The main purpose of these programmes, however, is not just to produce ASAT weapons but also to investigate the feasibility of developing defences against ballistic missiles.

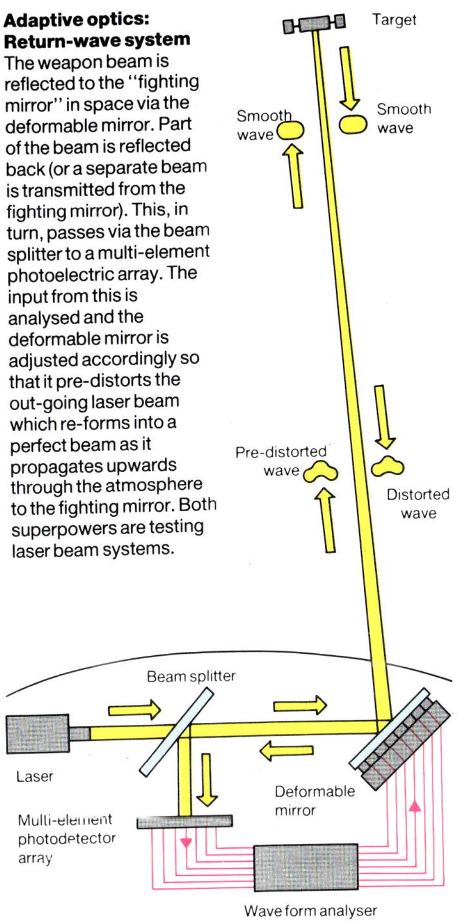


**Adaptive optics: Return-wave system**
The weapon beam is reflected to the "fighting mirror" in space via the deformable mirror. Part of the beam is reflected back (or a separate beam is transmitted from the fighting mirror). This, in turn, passes via the beam splitter to a multi-element photoelectric array. The input from this is analysed and the deformable mirror is adjusted accordingly so that it pre-distorts the out-going laser beam which re-forms into a perfect beam as it propagates upwards through the atmosphere to the fighting mirror. Both superpowers are testing laser beam systems.

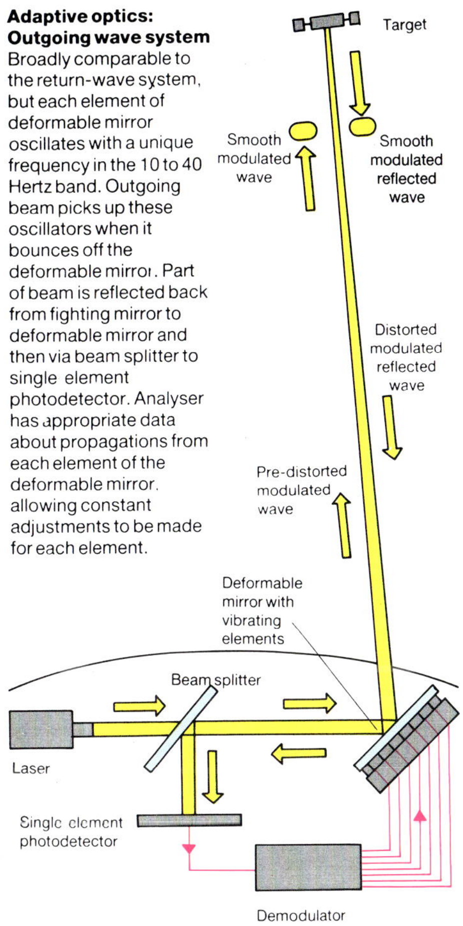


**Adaptive optics: Outgoing wave system**
Broadly comparable to the return-wave system, but each element of deformable mirror oscillates with a unique frequency in the 10 to 40 Hertz band. Outgoing beam picks up these oscillators when it bounces off the deformable mirror. Part of beam is reflected back from fighting mirror to deformable mirror and then via beam splitter to single element photodetector. Analyser has appropriate data about propagations from each element of the deformable mirror, allowing constant adjustments to be made for each element.

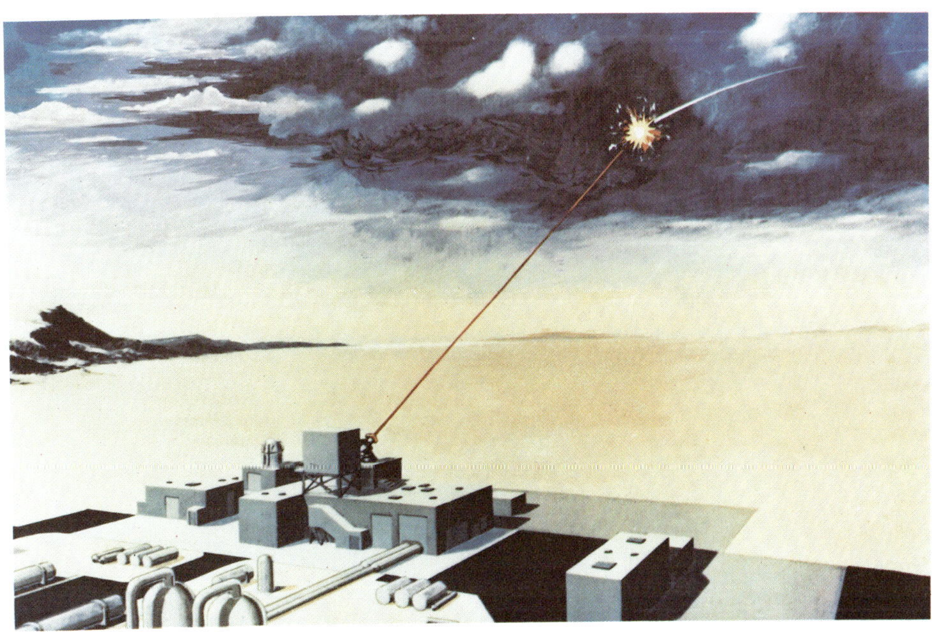

**Left: An artist's impression of the High Energy Laser Systems Test Facility (HELSTF) currently under construction at the White Sands Missile Range, New Mexico. The laser is depicted destroying a re-entry vehicle from a ballistic missile.**

The notion of constructing defences against ballistic missiles is not new. In the early 1970s both the United States and the Soviet Union deployed anti-ballistic missile systems consisting of nuclear-armed missiles controlled from the ground under the direction of large radars. The American "Safeguard" system, using Sprint and Spartan missiles, was sited in North Dakota to protect ICBM fields. The Soviet "Galosh" system was sited around Moscow.

Many doubts were expressed about the effectiveness of these systems and there was particular concern about the operation of radars and control links in a nuclear environment. Consequently, the American system was dismantled in 1976, though the Soviet system continues to be maintained and up-graded.

Research into ballistic missile defence (BMD) technology continued though, and throughout the late 1970s and early 1980s reports appeared suggesting that new technology was appearing which improved the prospects for constructing effective ballistic missile defences. According to some experts, new technology could not only overcome the problems encountered by terminal defences such as Safeguard and Galosh but might also provide the means for constructing much more comprehensive BMD systems.

In the West, matters came to a head in March 1983 when President Reagan made a major speech in which he called upon America's scientists to provide the means to make nuclear weapons obsolete. "What," he asked, "if free people could live secure in the knowledge that we could intercept and destroy strategic ballistic missiles before they reached our own soil or that of our allies?"

Soon afterwards, the United States re-organised its BMD research programmes and placed them under the heading of the Strategic Defence Initiative (SDI). Research funds were doubled compared with the previously projected levels so that the SDI's funding from Fiscal Years 1985 to 1989 stands at 26,000 million dollars.

The Soviet Union, of course, does not provide details of its BMD programmes, but a broadly similar effort to the SDI is in progress, possibly on a larger scale than its American counterpart.

The SDI is not geared towards constructing BMD weapons: its purpose – described as "technology definition" – is to investigate a range of BMD-related technologies to assess their potential. By the end of the decade, the knowledge gained will be used to identify the most promising technologies so that development of BMD weapons could begin if deemed technically feasible and politically desirable.

The research projects in the SDI are grouped into five major categories: attack monitoring, directed energy weapons, kinetic energy weapons, systems analysis and support programmes.

## Attack monitoring

A ballistic missile attack could consist of a thousand or more missiles launched, not only from Soviet ICBM sites, but also from Soviet submarines in virtually any of the world's oceans. There is a considerable premium on destroying the missiles as soon as possible after launch, ideally during the boost phase. After that – only a few minutes after launch – the missile starts to dispense its warheads and decoys, causing perhaps a ten-fold increase in the number of targets for defences to deal with. For boost-phase interception to take place, all missile launches must be detected and pin-pointed extremely rapidly.

Launch detection is already a mature technology and the demands of a surveillance system for boost-phase interception can be largely met by the improvements planned for early-warning satellites. Even so, additional systems will probably be required because early-warning infra-red sensors actually lock on to a missile's exhaust plume rather than the missile itself. Infra-red sensors will probably be suitable for initial detection but, to meet the demands of precise tracking and weapons-pointing, laser radars are believed to be most suitable. This type of technology is being developed in the US in a programme named Talon Gold.

Tracking during the post-boost and mid-course phases will be complicated by the presence of Soviet penetration aids such as dummy warheads, chaff and aerosol masking-clouds. Target discrimination will be enormously difficult but many active and passive sensors which hold promise are being studied. Examples include Fire Pond, which employs long wavelength infra-red and low-light-level optical sensors. Phased-array microwave radars, ultra violet laser radars, optical telescopes and various infra-red detectors are also

**Above: An experimental particle accelerator intended to demonstrate the more efficient production of high-energy particle beams.**

**War in space**
The diagram shows how war in space may proceed. At the outset each side tries to blind the other by hitting early-warning satellites and communications satellites. Air-launched ASATs (**1**) and co-orbital ASATs (**2**) may be used for this purpose. Land-based missile launches (**3**) are accompanied by missiles from submarines (**4**). Early-warning satellites (**5**) monitor these and tracking data is relayed via communications satellites (**6**) to ground control (**7**). Engagement of the missile starts as soon as possible. A submarine-launched, nuclear-pumped X-ray laser (**8**) and orbiting laser battle stations (**9**) start to destroy the attacking missiles. Surviving missiles dispense their warheads and decoys (**10**). A ground-based laser (**11**) engages the warheads through orbiting mirrors (**12**). Electromagnetic rail-guns (**13**) and satellites armed with small rocket interceptors (**14**) also attack surviving warheads. Remaining warheads are tracked by airborne (**15**) and ground-based (**16**) sensors. Ground-based interceptors (**17**) are launched to destroy warheads in various ways. Finally as warheads approach their targets, they encounter terminal "swarmjet" defences (**18**). Each side also tries to disrupt the other's space defences using mines (**19**) and other ASAT weaponry (**20**).

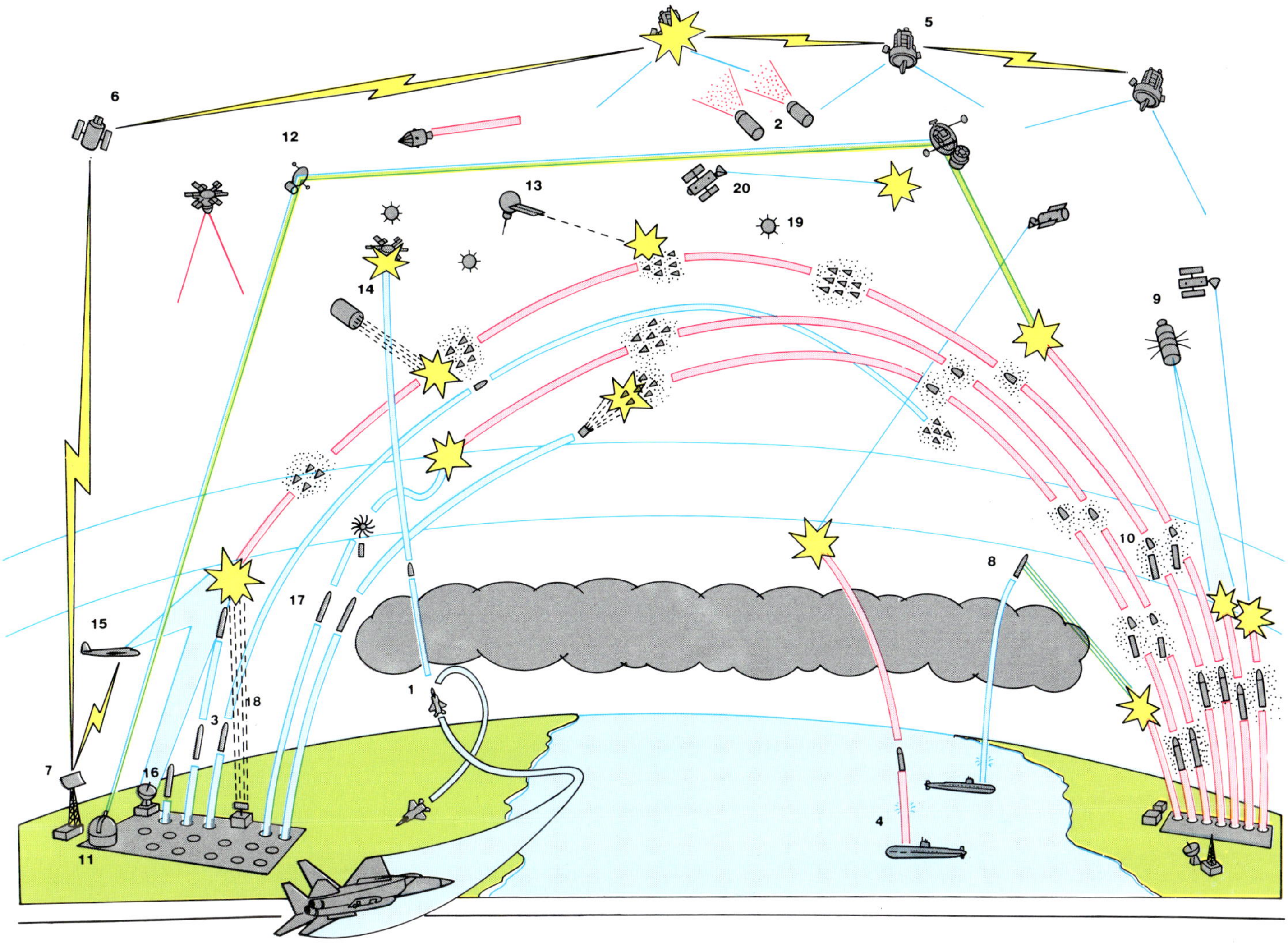

**Left: As part of the SDI, the US Navy's high energy laser beam pointing and tracking system is designed to track and destroy targets in flight.**

**Below: An artist's impression of an orbiting laser battle station. Specific designs for space-based laser weaponry are still a distant prospect.**

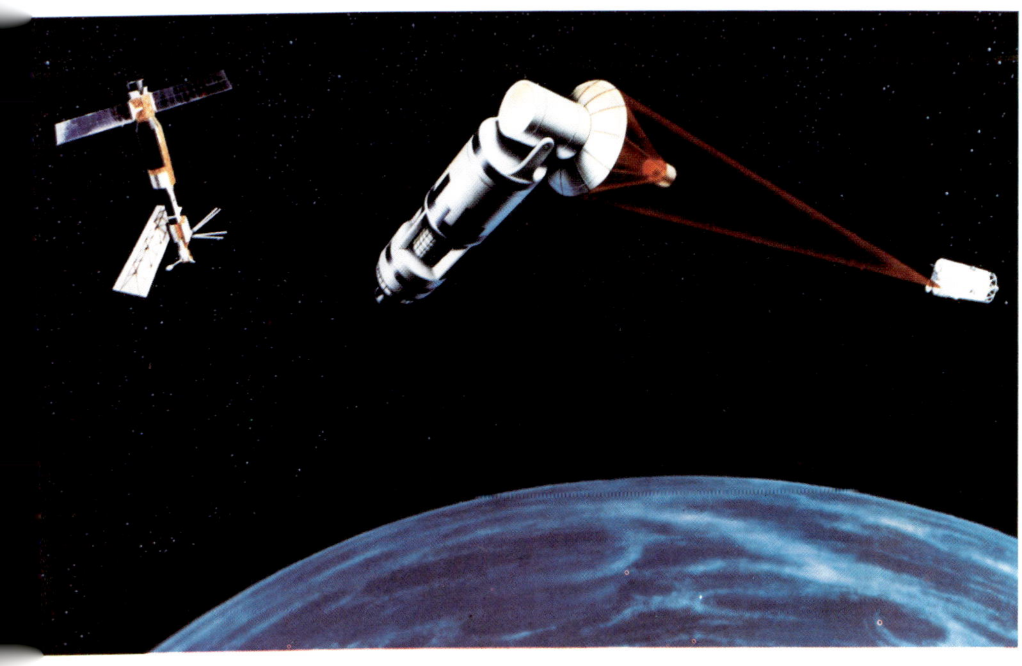

being investigated. Depending on type, these sensors could be based in space, on aircraft, or on the ground.

Other proposed detection techniques, which might be particularly good for target discrimination but which are more distant prospects, include lasers or particle beams powerful enough to induce variations in temperature or velocity which would reveal an object's nature.

During the re-entry phase, target discrimination problems decrease because debris and lightweight decoys heat up or decelerate more rapidly than warheads. Target acquisition and tracking can again be performed by many types of sensor but the main candidates are airborne infra-red and optical sensors, and ground-based radar. Once acquired by these detectors, tracking data could be passed on to an interceptor's own sensors.

## Directed energy weapons

Two types of directed energy weapons (DEWs) are being investigated within the American SDI programme: lasers and particle beams. At present, lasers are receiving most attention because they offer more immediate weapons applications. Several varieties of laser are being studied, the main types being chemical, eximer, free-electron and X-ray lasers.

Chemical lasers are powered by the reaction between two gases which also act as a lasing medium. In a programme named ALPHA, a 2MW hydrogen-fluorine laser is being developed to demonstrate the feasibility of constructing a 25MW chemical laser. Oxygen-iodine and deuterium-fluorine lasers are also being studied.

Chemical lasers could probably be used in DEW systems before any other type but other candidates are judged to hold better, long-term promise. This is because chemical lasers operate at relatively long wavelengths and long-wavelength beams disperse more and are not absorbed as well as short-wavelength beams.

Consequently, the SDI places greater emphasis on eximer and free-electron lasers. Eximer is short for "excited dimer", a dimer being a molecule consisting of two atoms. In this application, the dimer is made up of a noble gas and a halogen. An eximer laser operates by ionising a noble gas (usually xenon) with an electron bombardment so that it will react with a halogen (usually chlorine or fluorine). The dimers created by the reaction are unstable and when they split they emit light in the visible to ultra-violet range. The beam tends to be emitted in pulses rather than in a continuous wave.

Eximer lasers are less efficient than chemical lasers in terms of energy conversion but a device known as a Raman cell allows arrays of eximer lasers to be coupled, producing one very powerful beam at a slightly longer wavelength than the input beams.

The free-electron laser works by passing a beam of electrons through a specially-tailored magnetic field. This causes the electrons to oscillate which in turn causes them to emit photons of electromagnetic radiation. By adjusting the magnetic field, the emissions can be tuned to virtually any wavelength. Although, at present, free-electron lasers convert only a small percentage of their energy input into laser light, theoretical predictions suggest that conversion efficiencies of 25 per cent are possible.

A drawback with weapons using eximer and free-electron lasers is that they would probably be too large and too power-consuming to be based in space. The solution to this problem would be to construct just a few, very powerful, ground-based lasers which would engage their targets through a series of orbiting mirrors. Fabricating large, high quality mirrors and preventing vibration and distortion is a major task in itself, and placing the lasers on the ground raises the additional problem of atmospheric distortion. This problem, however, can be solved using a technique known as adaptive optics, which entails transmitting a low power laser beam from a beacon in space to a sensor near the weapon laser. The sensor measures the distortion in the beacon's beam and the weapon-beam is pre-distorted accordingly so that as it passes through the atmosphere it reforms.

The X-ray laser is radically different from the others in the SDI research programme. This weapon would operate by placing perhaps 50 lasing rods around a low-yield, nuclear explosive. The rods would be pointed at hostile missiles and, when the nuclear explosive was detonated, the rods would emit intense pulses of X-rays towards their targets. The missiles would be destroyed either by a shock wave caused by the action of X-rays on the missiles' skin, or by the effect of X-rays on the missile's electronics (the precise effect would depend on the frequency and intensity of the X-rays).

The exact nature of the X-ray laser is a closely-guarded secret but at least two configurations have been suggested. Both have a nuclear weapon surrounded by lasing rods. In one configuration the lasing rods would be dense, metal rods between 3 and 8ft (1 and 2.5m) long. In another configuration, the rods would resemble carbon-fibre hairs a centimetre long and one ten-thousandth that in diameter. Whether these descriptions are accurate or not, experiments conducted at the Nevada underground nuclear-test site have apparently been very encouraging.

Acting as "understudies" for lasers are particle beams. Various American projects are proceeding, examining different sorts of beam. For space-based applications neutral particle beams are the most promising, but generating and steering the beams are both difficult problems to solve. A project known as White Horse is intended to establish much of the basic technology.

Charged particle beams are also being investigated but, unlike neutral beams, they are deflected by the Earth's magnetic field. Charged beams also tend to disperse relatively rapidly due to mutual repulsion of the particles in the beam. At present, this restricts particle beams to short-range applications such as terminal defence. Electron beams in particular show promise as terminal defence weapons since they deposit a lot of energy on their targets. Propagation in the atmosphere is something of a problem but it appears that this can be solved by generating the beams in a series of extremely rapid pulses. Each

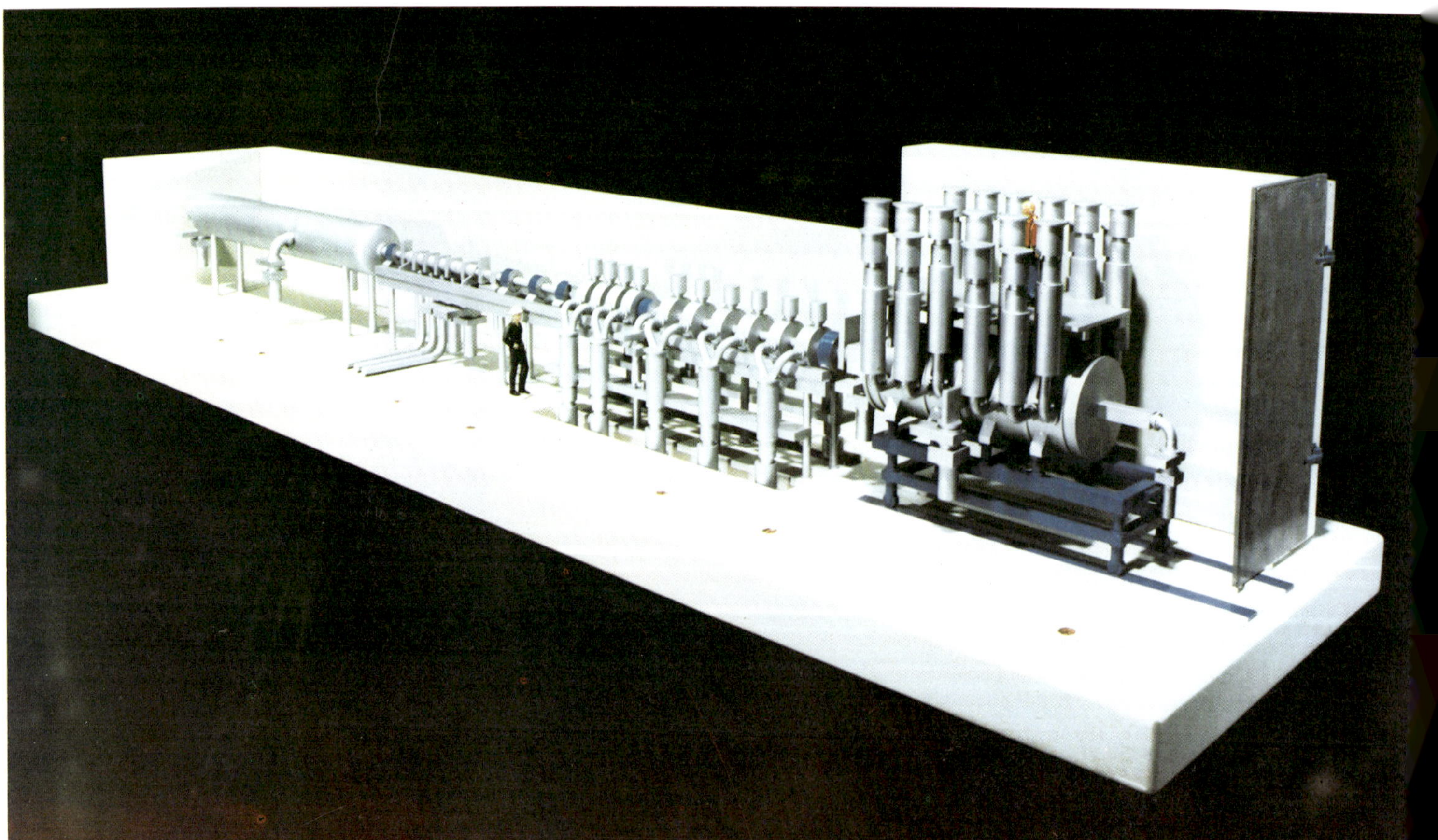

pulse effectively bores a hole through which the next can travel and the technique seems capable of working over respectable ranges.

## Kinetic energy weapons

Kinetic energy devices are intended to destroy missiles by collision, that is using kinetic energy of motion. Many kinetic energy interceptors are under investigation within the SDI programme. Some are intended to impact directly with their target and others would dispense a cloud of pellets into their target's path. At the very high speeds involved, the impact with even a small body should impart enough energy to destroy an incoming warhead.

Kinetic energy weapons of various kinds could be employed to intercept missiles from the boost phase right until just a few moments before the warheads reach their targets.

A device receiving a great deal of attention is the electromagnetic rail gun. This works by passing a powerful electrical current - up to several mega-amps - through two parallel rails. A projectile fits between the rails and makes an electrical contact across them. As the current flows, Lorentz forces act to propel the projectile along the rails. Experimental rail guns have been used to accelerate projectiles weighing 3.5 ounces (2.5 grams) up to velocities of 5 miles (8.6km) per second, and these have penetrated steel plates over half a centimetre thick. Projectiles weighing 10.5 ounces (300g) have attained speeds of 2.6 miles per second (4.2km/sec). It is believed that projectile velocity could be increased to 62 miles per second (100km/sec) with rates of fire up to 60 shots per second.

Two basic modes of operation are possible, one firing many unguided projectiles, the other firing fewer, heavier projectiles equipped with guidance systems. Both modes could be applied to interception from the boost phase through to the terminal phase. Rail guns suitable for BMD applications will probably be available well before directed energy weapons, and the Department of Defense is already developing an orbital gun system to defend satellites from ASAT weapons. Projectile velocity would be about 6 miles a second (10km/sec), and a ground-based demonstration should take place around 1990.

Another kinetic energy BMD concept being studied by American scientists involves a series of orbiting satellites, each fitted with many kinetic energy intercept mechanisms. The intercept mechanisms would operate along the lines of the miniature homing vehicle and each would be fitted with propulsion devices. Acquisition sensors would be fitted either to the parent satellites or to remote "battle managing" satellites which would coordinate the system's operations. The number of satellites needed and the number of intercept mechanisms on each satellite would depend on the capability of the interceptors – range, weight, kill probability, etc. – and on the degree of protection desired. One proposed scheme based on this concept is known as "High Frontier" and would be composed of about 400 satellites each fitted with up to 50 interceptors.

For terminal defence, rockets equipped with homing vehicles could be used. Studies of a variety of ground-based, antiballistic missile schemes are in hand. Some involve relatively short-range missiles for interception within the atmosphere, while others are looking at longer-range missiles intended for interception beyond the atmosphere. The principles are similar but interception inside and outside the atmosphere calls for different types of rockets,

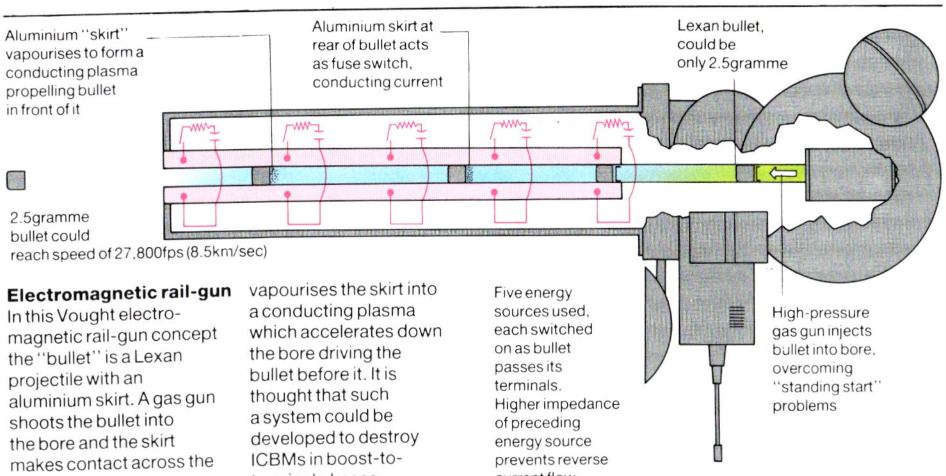


**Electromagnetic rail-gun**
In this Vought electromagnetic rail-gun concept the "bullet" is a Lexan projectile with an aluminium skirt. A gas gun shoots the bullet into the bore and the skirt makes contact across the rails. The current vapourises the skirt into a conducting plasma which accelerates down the bore driving the bullet before it. It is thought that such a system could be developed to destroy ICBMs in boost-to-terminal phases.

**Left: A model of the Experimental Test Accelerator at the Lawrence Livermore National Laboratory. On the support stand to the right is a 2 MeV electron gun.**

**Below: A 7-gramme Lexan projectile (as shown) hit this aluminium block at 4km per second, showing how lethal small, high-velocity projectiles can be.**

**Right: Artist's impression of a space-based electromagnetic rail-gun, firing hypervelocity projectiles whose speed may exceed 62 miles (100km) per second.**

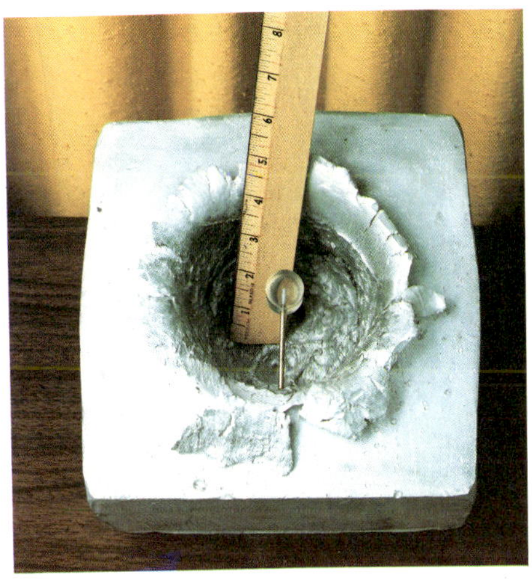

intercept mechanisms and sensors.

Advances in sensor and guidance technology mean that such interceptors would operate much more autonomously than older, superficially similar systems such as Spartan and Sprint. Indeed, progress has been so great that the interceptors should not need nuclear warheads to destroy their targets but should be able to impact directly with them.

A vivid illustration of this was provided in June 1983 when a ground-launched interceptor missile successfully destroyed a re-entry vehicle from a Minuteman ballistic missile by colliding with it. Just before impact, the interceptor unfolded weighted, umbrella-type arms about 8ft (2.5m) long to increase its destructive radius and increase its chances of destroying the target. The interception - which occurred at an altitude of around 90 miles (150km) with a closing speed of 3.7 miles per second (6km/sec) - was intended primarily to validate a long-wavelength infra-red sensor for exoatmospheric use; many of the interceptor's parts were modified, "off-the-shelf" systems. (It is interesting to note that guidance errors on ABMs of the 1970s were expected to be so large that the missiles carried 5 megaton warheads.)

For "last-ditch" defence close to high-value targets, radar-controlled, rapid-fire machine guns and cannon-launched "swarms" of up to 1,000 small projectiles are among the ideas being studied. At most, the effective range of these systems would be a few miles.

## Systems analysis and mission support

One of the major tasks for systems analysis is research into battle management systems which could control and co-ordinate a defensive network. Remarkable data-processing and communications capabilities will be needed to ensure the efficient

**Above: The intercept mechanism used in the Homing Overlay Experiment. The 6ft (2m) metal ribs unfurl seconds before impact with the target re-entry vehicle.**

**Below: The Homing Overlay Experiment rocket streaks skyward. Minutes later it destroyed a re-entry vehicle from an ICBM launched from Vandenberg AFB.**

**Below: A photograph of the rocket plume from the Homing Overlay Experiment homing-and-kill vehicle just before impact with the target re-entry vehicle.**

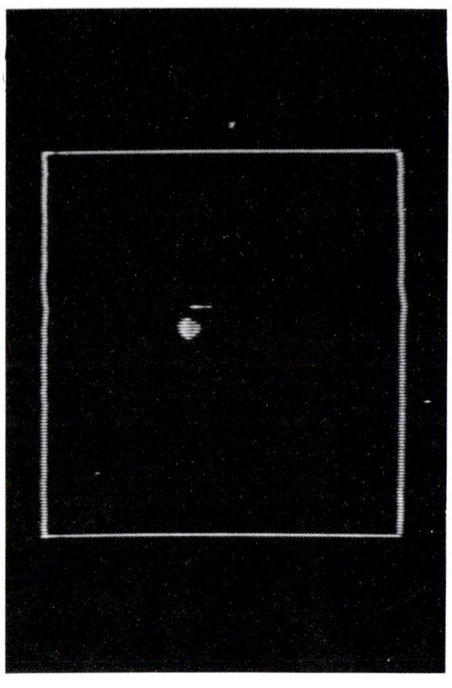

engagement of many thousands of objects, each travelling at several miles per second. Indeed, despite the advances made in computer technology in recent years, battle management components with the necessary combination of size, speed and power will require further, major progress. Indeed, some experts see this area as one of the greatest technical challenges faced by the SDI.

Systems analysis also includes research into the overall architecture of a BMD network.

Mission support includes many research projects in diverse fields. One key area is the investigation of power sources for orbiting systems. These range from using nuclear reactors to conventional explosives. Other subjects include studies into damping vibration on large space structures and examining the feasibility of remotely-controlled, long-duration aircraft to carry sensors. Attention is also being devoted to the protection of space systems and to the ways in which Soviet missiles might be modified to reduce their vulnerability to defensive systems.

**Above: Artist's impression of a new Soviet silo-based ABM. This high-acceleration interceptor is designed to engage targets within the atmosphere.**

## Soviet BMD programmes

Although it is very difficult to obtain details of Soviet BMD research, an effort similar to the SDI is believed to be under way. In some areas, the Soviet effort is greater and more advanced. Research into directed energy weapons, for instance, is believed to be on a larger scale and at least six, large DEW research facilities appear to have been built. At Saryshagan, an area long-associated with BMD work, there are three DEW complexes, each exploring different types of laser.

There have already been suggestions that the Soviet Union has used lasers to blind American satellites and as mentioned earlier, some reports have asserted that an anti-satellite laser weapon will be placed into orbit before the end of the decade.

Particle beam research is believed to be at least as advanced in the Soviet Union as in the United States. In one instance, American research was, in fact, aided by some sketchy details of a neutron beam device which appeared in Soviet technical journals.

A source of particular concern in the West is that the Soviet Union may be assembling the components for a widespread BMD system based on ground-based interceptors. The Galosh ABM system around Moscow is being up-graded and a new, mobile system designated ABM-X-3 has also been tested. This consists of SH-04 exoatmospheric interceptors and SH-08 endoatmospheric interceptors, plus Flat Twin and Pawn Shop radars. Some Soviet air-defence missiles also have an ABM capability, in particular the SA-12 which has been tested against ballistic missile re-entry vehicles.

In addition, the Soviet Union has built a new, phased-array radar near Abalakova, 500 miles (800km) from Mongolia and 1,865 miles (3,000km) from the Pacific. This covers the trajectory that American SLBMs would follow if directed at the Soviet southern missile fields, leading to suggestions that its purpose is ABM battle management though, as mentioned earlier, the Soviet Union maintains that the radar is to be used for tracking satellites.

**Soviet missile tracking**
The current Soviet launch-detection satellite network provides about 30 minutes warning of any US ICBM launch. This is complemented by two over-the-horizon radars directed at US ICBM fields. Eleven Hen House radars at six locations on the USSR's periphery give more detailed information and could provide target-tracking data to support ABM systems. Five phased-array radars are already in operation with a sixth being built in Siberia, probably in violation of the 1972 ABM Treaty. The USSR may be preparing to deploy large-scale missile defences.

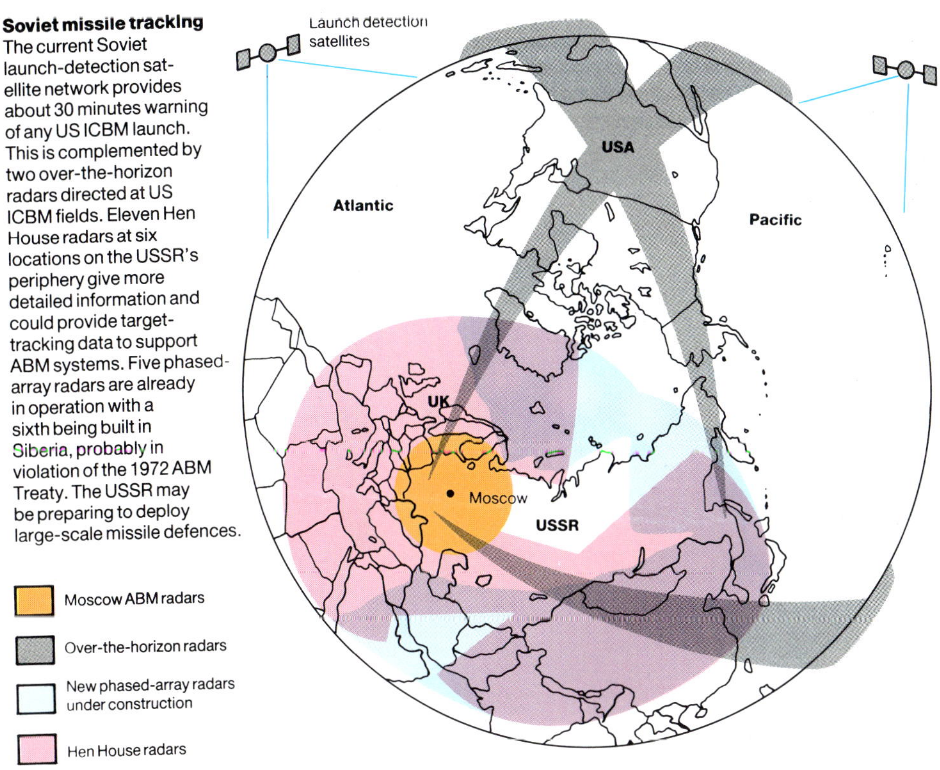


**Moscow's ABM defences**
The only operational ballistic missile defences in the world are deployed around Moscow. Until recently, only 64 defensive missiles were deployed but the number is now being increased to 100, the limit imposed by the 1972 ABM Treaty. The new Moscow defences will have two "layers", one each for engagements inside and outside the atmosphere. New radars are also bieng built. These new defensive systems could be fully operational by 1987.

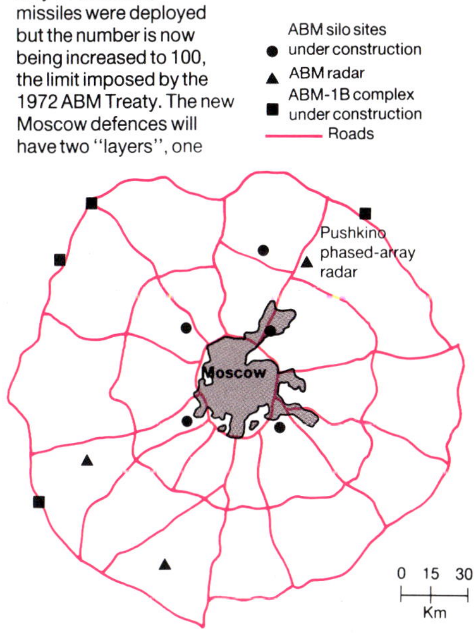

If space-based ballistic missile defences are ever constructed, manned space missions will probably be indispensable for assembly and maintenance. Even without such defences, manned missions will be a major, and probably essential, part of the growing military exploitation of space.

The American Space Shuttle was envisaged from its inception as an integral part of United States military space activity. Roughly 25 per cent of Shuttle missions will be devoted exclusively to military purposes and many other missions will include military payloads in their cargo.

The most obvious use of the Shuttle is to launch military satellites but the Shuttle is much more than a manned rocket booster. It can serve as a platform for testing sensors such as Validator and it offers capabilities unmatched by unmanned systems. The ability to retrieve satellites for repair or refurbishment – and to perform in-orbit maintenance – has already been demonstrated, and in view of the enormous costs of military satellites this ability will be extremely valuable. Studies are already being conducted on the feasibility of modifying reconnaissance satellites for refuelling and replenishment in space.

Concern has been expressed by the Soviet Union about the Shuttle's potential to pluck satellites from orbit. While this form of ASAT capability is theoretically possible, it would be a dangerous enterprise since booby-trapping satellites would be very simple. However, the Shuttle could be used to inspect foreign satellites - from a safe distance – and it could carry ASAT weaponry although cheaper, unmanned systems would be more suitable.

Various supplements and successors to the Shuttle are being put forward, though which ones will emerge as serious development possibilities remains a matter for conjecture.

**Above: The Space Shuttle may prove to be the forerunner of a variety of more flexible manned space vehicles with military and civilian applications.**

Small, single or twin seat space cruisers have been proposed for carrying out reconnaissance missions, placing satellites into orbit, repairing satellites, and for rotating crews on permanent space platforms. Space cruisers might be launched by the Shuttle, by MX-type rockets or even from the back of a Boeing 747. Space cruiser proposals suggest a capability for higher altitude operations than the Shuttle can achieve so the vehicles would undoubtedly be useful, but whether they would be cost-effective is more open to question.

Other concepts envisage a variety of Shuttle successors, including spaceplanes which would be able to take off and land using conventional runways.

A project much closer to realisation is the American Space Station. This will be a permanently-manned, orbiting facility due to be in place in the early 1990s. The Space Station is an international venture and the Department of Defense has no part in it. In the future though, particularly if large-scale BMD systems are deployed, the military may decide to acquire a similar facility to provide – among other things – in-orbit repair and maintenance for other military space systems.

As mentioned earlier, the Soviet Union already uses its Salyut space stations for military purposes, and a space orbiter similar to the American system is in the advanced stages of preparation. The Soviet shuttle will have a larger payload capacity than its American counterpart and can be expected to perform similar functions. A sub-scale model of a smaller space "shuttle" has been tested but it is unclear whether the model is merely for engineering test purposes or whether it precedes full-scale development.

The Soviet Union has also developed a rocket booster capable of orbiting five times the payload of the American Shuttle. This may be used to launch space stations or laser ASAT devices.

## Conclusion

The scene is set for a dramatic expansion of military space activity. The capability of military satellites is increasing and terrestrial military forces are becoming ever more dependent upon them. The growing militarisation of space is frequently denounced by critics but blanket condemnation is unjustified because many military

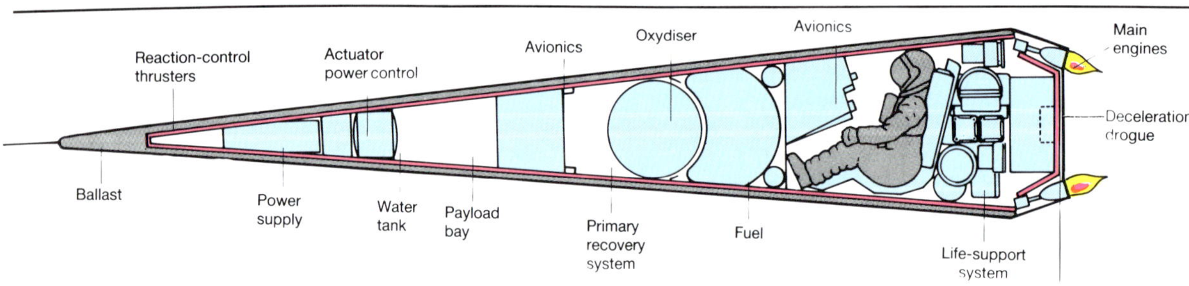


**Advanced space plane**
In this concept of a manned space plane the pilot remains suited to avoid excess weight and complexity in the space plane's design. The nose folds back to give access to the payload bay. Endurance is about 24 hours.

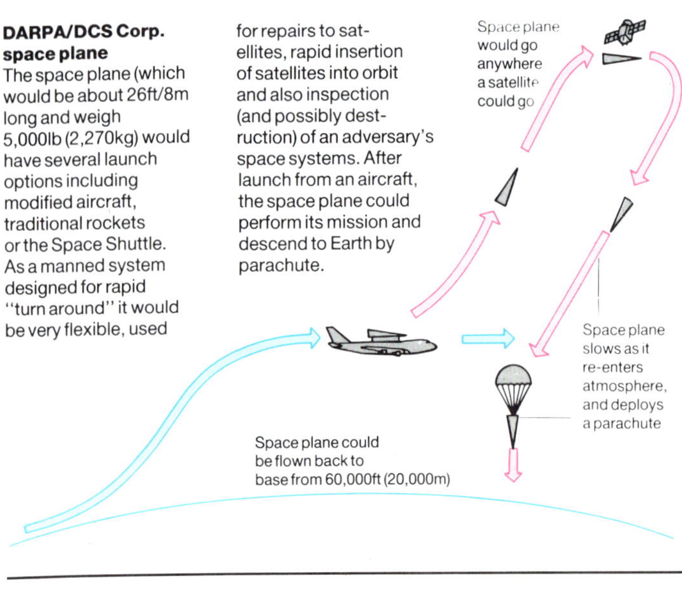


**DARPA/DCS Corp. space plane**
The space plane (which would be about 26ft/8m long and weigh 5,000lb (2,270kg) would have several launch options including modified aircraft, traditional rockets or the Space Shuttle. As a manned system designed for rapid "turn around" it would be very flexible, used for repairs to satellites, rapid insertion of satellites into orbit and also inspection (and possibly destruction) of an adversary's space systems. After launch from an aircraft, the space plane could perform its mission and descend to Earth by parachute.

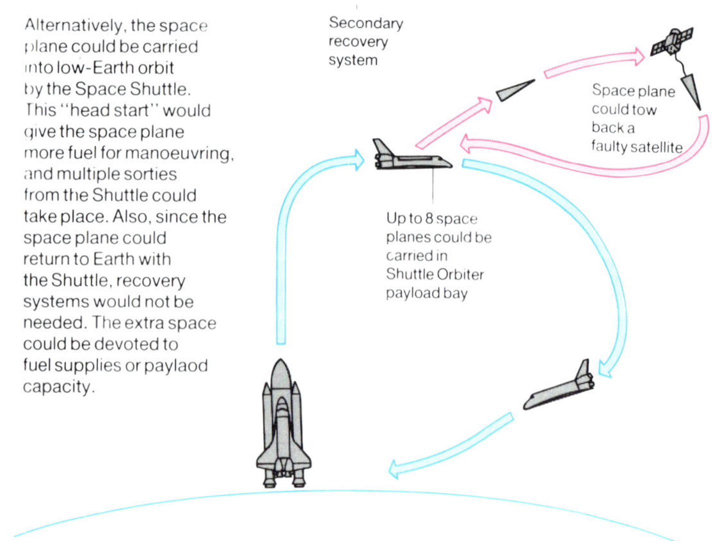


Alternatively, the space plane could be carried into low-Earth orbit by the Space Shuttle. This "head start" would give the space plane more fuel for manoeuvring, and multiple sorties from the Shuttle could take place. Also, since the space plane could return to Earth with the Shuttle, recovery systems would not be needed. The extra space could be devoted to fuel supplies or paylaod capacity.

Above: **The utility of manned space systems has been demonstrated several times by Shuttle crews who have repaired or retrieved ailing satellites.**

satellites are useful for strategic stability and arms control. ELINT satellites can provide advanced warning of increased military activity, reconnaissance satellites can establish force structure and numbers, and can also monitor arms-control agreements. All this information is useful for building confidence and for preventing overreactions based on prudently pessimistic appraisals of a potential adversary's actions and capabilities. Consequently, although "spies in the sky" seem sinister, those spies help to keep the superpowers both honest and calm.

The "weaponisation" of space is a different matter. Controversy rages about the effects of ASAT on stability. Most people would agree that strategic stability would be enhanced if ASAT weapons were not deployed. Unfortunately, technical realities preclude a simple superpower agreement not to deploy such systems. The problem is that many ASAT systems – lasers, interceptors, ECM, etc. – would be virtually undetectable until they were actually used, and an ASAT treaty would be unverifiable because neither side might be willing to trust the other's word.

So, although a negotiated agreement on ASAT systems would be desirable - and both superpowers can see the advantage of preserving their satellites' survivability – it would not be surprising if ASAT development continued. It could be argued that the best way to protect satellites from hostile action is by posing a counter-threat to an opponent's satellites.

The situation regarding ballistic missile defences is even more complicated. The emotional appeal of constructing such defences is obvious but controversy surrounds both their technical feasibility and their strategic desirability. On both issues, the answers are by no means clear. Research programmes now in progress will eventually resolve the technical debate one way or the other but the strategic questions will remain, simply because there are no definitive answers.

One firm conclusion *can* be made, however. Both superpowers are vigorously exploring BMD technology, yet, because it is more public, the American programme has attracted most criticism. But critics should bear in mind that, despite understandable reservations about actual *deployment* of BMD systems, while Soviet BMD research continues American research should also continue as a hedge against future Soviet deployments, or as a bargaining chip. If the Soviet Union maintains its BMD research programmes, it would be foolhardy for the United States not to do the same.

**Above: Current US space station plans are purely civil but if extensive space defences are constructed, military space stations may perform support activities.**

**Below: A Soviet model "mini-shuttle" is retrieved after splashdown in the Indian Ocean. Subsequent tests have splashed down in the Black Sea.**

**Below: The Soviet Union has already launched several Salyut space stations. This concept shows a large space station made up mainly of Salyut-type modules.**

# AIR WARFARE

Historically the first aerial warfare mission was reconnaissance. From the Battle of Fleurus in 1794, observers carried aloft in balloons, dirigibles, aeroplanes and helicopters have played an increasingly important role in deciding the outcome of land and sea battles. Today human eyes have largely been replaced by optical cameras (using clever films that defeat camouflage effective only at visible optical wavelengths), radars, IR linescan and EO sensors, while the carrier vehicles have expanded to encompass satellites and unmanned RPVs.

Combat aircraft today do not necessarily have a human aboard. The development of radio-controlled aeroplanes dates from 1915, and in World War I such machines were built in numbers to carry bombs or explosive charges, torpedoes and even cameras (but the technology was unreliable). In the 1950s the development of supersonic cruise missiles and targets with sophisticated guidance systems appeared to presage a gradual elimination of human crews from combat aircraft, and in Britain the official view went so far as to announce, in April 1957, that the RAF would be "unlikely to require" any more fighters or bombers! With the benefit of hindsight it can now be seen that this belief was an error on a gigantic scale, which has had extremely damaging consequences to Britain. In a later subsection the prospects are discussed for eliminating the human crew in the near future.

Perhaps the most difficult mission to automate is that of the fighter. Here the pilot may well take off with only a hazy idea of the threat type, numbers and even height or direction. In close combat it is very difficult to duplicate the judgements and skills of a human pilot, but in the plain interceptor mission the situation is likely to be different. In peacetime it helps to have at least one human aboard in order to effect positive identification of intruders in a way that an RPV would be hard-pressed to emulate. A backseater in a Tornado F.2 can shine a torch or searchlight at a "Zombie" and say, *"It's a Bear . . . a Bear-F . . . it has a long pod projecting aft of the fin tip, and there are no pods on the tips of the tailplane."* An RPI (remotely piloted interceptor) would find this difficult, and a Soviet Bear crew might get jumpy to have such a machine in tight formation. But in wartime the interception mission would be far more cut-and-dried, masterminded by the national electronic defence network. This could fire off pilotless interceptors just as easily as it could fire SAMs, the difference being that the interceptor would be reusable.

Likewise the attack or transport missions appear ready candidates for automation, perhaps even where the target is mobile. With modern sensors and weapon-delivery systems it is possible for an unarmed or remotely piloted vehicle to discover, pinpoint and then destroy moving targets such as a ship, tank or even a hostile aircraft. As long ago as September 1947 a C-45 Skymaster flew from Newfoundland to England without a human pilot at any time touching the controls. In transport

**Above: This photograph was taken from a Western interceptor which has already reported the intruder as a Soviet "Bear". The interceptor is itself watched impassively by the Soviet rear gunner.**

**Below: Taken aboard an F-18, this photo emphasizes the colossal versatility of human capabilities, some of which can be far surpassed by tiny boxes, but it is hard to duplicate them all.**

**Air warfare contains more advanced technology than any other kind of warfare, if capital expenditures can be considered a fair indication. Moreover, because our atmosphere covers our planet, air warfare has a direct impact on combat on land, at sea and under the sea. Indeed, today "air warfare" also stretches into outer space.**

aircraft, however, the advantages of a human crew are considerable, and the drawbacks relatively small. The same applies to the medevac/casevac mission, which is basically a transport operation.

A particular form of casevac is the SAR and aircrew rescue mission, and here at least it can confidently be predicted that human crews will be needed throughout the foreseeable future. Such missions began with small helicopters in Korea more than 30 years ago, and in Vietnam often assumed the proportions of local battles in their own right, involving large numbers of surrounding troops trying to thwart the rescue efforts of up to a dozen helicopters and aeroplanes to pick up a single downed pilot. Today the SAR helicopter fulfils a vital role in saving (usually civilian) lives in all advanced countries, but this is not warfare. In war, the same helicopters would have the grave disadvantage of high vulnerability. Even such battleworthy machines as the USAF's HH-60 Night Hawk would find survival difficult over tomorrow's battlefield.

In general it can be stated that, the closer one approaches to a future land battle, the more difficult it will be for *any* aerial vehicle to survive. Already front-line troops have SAM systems which can destroy small supersonic missiles and even artillery shells, so what chance does an aeroplane or helicopter have? Of course, the apparently near-zero chance may well be modified by devices or techniques that are classified, and it is difficult for the impartial observer to form a fair objective judgement. For example, the fact that on several occasions small RPVs flying at only about 100mph (160km/h) have loitered at about 10,000ft (3,050m) for over an hour despite frantic efforts by air defence troops to destroy them should not necessarily be taken as proof of survivability. In just the same way, claims of the effectiveness of anti-aircraft defence systems under test conditions can seldom be taken at face value. With deeper knowledge one often finds that the target was held in straight and level flight and laden with devices to increase its IR or radar cross-section!

**Above: Warfare involves a mixture of technologies: here Afghan villagers study a Soviet Frontal Aviation bomb that failed to go off. People like this have been chief targets in recent wars.**

**Below: This small RPV was photographed in December 1984 testing a low-cost autopilot which could be used in tactical missiles. Known as the USAF's XBQM-106, it typifies the way technology can replace humans.**

Most of this chapter addresses itself to what might happen in a full-scale war between superpowers. We all hope this is an "unthinkable" scenario, but it is the one that throws up the greatest number of interesting problems and possible solutions. So perilous would all defended airspace become in such a conflict that aircraft manufacturers might well think they were in the wrong business, and that military customers would simply cease buying such costly, frail and totally vulnerable toys. In any future full-scale war airborne aircraft will be lucky indeed if they escape detection and destruction; but in the air they are at their least vulnerable. On the ground they have no chance whatsoever (if the enemy can find them).

But history since 1945 shows that the wars people are more likely to have to fight are of the so-called limited variety. Even if a superpower is involved, the weapons of mass destruction – nuclear, chemical and biological – are unlikely to be used. Many missions, even as recent as the Black Buck bombing of Port Stanley airfield in May 1982, have been at a level of technology scarcely any more advanced than that available in World War II. Prolonged warfare in the Middle East, Afghanistan and between Iran and Iraq, has been marked by plenty of bangs and bloodshed but a near-absence of new technologies.

The world's planemakers can take heart from such limited wars, in which most of the pilots can sleep in their barracks, houses or tents fairly secure in the knowledge they will wake up next day. The worst that can happen to their airfields is likely to be a string of cratering weapons along the runway. But this is yesterday's kind of war, and since it is easy to counter there are plenty of available solutions. The kind of war that would ensue if "the balloon were to go up" in Central Europe could hardly be more different. Only the most technically advanced, dispersed, stealthy and lucky aircraft would see Day 1.

Most countries have no air-defence infrastructure. They may have fighters, and probably mobile triple-A (anti-aircraft) or SAMs forming part of the army, but unless they are actually engaged in protracted warfare they very rarely possess a comprehensive nationwide system. So far as is known not even Iran and Iraq possess such systems, though they are clearly crucial to any effective air defence of more than a very small area, such as a single airfield, small town, refinery or similar-size target.

Different kinds of AA defence have very different lethalities against different kinds of target, and in particular against targets flying at different heights. In the period immediately after World War II it became obvious that continued improvement of radars, AAA and SAM systems would make it not only increasingly dangerous to intruding aircraft at traditional altitudes (say up to 30,000ft/9,150m) but would soon extend the height of effective AA defence above the level at which practical military aircraft can operate. For example, in January 1958 the US Army began to operate the Nike Hercules SAM (still in use by European NATO nations) which is effective to a height of 150,000ft (45,750m). This introduced a new situation in which no increase in bomber engine power or wing area, to gain extra over-target height, had any effect in reducing vulnerability. Even dramatic increases in target speed, as exemplified by the B-70 and SR-71, do not appear to be of much advantage except to force the hostile defence system to respond quickly.

By 1952 it was accepted that to survive in defended hostile airspace there was no alternative to flying at the lowest possible level, as explained in the next subsection. In turn this focussed attention on low-level air defence systems. It became general practice to site surveillance radars on mountain peaks, and to locate all AA defence installations at the highest points in each locality in order to have the longest possible warning time and best possible view, both to see enemy aircraft and communicate between sites by line-of-sight microwave link. It was at about this time that it was also recognised that 1st-generation air defence systems involving "croupier" girls pushing model aircraft about on a giant map, and with the information passed on by telephone (often over the public system) had to be swept away entirely.

Most advanced countries contracted for nationwide systems, and the earlier they were started the quicker they became obsolescent. Biggest and earliest was the gigantic SAGE system (semi-automatic ground environment) of the USA, later augmented by BUIC (back-up interceptor control). This involves house-sized computers and masses of other electronics which make a significant difference to the national consumption of electricity.

The European NATO nations fortunately spent so much time arguing that their Nadge (NATO air-defence ground environment) was not designed until the late 1960s, and by the time it was being built in the early 1970s it was about two generations newer than SAGE/BUIC. Nadge broke new ground because, to make sense, it had to be a single system stretching 3,000 miles (4,800km) from North Cape down through Norway, Central Europe and Italy across to Greece and then along the thousand miles of Turkey. Its 84 sites were designed and built as a single organism linked by computers and instantaneous unjammable radio and microwave beams, so that, should a single fast jet travelling at about the speed of sound cross over into Western airspace, at any height and following the most "undetectable" track (for example, along gorges and valleys), it would immediately be detected, recognised and tracked. But it was then up to local interceptor or SAM forces to do something about it.

Subsequently the defence possibilities have been multiplied by AWACS (airborne warning and control system) type aircraft, which by lifting a powerful multimode radar to a height of some 28,000ft (8,500m) for the Boeing E-3 series (considerably higher for the Nimrod AEW.3) changes the geometry of its range of vision. The limit of good vision is extended to at least 230 miles (370km) against targets flying close to the ground; against high-flying targets the effective radius is much greater. By thus opening out the radius of vision such a radar can simultaneously track all air

**Below: Empty cases stream from the twin Bofors L/70 40mm guns of the US Army's Ford Gunfighter (Sgt York), just one of hundreds of "triple-A" systems.**

**Above: An ADATS (air defence anti-tank system) missile zips from its launcher during guided prototype tests in 1982. This dual-role missile rides a laser beam.**

**Below: Certainly the most numerous, and possibly the most respected, of all mobile triple-A systems, the ZSU-23/4 is seen here in service with the East German army.**

vehicles, including even small missiles, within a vast block of airspace. If the carrier platform is big and capable enough it can simultaneously store in its memory billions of items of information regarding friendly and hostile forces, types, locations, serviceability states, radar and other emitter signatures, numbers and many other quickly changing variables. It can also provide a vital command and control function, not only in managing local air defence but also friendly tactical attack forces. This is referred to again later in this section.

In general, the more sophisticated the defence or attack system, the greater are the opportunities for defeating it. To every measure there has so far been discovered a countermeasure, so that electronic countermeasures (ECM), infra-red countermeasures (IRCM), and electro-optical countermeasures (EOCM) are commonly found in the list of systems or equipment items carried aboard modern combat aircraft.

Occasionally a system presents peculiar difficulties to the designers of countermeasures. For example a missile that rides a laser beam may have aft-facing receivers that ignore all signals except those coming from the launcher, or at least from within a cone centred on the launcher and with such a small semi-angle as to preclude all attempts by hostile forces to make the missile break-lock. But how is the laser held on target? If it is aimed by a human, such as an infantryman, this at once introduces both unreliability and inaccuracy; if it is automatically locked-on the target, this opens the way to confusing or distracting the beam by providing spurious targets that the system finds more attractive.

Of course, all forms of missile that home automatically on the target inevitably are vulnerable to jammers, decoys and other false sources that either lure the missile away or, at least, remove the source of homing signals. Yet what can the cleverest attack aircraft do against crude weapons such as a gun aimed by a man who can see the aircraft and has rudimentary knowledge of how to aim just ahead of the speeding target? In the Falklands the AA defence of British ships, including warships, was so poor that it was standard practice for men from all services to pick up GPMGs and other infantry weapons and fire them at oncoming aircraft. Such weapons lack punch, but in fact are disliked by modern aircrew because they are immune to countermeasures – except, perhaps, for a few cannon shells or bombs in return.

This enhanced importance of crude weapon systems has a parallel in the fact that, because of the lethality of modern SAM systems, virtually all attack aircraft fly as low as possible to escape them. For at least 25 years this has thrown the emphasis on weapon systems to defeat the low-level attacker. With modern rapid fire triple-A in calibres from 20 to 40mm it is possible to engage any aircraft and almost any tactical missile at a range of 2 to 5 miles (3 to 8km) with rapid-fire ammunition having extremely high lethality.

Many countries and service staffs have tended to fall into the error of thinking that each new technology automatically invalidates its predecessors. The advent of SAMs, as in the case of jet engines and many other new inventions, has often resulted in not only discontinuance of purchasing anything seeming to use earlier technology but also premature withdrawal of hardware already in service. For example, Britain's RAF Regiment long ago discarded all its triple-A, but is now learning the error of its ways. Money for new guns might not have been forthcoming, but the AEW Nimrod base at Waddington is now defended by a new squadron formed with 12 twin-35mm guns with Skyguard fire control all captured during the 1982 Falklands War!

**Above: Potentially the best surveillance aircraft in the world, this Nimrod AEW.3 is one of 11 whose development costs are expected to exceed £1,200 million.**

**Below: Wallop Industries Rampart combines smoke, chaff/IR decoy flares and also rocket-fired Skysnare tethered balloons to catch low-flying attacking aircraft.**

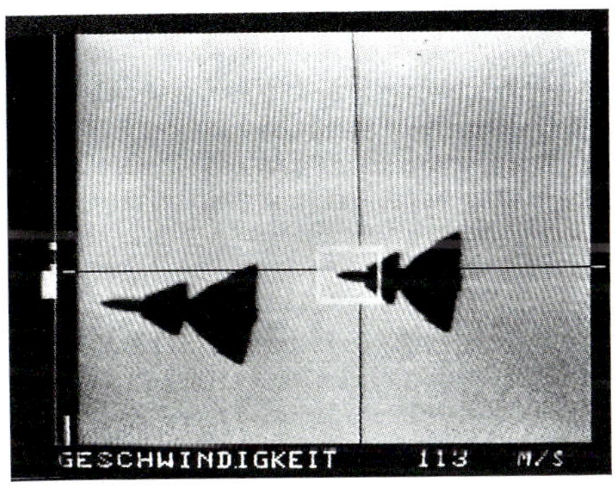


**Above: A Swedish Viggen caught in the TV/radar video imagery of the Swiss Contraves Skyguard fire control (see below). Note readout of target speed (metres/sec).**

**Below: Contraves Skyguard is a track-while-scan radar fire control seen here linked with twin Oerlikon 35mm triple-A. It can also control SAM batteries.**

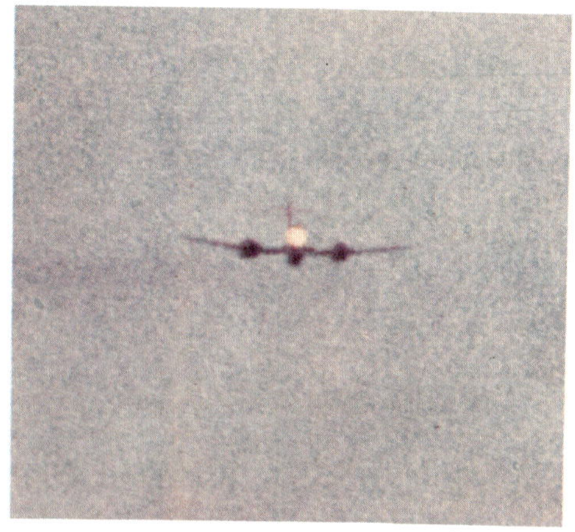

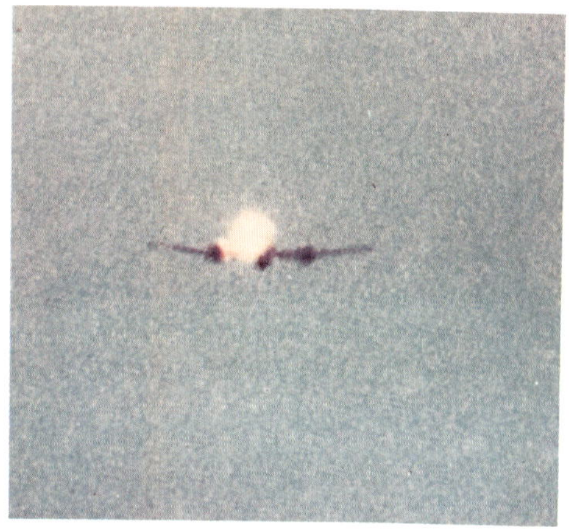

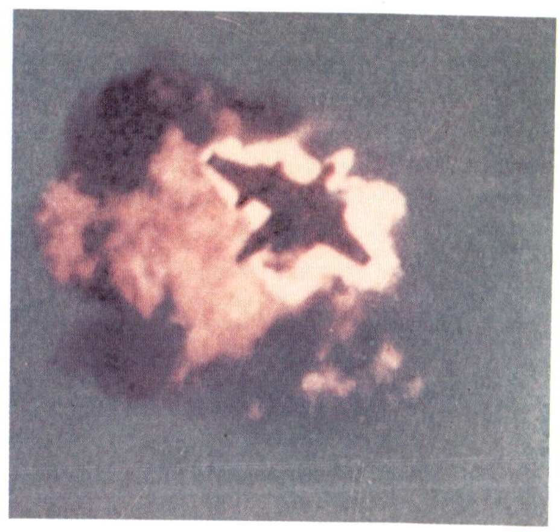

Among the advantages of triple-A are short flight-time to the target, reliability, virtual immunity to countermeasures and very quick reaction time. How many pilots would like to face a single Bofors Trinity? This is a direct descendent of the World War II Bofors, and retains the single barrel of 40mm calibre, but there resemblance ceases. Trinity can fire a burst of 10 individually aimed rounds in less than 2 seconds, each round having a devastating proximity-fused warhead with more than 1,000 lethal pellets, forming a computer-planned compact group surrounding the aircraft or oncoming missile. Even the precise details of fusing for each round are inserted by computer a split second before firing, and the bull's-eye accuracy is enhanced by such factors as radar/optical/TB sighting, laser ranging, computerised meteorological input and extremely high muzzle velocity. Anyone who watches this system must wonder why bother to fire ten rounds when one will do!

What is the inevitable result of such deadly low-level defences as Trinity, and the highly accurate British Rapier SAM? It is bound to be a renewal of interest in flight at over 30,000ft (9,150m), beyond the effective reach of these systems. At higher altitudes the population of defence systems thins out greatly, and all the simple manual systems, which cannot be countered by countermeasures (short of making aircraft invisible, of which more later), are eliminated. One is left with ponderous large-scale systems such as interceptors and a very few large SAMs such as Patriot (MIM-104), SA-4 "Ganef", SA-6 "Gainful" and, less assuredly, Improved Hawk (MIM-23B).

There is justification for believing that a pilot penetrating hostile airspace would

**Above: A Harrier prepares to land at an airfield defended by an RAF Regiment Rapier unit. From the right, optical tracker, quad launcher and blindfire tracking radar.**

have a better chance at high altitude using the maximum of well-programmed countermeasures (and, of course, if possible stealth techniques) than at a height between 200 and 500ft (60 and 150m). This is a matter for much argument, but if even small attack aircraft are not viable at 30,000ft, what price the E-3 Sentry or TR-1, both repeatedly described as "survivable", at heights well within the engagement envelopes of the Soviet SAM systems just mentioned?

This alleged survivability of these large, slow and ponderous aircraft at medium to high altitudes, where their contrails can be watched with the naked eye from enemy territory, is something Western taxpayers have to take on trust. At first glance it seems to be arrant nonsense; no aeroplane with ordinary jet engines and a radar cross-section bigger than the proverbial barn door is going to divert the cannon shells and missiles by clever electronics or other kinds of countermeasure – or is it? Could it just be that the USAF and other operators possess a secret defence that renders them invisible at all EM wavelengths?

Little purpose is served by speculation, but it is worth also noting that, in their recent (October 1984) triumphant participation in "Prairie Vortex", the annual USAF SAC bombing competition, RAF Tornados from 617 Sqn not only achieved unprecedented scores in bombing, navigation and precision timing but also, uniquely, completed the entire exercise without a single aircraft being "shot down". These competitions are made as

**Above: Few SAM systems have been as well tested as BAe Dynamics Rapier. This Meteor, approaching head on at full throttle, was intercepted in one of the very first guided trials. The missile carried a telemetry pack in place of a warhead, the target being destroyed by direct impact. Rapier is designed to strike its target, the warhead detonating inside it.**

Above: Looking up at a Luftwaffe Tornado dispensing two types of munition from its MW-1 multipurpose dispenser. MW-1 comprises up to four large boxes each containing 28 double-ended tubes from which different submunitions can be fired to left and right in an exact pattern, the munitions being selected according to the target. Tornado can attack at full speed below 100ft (30m).

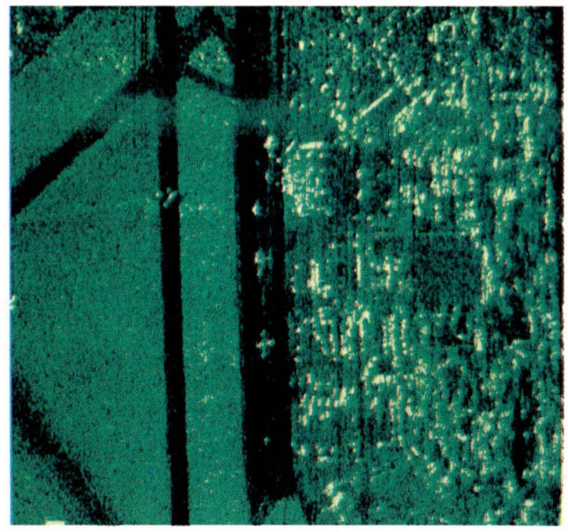

Above: Two parked USAF C-9As and a One-Eleven on the runway show up clearly in the radar image (top) of Scott AFB taken by a modified (doppler beam sharpened) APG-63 radar in a special F-15E prototype.

realistic as possible, and the defences use real radars, and the most realistic simulations of SAM belts, as well as crack USAF interceptor squadrons. Yet the Tornados repeatedly penetrated the SAM belts unscathed, and not one was ever detected from the air. This can hardly be attributed to mere luck, and one factor is certainly the Tornado's unequalled low-level attack capability, in that if flies faster than the defences expect and at a lower height. This is discussed further in the final part of this section.

With targets very close to the ground MTI (moving-target indication) becomes an important technique in separating the targets from the background. Without it there are problems in eliminating the radar return (reflection, or echo) from the ground or sea surface, which would otherwise obliterate any useful information from an interceptor's cockpit display. If the Earth's surface were still (relative to the local part of the planet, that is) then MTI would be simple; the rule would be "if it moves, the radar sees it". But in the real world we have leaves fluttering or blown away by the wind, waving crops and other vegetation, waves, wind-blown spray and spume often moving at about 60mph (100km/h). We also have birds, animals, and man's vehicles. It is essential to set the interceptor's MTI circuits to eliminate all unwanted (slow) moving targets and respond only to those travelling across the Earth's surface at a chosen speed, say 100mph (160km/h). But when they flew near local autobahns (motorways) the F-15 pilots of the USAF 36th TFW at Bitburg in Germany found their displays choked with speeding Mercedes and BMW cars. They had to set the MIT circuits to a still-higher speed; so does that mean the Russian clattering across Germany in his Po-2 biplane at 80 knots could not be detected?

The helicopter's vulnerability over the modern battlefield has always been questioned. Though modern types are designed to survive direct hits by bullets and shells up to (in the case of the AH-64A Apache) 23m calibre, nothing can be done about the bullet that hits the pilot in the head; and not much can be done about the SAM warhead, or an HE round fired from the main armament of a tank or SP gun.

But is not a modern attack helicopter seen head-on a very small and elusive target, merely popping up occasionally from behind cover to loose off and guide a missile? Though no such development has been announced, it should be relatively simple to provide front-line army air-defence radars with circuits tuned to recognise the blade-passing frequency of known enemy types of helicopter (this can be done by measuring doppler frequencies). Such a radar would instantly detect the rotor blades of any enemy helicopter, even from a great distance and with the rest of the machine "hull down" and out of sight (so long as the rotor disc was viewed approximately edge-on, as usually would be the case). Shall we later see anti-tank helicopters whose rotors are *under* the fuselage?

**Soviet SAM coverage** Some possible locations of USAF AWACS, ECM jamming and tactical surveillance aircraft are shown in relation to the FLOT (forward line of troops), previously called the FEBA (forward edge of battle area), and to sky volumes covered by Soviet front-line mobile SAM systems. SA-4 has been in use many years but has been repeatedly updated to counter the latest NATO counter-measure systems. SA-10 was first deployed in 1980 (mobile version 1985). SA-12 is coming into service. It can intercept Pershing missiles and range might be 340 miles (550km).

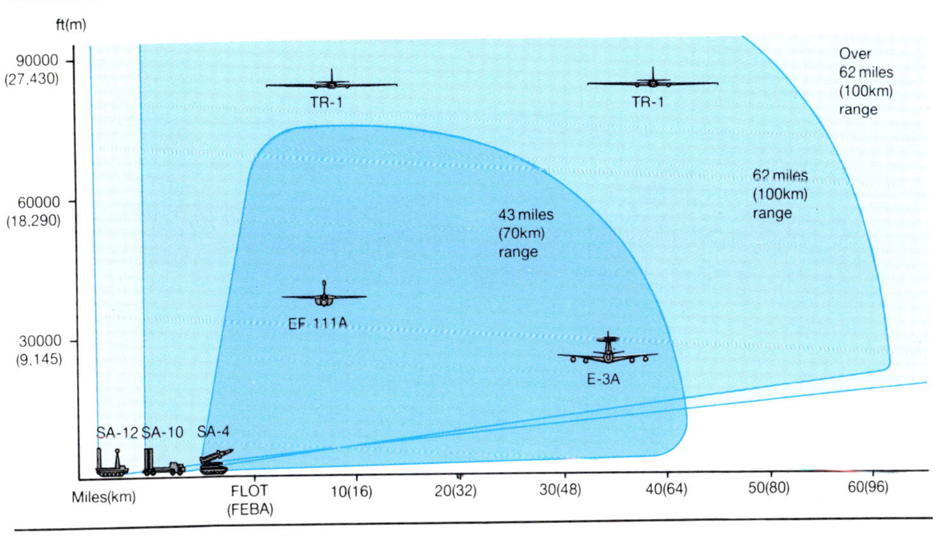

It will already be obvious to the reader that the viability of any kind of aircraft in hostile airspace has little to do with its speed, manoeuvrability or armour protection, but a very great deal to do with its ability to escape being seen. A half-century ago a Russian designer tried to build an invisible (ie, transparent) aeroplane; of course he was defeated by such things as the engine, tyres and even the fuel. At the time such an objective seemed laughable, but today the search for invisibility has become a matter of life or death.

Invisibility today has to be sought not only at the wavelengths of visible light but also throughout all parts of the electromagnetic (EM) spectrum that are used by military sensors. Lasers almost universally operate at a wavelength of 1.06 micrometres, at the extreme red (long-wave) end of the visible spectrum. Still longer waves are called IR, or heat, the usual sensor bandwidth being 8 and 13 micrometres. The far IR region merges into the vast field of radars at a wavelength of about 1mm (1,000 micrometres). There are not many millimetric radars, though one rides in the nose of each Hellfire anti-armour missile. Most radars have wavelengths measured in centimetres, though a few operate at a wavelength of almost a metre. When we set out to make an aircraft invisible we have to try to achieve this from one end of the hostile sensor spectrum to the other, and this is probably one of the very few tasks (short of flouting laws of physics) that can truthfully be described as impossible.

Even the simplest possible form of invisibility, by use of camouflage paint, is easily defeated by a small change in the sensor wavelength. By, as it were, "Looking through rose-coloured glasses", objects normally invisible can be made to stand out from the background, but the accepted technique is to use so-called "false colour" photographic film. This makes use of different parts of the EM spectrum, and sidesteps attempts to render an object invisible in the purely visual range of wavelengths.

In any case what is the background? An F-15 painted in air-superiority grey may be hard to see from a distance in most air-to-air encounters but it will show up clearly against a green field or other surface backgrounds. Lightning F.2As in RAF Germany spent so much time at low level, or on the ground, that they were sprayed dark green over all upper surfaces. Artist Keith Ferris carried out much research for the US Navy on how paint can be used either to make a fighter invisible or, at least, very hard to see clearly. Like the "dazzling painting" of ships in World War I, Ferris sought to break up the characteristic outlines so that an enemy, even if he saw the aircraft, would be unable quickly to determine its attitude or direction of travel.

OCM (optical countermeasures) remains

**Above: Four F-15s of the 49th TFW, seen over Germany on exercise Reforger in 1983, could hardly be more conspicuous. What is needed is an adaptive coating that, like a chameleon, matches the background.**

**Below: Artist Keith Ferris worked for the USAF and the Navy, and one of his full-scale trial paint schemes was applied to the sixth USAF F-4C Phantom. In this case national markings were omitted.**

Above: Different conditions and these F-15s are not quite so conspicuous as they just clear the towers of a fog-girt schloss on the Austrian border. But look-down radars would see them.

Below: Contrails formed by vortices as intense as those generated by a manoeuvring F-16 show on radar, though they are usually much shorter than those formed by freezing water vapour in the propulsive jet.

an imperfect and often frustrating art. One has sympathy with people who try a more direct approach. In Sweden the FFV company is marketing (initially for hovering helicopters) a self-screening launcher which can fire ten rockets over an arc of 120° which burst 330ft (100m) ahead to generate an instant smokescreen good from 0.4 to 14 micrometres.

It goes without saying that, to survive, an aircraft must neither make a visible contrail nor pump out a smoky jet. At low levels it is not difficult to eliminate any kind of visible wake or trail, but in the stratosphere the formation of a contrail is difficult to avoid. All engines that burn hydrocarbon fuels pump out exhaust containing a high proportion of water vapour, and if the atmosphere is cold enough this quickly freezes to form visible ice crystals. In a European winter a contrail of this type could easily be formed at low levels, and there is no simple way to avoid this.

So far we have discussed invisibility at optical wavelengths, and come to the conclusion that this objective is unattainable. But modern air defence systems must not wait until the enemy aircraft can be seen; that would be a last-ditch situation! In the first instance aircraft survival depends upon invisibility to defending radars, and here we are dealing with a very much longer wavelength of EM radiation. Moreover, as in the case of visible light, radars operate over a wide band of wavelengths – which could be thought of as different "colours" of radiation. Thus, we might think of a surveillance radar as using red radiation, a SAM tracking set as yellow and a fighter radar as indigo-blue. So to survive we have to try to look invisible at all these wavelengths as well, and again we are confronted by impossible requirements. The best we can do is, first, to shape our aircraft so that it reflects hardly any radiation back along its original path, and second, to cover important parts of the aircraft with the best RAM (radar-absorbent material) we can invent.

Like Britain's Chobham armour, RAM is supposedly a classified subject. Not much has escaped even into the US technical press, but a few conclusions appear to be obvious. First, the RAM has to be applied in the form of a coating sufficiently thin neither to alter the aircraft nor significantly increase the structure weight. Second, it has to stay on, despite prolonged encounters with rain or hail at speeds around Mach 1, and (almost certainly) despite the existence of forward-facing edges where the coating comes to an end. Third, as a "variable-geometry" or inflight-reprogrammable RAM coating seems to be too much to ask, the existing coat must counter as many as possible of the known hostile radars. We have already seen how wide a range of wavebands these cover, but what is not known is the success that has been achieved in designing RAM coatings that are highly effective over a spread of wavelengths. Put another way, if they depended on the physical dimensions of their structure or thickness, such coatings would probably be of little use except over a narrow waveband, or even at discrete wavelengths.

The way a solid structure reflects, or reradiates, radar signals is extremely complex. To minimise the echo it is necessary to pay attention to both the type of material used and the structure's shape. A non-metallic structure has an inherently smaller RCS (radar cross-section) than a conductive metal one, yet conversely a conductive paint can even out surface potentials (voltages) and reduce RCS significantly in comparison with a paint that is an insulator.

The Lockheed U-2 was used to pioneer conductive paint, containing astronomic numbers of microscopic iron balls. It was explained that this near-black paint was used to radiate heat efficiently from the highly supersonic SR-71A, but such an explanation is less convincing for the 400mph (644km/h) TR-1. What is particularly interesting is the willingness of the DoD/USAF/CIA to trade enhanced visibility at optical wavelengths in return for much-reduced visibility at radar wavelengths. There seems to be no physical law against "iron ball" paint that at optical wavelengths looks pale grey. It has been stated that plain grey is one of the colour schemes being considered for the B-1B. From the visual aspect this is obviously preferable to black.

The point has also been made that thickness of paint has to be tightly controlled to avoid significant escalation in aircraft empty weight, but little is publicly known about the effect varying paint thickness has on RCS at different wavelengths.

Airframe materials exert a large influence on RCS. Structures made wholly of glass-reinforced plastics, and to a lesser degree carbon-fibre composites, are much better for stealth aircraft than metal structures. Except for the main pure-delta wing box of B-120 titanium alloy, the entire wing and elevons of the SR-71A is made of a classified plastics-honeycomb material which has to meet extremely severe operating stresses and temperatures. The internal ribs are arranged at small angles to each other, in a fine zig-zag all round the wing. Radar signals are reflected to and from between each funnel-like gap, losing energy at each reflection, in the same way that RF energy is absorbed by the inward-pointing pyramids of an anechoic chamber.

Of course it would help if we could cover the whole surface of the aircraft with such spiky pyramids, because the RCS would then be close to zero no matter what shape the underlying structure was. Unfortunately this kind of flying porcupine would have high drag, so we have to use a smooth skin but shape it with extreme care to minimise reflectivity. We must avoid any large surface likely to be more or less perpendicular to the radar beam direction. If we are going to penetrate the defences at low level, this means that we must avoid vertical faces such as fins or fuselage sides of the kind seen in the F-15.

It is easy to avoid flat sides – look at the SR-71 – but how do we avoid fins? A Harrier-type reaction-jet control system could be used instead. An alternative could be split wingtip ailerons of the kind that Northrop, under the patented name Dece-

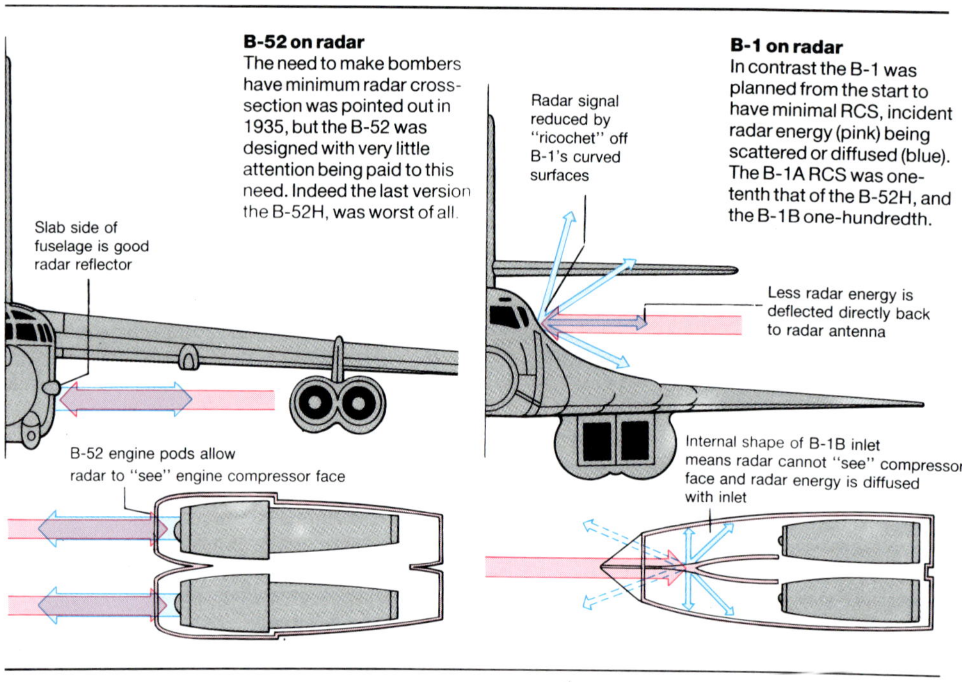

**Above: Even without putting the flaps down the B-52 (this is a turbojet-engined B-52G, not quite as bad as an H) has a radar cross section "like a barn door".**

**Below: Though it does a much bigger job, the Rockwell B-1B is much smaller than a B-52 and represents a 30-year difference in basic technology.**

**Above: Why are the U-2R and TR-1A black? U-2R 68-10336, with extended SLAR nose, shows the dull, smooth surface of its electrically conductive "iron ball" paint.**

**Left: Lockheed's amazing SR-71 was the first aircraft deliberately designed with stealth characteristics, though some features had to be considered compromises.**

leron, built into the YB-49 eight-jet flying wing bomber. This huge aircraft lost out to the B-36, but today it can be seen as potentially a fantastic basis for a stealth aircraft. The USAF probably carried out trials on B-49 radar signature but, because in those days bombers flew at over 40,000ft (12,200m) and the radar would have looked at the flying wing almost square-on, the results were probably discouraging. But bring a B-49 in at tree-top height and the RCS would be very different.

In most aircraft the most reflective single part is the front of the engine. In many fighters this cannot be seen, but significant radar energy can still be pumped out of the inlet duct. One of the worst aircraft ever built for engine RCS was the B-52 bomber, with eight large turbojets facing straight at the enemy. The final model, the B-52H, made matters worse by enlarging the engine diameter from 39in (990mm) to over 53in (1,350mm). The wish by Pratt & Whitney to put its new PW2037 engine – urgently in need of orders – into the B-52 would halve the number of engines but actually make matters worse still because the diameter of that engine is 85in (2.16m).

In the B-1B, Backfire and Blackjack the engines are at the far end of long inlet ducts which are specially curved and equipped to minimise radar signals reaching the engines or getting out again. In fact the USAF claims that the RCS of the B-1B is one hundred times less than that of a B-52 (model not specified).

Of course, one problem with carefully shaping an aircraft to have minimum RCS is that the stealth qualities depend on the direction of the incident radiation, in other words on how the bomber looks to the defending radar. To some degree the RCS from different aspects parallels the visual appearance. For example, a B-49 looked enormous from below but small from the side. In the same way a low-level penetrator configured to have minimal RCS from the front or side might present a juicy target on the radar of an interceptor looking down from above. Of course, this is mainly because of the presence of the wing. If we could cruise at Mach 1.5 or more we need not have a wing but instead could rely on body lift.

Many missiles do this already, including the Asraam and ASMP. The reason today's AGM-86B Tomahawk cruise missiles need retractable wings is that they cruise slowly, at 500mph (805km/h) or less. A wingless vehicle naturally has smaller RCS from most aspects than one with wings. Indeed it is possible today to build a V/STOL vectored-thrust combat aircraft having the classically simple form of a tube, with an ogival nose and pointed tail. There is no need for it to have any aerodynamic surfaces; it would take off vertically, accelerate to supersonic speed and both fly level and manoeuvre on body lift alone. Its massive advantages would include the ability to move sideways or vertically in any direction without the prior need to roll, the extreme simplicity of the structure and absence of concentrated loads, and a RCS much smaller than anything yet achieved in aircraft of similar weight and capability. Of course its ability to hide in forests again would be unrivalled.

The absolute need to get away from airfields is now so self-evident it is even becoming accepted, albeit reluctantly, by the USAF. This factor is referred to in the next section, in discussing airbase requirements.

Now we can do wonders with combat aircraft in the form of vectored-thrust aircraft looking like bits of oil pipeline, but are we then going to ruin everything by making it behave like a mobile radio station? Modern attack aircraft typically have 12 separate aerials (antennas) from which they pump out EM radiation of distinctive kinds, all of them known to and recorded by the enemy. There is little point in building a nearly perfect stealth aircraft if it is going to broadcast its presence and provide ideal radiation on which enemy

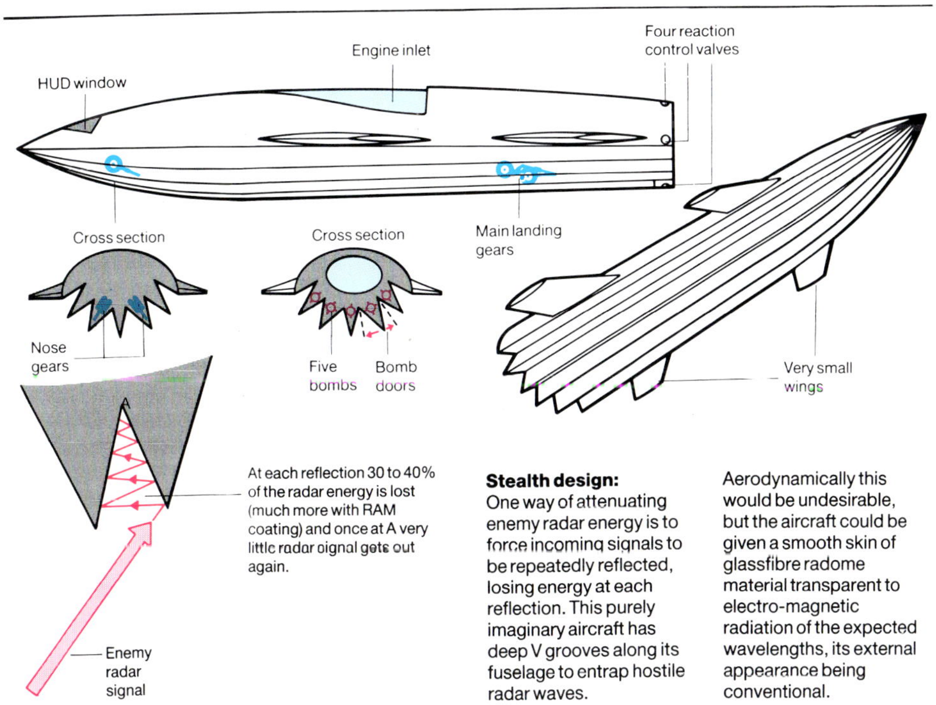


**Stealth design:** One way of attenuating enemy radar energy is to force incoming signals to be repeatedly reflected, losing energy at each reflection. This purely imaginary aircraft has deep V grooves along its fuselage to entrap hostile radar waves. Aerodynamically this would be undesirable, but the aircraft could be given a smooth skin of glassfibre radome material transparent to electro-magnetic radiation of the expected wavelengths, its external appearance being conventional.

missiles can home. And, even if we can make our penetrator electronically "quiet", or passive, it is difficult or impossible to propel it at high speed without simultaneously generating large quantities of IR (heat). The latter alone is all that many anti-aircraft missiles need as a source on which to home.

Clearly, the best way to avoid having one's emissions detected is not to emit any. Much can be done to render one's own emissions less suicidal (see the Electronic Warfare of this book), but it is clearly impossible (yet) to create an effective combat aircraft that remains electronically silent throughout its mission. Even the presence of aerials for radio communication, doppler navigation, radar altimetry or any other radar function ensures that they will pick up and retransmit an enemy's radar signals.

If the enemy's surveillance and other systems do use radar wavelengths then it is possible to minimise response of one's own aerials by making the latter work at very different wavelengths, an obvious choice being to use lasers. These may be more rapidly attenuated by atmospheric particles or precipitation, but have a further advantage in being propagated in the form of extremely narrow beams detected by the enemy only if they happen to be aimed at him.

As most readers will know, a manned penetator cannot safely fly manually at very low level, but instead must be controlled by a reliable and instantly reacting TFR (terrain-following radar). This projects what has been likened to a ski ahead of the aircraft, which rides smoothly over (or sometimes around) all hills, radio masts, buildings and other obstructions.

A drawback of the TFR is that it provides an excellent warning to the enemy of the oncoming aircraft, but it should be possible to replace it by systems that are less of a hazard to the aircraft that carries them. To obtain a visual picture a FLIR can be used, operating passively and as effective by night as by day (though with performance usually degraded by precipitation, such as heavy snow). Distance to the obstruction can be obtained by a pulsed laser.

There is no difficulty in linking these virtually undetectable sensors though a computer and thence to the aircraft flight-control system, but at present it is difficult to achieve such guaranteed safe flight along the NOE (nap of the Earth) as is possible with a TFR. A laser altimeter is no problem, and indeed several exist. Doppler navigation could be achieved with very narrow laser beams, but a self-contained INS (inertial navigation system) updated by other means, such as astro star-tracking or satellite reception, seems preferable. The difficulty lies not in navigation but in flying very low with assurance that obstructions will be avoided, and this requires some form of coupling, at a chosen electromagnetic wavelength, between the speeding aircraft and its environment.

## Aircraft heat emissions

Turning to the vexed question of IR, there is no way that a combat aircraft can be

**Above: All versions of the F/A-18 Hornet can carry an AAS-38 FLIR pod on the left body pylon, in place of a Sparrow or other weapons. This aircraft already had a very versatile multirole radar in the nose.**

**Right: A house at Piney Point, Maryland, as seen by an AAS-38 FLIR. In low-level attack at night FLIR imagery is better than any other wavelength, especially against targets emitting significant heat.**

**Below: Low-level training in a two-seat Jaguar attack aircraft, with a HUD display in each cockpit. With the HUD, radar altimeter and laser the absence of radar in most Jaguars is unimportant.**

**Above: Two afterburning turbofans make this F-111 a target IR-seeking missiles dream of. On a clear night a sensitive IR seeker could lock-on this aircraft at a range of more than 40 miles (65km).**

designed without emitting heat. Even a man-powered aircraft emits heat, from the operator's body and breath; and kinetic friction results in (very small) rise in temperature of the airframe. With our supercruise stealth aircraft the heat emission is likely to be a few million times greater, and it poses very large problems. Gone are the days when IR-homing missiles were distracted by the sun, or its reflection in a friendly greenhouse: today's IR seekers are tuned to ignore everything but the hostile aircraft for which they are looking; and in most AAM systems using this form of guidance the seeker head is locked-on before launch.

IRCM has thus swiftly grown into a science – or dark art – rivalling that of ECM. There is one obvious way to try to survive against such weapons: cut down one's IR emissions and/or change the wavelength. The former is not easily done. Engine designers the world over are striving to increase the thermal efficiency of aircraft propulsion in order to reduce the consumption of costly fuel, and their efforts automatically cut down the emission of waste energy. Barring an unforeseen technical breakthrough, there is nothing to be gained beyond one or two percentage points in trying to reduce emissions by making more efficient engines. Put another way, the propulsion system of any modern jet aircraft cannot avoid pumping out heat, and because of the need for propulsive efficiency there is little that can be done to change the frequency/wavelength. All that can be achieved easily is to extend or shroud the jetpipe to make life harder for missiles that home in on hot metal.

A jet engine, no matter whether it is a turbojet or turbofan, presents an ideal IR source if it is operating in the afterburner or augmented mode. In this case there is so much IR and so little prospect of altering or hiding it that one may as well forget about any thoughts of stealth design. Thus, future combat aircraft will have to avoid afterburners or have very clever IRCM protective systems – or have very short lives indeed.

How do you provide IRCM? Assuming that you cannot simply screen or mask the source, the simplest way is to offer the IR-homing missile an alternative target. This involves creating a second IR source. Clearly, for the best results this must be matched exactly to the frequency/wavelength at which the enemy seeker head is most efficient.

In its simplest form an IRCM source is merely a juicy IR emitter which operates (burns) continuously and with sufficient power for it to be preferred by the enemy missile to one's own aircraft. This kind of decoy serves little purpose if it is fixed to one's aircraft, so to offer protection it has to be thrown as far away as possible. Moreover, for obvious reasons, the timing of the throw has to have split-second accuracy to have maximum decoy effect. If it is ejected too late the aircraft will still be within the lethal radius of the enemy warhead. Equally, there is no point in ejecting it until an enemy IR-homing missile is well on its way towards the aircraft, and fully locked-on. Thus, this kind of IRCM is linked with a reliable surveillance system, which can take any of several forms and which unfailingly triggers the IRCM at the correct moment to give protection.

This kind of IRCM uses what are called expendables. These are flares, packaged into (usually cylindrical) cartridges which are interchangeable with chaff cartridges and active RF jammer cartridges of the kind mentioned in the section on Electronic

**How target looks to anti-aircraft systems**

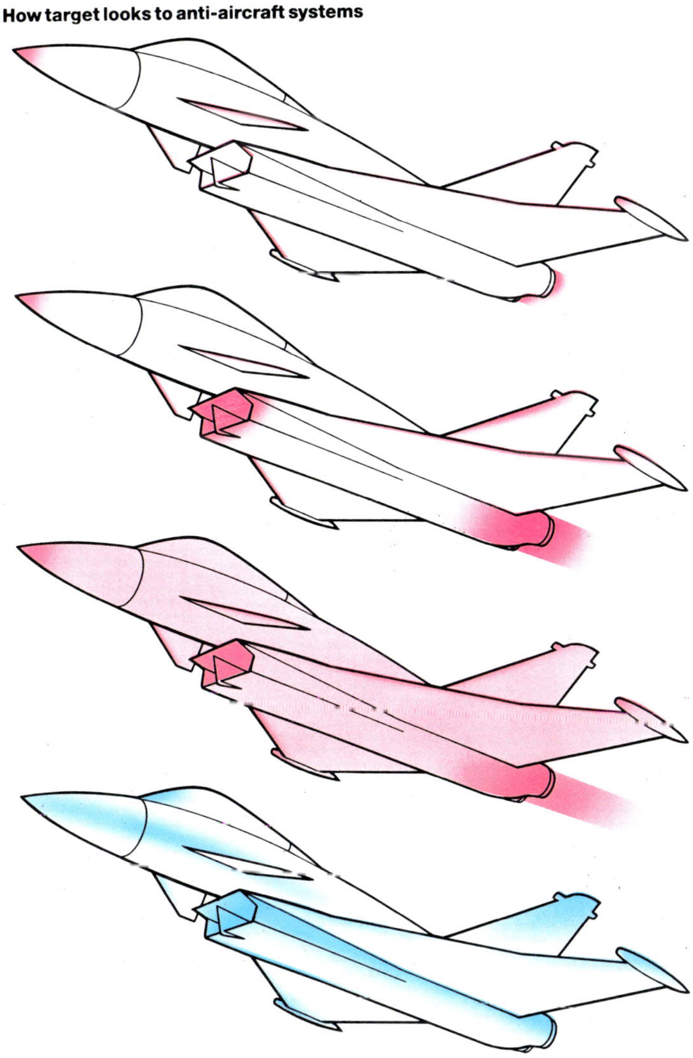

**Cruise**
To an IR-homing missile a target aircraft consists of just a few attractive local "hot spots". This fighter is shown as it would look in subsonic cruising flight to an AAM tuned to a wavelength of 2 to 3 microns, which "sees" high temperatures.

**Dash**
Here the same IR missile, seeing radiation of 2 to 3 microns wavelength, has locked-on to the same fighter in full afterburner making a supersonic dash. All returns are greatly strengthened, and the afterburner nozzle is absolutely dominant.

**Longer wavelength**
While some IR sensors, such as doped indium antimonide, see 2 to 3 micron radiation, germanium and some others work best at 10.6 microns. This longer wavelength sees cooler targets, taking in virtually every part of the target aircraft.

**Radar**
Operating at wavelengths roughly 10,000 times longer than IR seekers, the radar-guided missile can "see" better through clouds and rain. It receives the strongest echoes from flat surfaces, especially from junctions where flat surfaces meet each other. Stealth designers have this in mind.

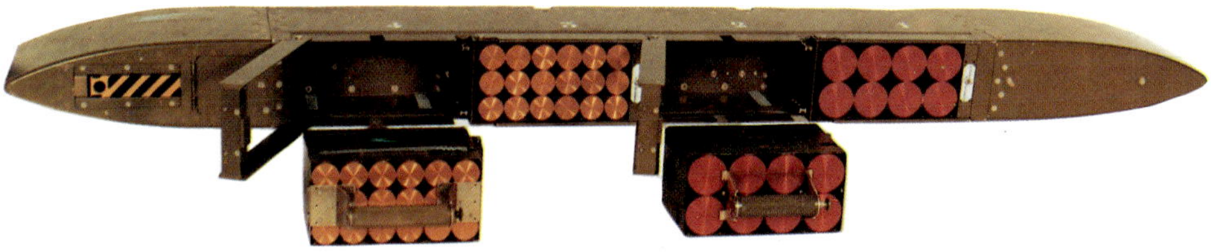

Above: Countermeasures dispensers can be mounted internally, or scabbed on the outside of tactical aircraft, or attached to a pylon like any other store. This French Alkan dispenser comes in the third category. It is of modular design and is shown here loaded with two magazines of 40mm chaff cartridges (ECM) and two magazines of 60mm flare cartridges (IRCM). This model flies on Jaguars and Mirages.

Below: Virtually all high-value military aircraft of Western air forces have at least the minimum self-protection kit of chaff and flares dispensers. In this exercise one MC-130E Hercules is following close behind another which is dispensing IRCM flares. So far 46 have been fired, in a time of about 4 seconds, from a dispenser installed on each side of the rear fuselage, firing in unison.

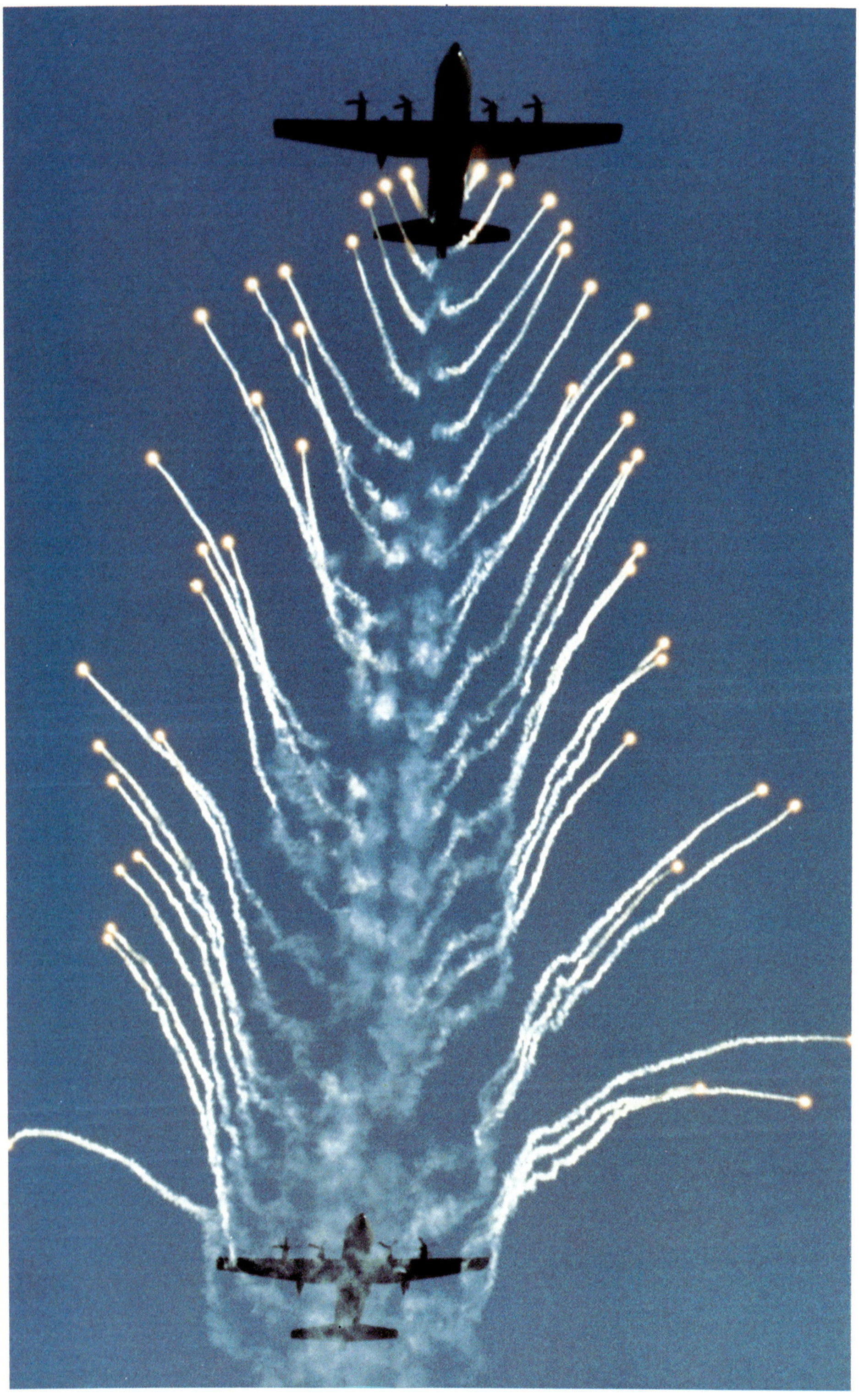

Warfare. The aircraft is permanently or semi-permanently fitted with a dispenser rather like an egg-box with anything up to 100 firing tubes. Each tube can be loaded with any of three common types of expendable: chaff, RF jammer or flare.

The flare, the IRCM expendable, is a cartridge filled with a composition which burns as long as possible at a suitable high temperature to achieve the desired intensity and wavelength. A common filling is finely divided magnesium metal plus TFE (tetrafluoroethylene), which gives a bright burn at about 2,200K (1927°K). The flare cartridges are usually fired in sequence, each being ejected by a small explosive charge. Small salvoes have been seen by millions on TV newsreel in 1983-84 when they were fired from Israeli fighter and attack aircraft over the Bekaa Valley and from Soviet aircraft over Afghanistan.

Some of the Soviet aircraft were attack helicopters. This is the only category of powered aircraft which can reduce its IR emissions close to zero. This is because its turboshaft engines extract as much energy as possible from the gas flow and put it into the shaft driving the rotor. The relatively cool gas can then be ejected through an IR suppressor where it is mixed with induced cold air and finally escapes at little above ambient temperature on a hot tropical day. Only the slow-flying helicopter can accept the bulk and aerodynamic bluffness of the suppressor, which has to be very large in order not to spoil engine performance by causing high back-pressure in the jetpipe. Such a suppressor cannot, with present technology, be fitted to a jet aircraft without causing unacceptable penalties in flight performance.

On the other hand, there is an IRCM technique which can be fitted to anything, even a supersonic missile. This is the IR pulsed emitter, a subject which is still classified, so that precise details of the relationship between the hostile missile threat and the IRCM emission cannot be given. Suffice to say that each installation

Below: Loral ALQ-123 pods on test before delivery. As described in the text, this uses windmill energy to pump out a coded sequence of IR emissions. ALQ-123 has been in service since about 1975.

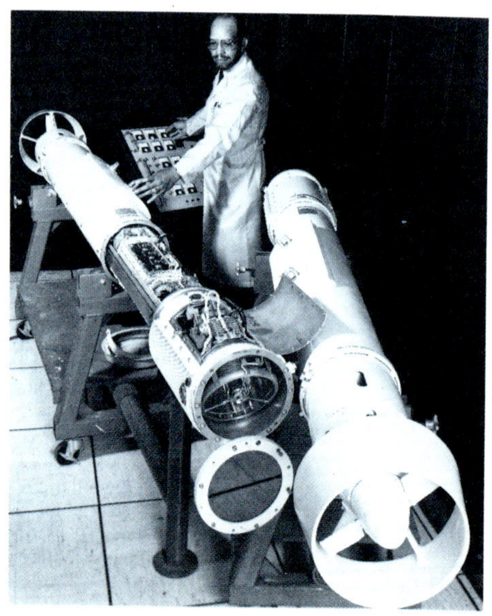

consists of a powerful IR lamp which converts electrical power into heat and whose emission can be "turned on or off" extremely suddenly, plus a computer control system. The words in quotes are used loosely, because it is not yet known if current pulsed systems use an actual pulsed lamp or a continuous lamp fitted with rapid-acting shutters. Each installation is tailored either into an external pod or for hard mounting on a particular type of aircraft.

Three leading suppliers of airborne IRCM systems are Loral, Sanders and Xerox. The Sanders ALQ-147 family of the "hot brick" devices are carried in external pods on aircraft with limited electrical power, the high-power emission being generated by a special ceramic source made white hot by a flame fed by jet fuel mixed with ram air. Such a source obviously has to be continuous, so modulation according to the desired codes is effected mechanically. The shutter system also has to mask or filter all visible radiation, so that nothing can be seen even at night.

Another pod that needs little from the carrier aircraft is Loral's ALQ-123, which is easily the most common IRCM on US and NATO fighter and attack aircraft. Its heavy electrical current is generated by a ram-air turbogenerator on the nose. Amidships is the modulator and at the rear is the IR lamp and lens. The ALQ-157 is a system for helicopters, with two or three transmitters carried on the sail, rotor pylon, nose or other locations to achieve 360° coverage. For high-value multi-engine aircraft such as an E-3A, Loral supplies a modular system, each transmitter taking an electrical load of 6kW, ideally one being installed for each engine.

All these pulsed IRCM systems pump out IR radiation controlled to a preprogrammed jamming code, which is often selectable by the flight crew during the mission to counter different kinds of threat. The code is chosen to make the enemy missile home first on the aircraft engine(s) and then on the IRCM source. When this is done correctly the oncoming missile totally breaks lock and thereafter flies in a roughly straight line, or even finds a different target entirely.

For over 30 years it has been appreciated that aircraft that penetrate hostile airspace ideally need support from specially equipped (so-called dedicated) aircraft whose sole purpose is to help confuse or defeat the defences. Not much has been done with dedicated IRCM aircraft, though this mission is bound to become increasingly important. At RF wavelengths the story is different. The dedicated RF jammer has long been an established type, and as in the Western world it tends to be US technology that gets accepted (for lack of competition) by the other NATO partners, the operative buzzword here is "Wild Weasel". This was the name of primitive F-105 conversions used in Vietnam to bother radar-based defence systems, and these in turn led to the F-4G Advanced Wild Weasel which packaged into the F-4 Phantom II airframe a useful RF reception system (mainly APR-38) and missiles which home automatically on enemy radars, starting with AGM-45 Shrike and AGM-88 Harm.

**Above: Another propeller-driven IRCM pod is Northrop's AAQ-8, which is automatically controlled to emit modulated IR signals to break the lock of oncoming missiles. The aircraft is a Buccaneer at the RAE.**

**Below: At a distance the F-4G could be mistaken for other Phantoms, but it is a dedicated anti-radar platform. Its internal APR-38 "Wild Weasel" set provides eyes and ears for anti-radar missiles, which here are a Standard ARM, a Maverick, a Harm and a Shrike.**

Back in 1959 the US Navy also took the trouble to think about the problem and eventually bought first the EA-6A Intruder and then the EA-6B Prowler, a true dedicated ECM platform which does its best to counter all known hostile anti-aircraft defence radars. It does this by receiving signals at the top of the fin, processing and analysing these and then jamming or otherwise countering them with emitters housed in up to five external pods, each drawing current from a nose turbogenerator, with systems management by two backseat crew-members. Almost the same system was repackaged into the USAF EF-111A Raven, but with the whole panoply of jammers contained in the fuselage, radiating via aerials (antennae) faired inside a ventral "canoe radome" and with no extra crew members.

The RAF must have realised what a good jamming platform the surplus Buccaneers would make, and the Luftwaffe (having no spare Buccaneers) has for almost three years been trying to define the specification for an ECR (electronic combat and reconnaissance) version of the Tornado. Any future NATO ECR platform is likely to carry the BAe Dynamics Alarm missile, which is by far the best weapon in the West for knocking out hostile radars. One of its operating modes is a slow parachute-retarded "loiter" in which it spends a long period waiting for the alerted enemy to switch on their inactivated radars. Of course an ARM (anti-radar missile) as clever as Alarm could transform the capability of even the most primitive air force.

Before going into detail on such topics as aerodynamics, structures and propulsion, the point has to be made that in any future war between the superpowers one may as well discount any aircraft that needs an airfield. No significant military asset whose location is geographically known is going to see the dawn of Day 2; this is self-evident from a simple count of Soviet missile-delivered warheads. Yet the continuing lunacy forced on NATO by Washington says, "There's no problem, our F-15s will be able to *take off between the craters!*"

Even if the Warsaw Pact decided not to fire its missiles, but instead sent wave after wave of aircraft to drop runway-cratering munitions, NATO's problem would not merely be how to take off between the craters: once in the air, our intrepid NATO aviators would at some future time have to try to get back. If you ask the purveyors of the official party-line, "How are we going to *land* between the craters?" they look rather pained and explain that that little problem is being studied at the moment!

Clearly it is much easier to design a modern combat aircraft if you can count on tearing along a smooth paved runway between the craters, than if you recognise from the start that no fixed military asset is in the slightest degree survivable. The preceding part of this section was concerned with survival whilst flying through enemy airspace, which is relatively simple. Survival of fragile aircraft, and fragile humans, whilst they are on a known airfield is today a totally unsolved problem. Nuclear weapons, giant FAE (fuel/air explosive) warheads, lethal or incapacitating chemicals, corrosive/incendiary chemicals, and various forms of sabotage are all obvious hazards more difficult to deal with than a crater in a runway.

## Aircraft shapes

Aerodynamically there is no problem in deciding on the "best shape" for each part

**Above: Setting out on a trans-US flight in 1949, the Northrop YB-49 was the greatest stealth bomber ever built. Today such an asset is better appreciated, and Northrop is prime contractor for the ATB – or "B-2".**

of a mission. The problem is that each part of the mission calls for a different shape. A quarter of a century ago the VG (variable-geometry) aeroplane was all the rage, though VG was interpreted merely as having pivoted wings whose sweep angle could be varied. Gradually this idea went out of fashion, because the decision-takers began to concentrate not on the obvious advantages but on the supposed penalties.

To some extent variable wing sweep was replaced by variable wing profile, this reaching its highest level of expression in the MAW (mission-adaptive wing) created by NASA and Boeing Military Airplane Co., and now flying on an F-111. It is probably safe to predict that for as far ahead as can be seen all combat aircraft will have at least some form of variable wing profile using hinged leading and/or trailing edges. The obvious demands for stealth qualities could lead to new forms of aircraft with sharply reduced span, and these are likely to rely on vectored engine thrust as much as on wing lift in the takeoff and landing portions of each mission.

The use of vectored thrust is made more obvious from the fact that the installed thrust at sea level is likely to be similar to, or even greater than, the aircraft weight. So far aircraft designers have barely scratched the surface of the technique of integrating a jet engine with an aeroplane. Merely to have two giant afterburning nozzles facing aft is an inadequate solution, but it happens to be easy and, in peacetime, efficient. The contrast between a modern fighter and a helicopter is roughly as great as that between an artillery rocket and a dragonfly, and any designer who takes a future war seriously cannot avoid having to bridge the gap.

If we recall the shapes that give stealth qualities we see that one is a vehicle

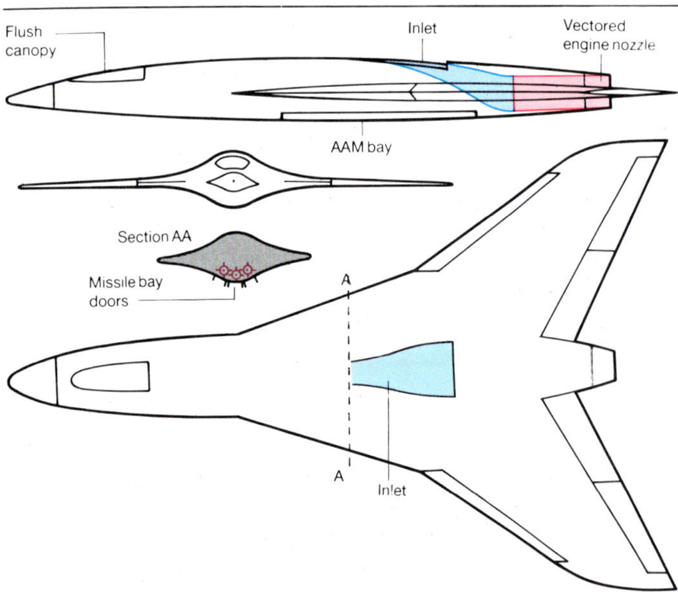


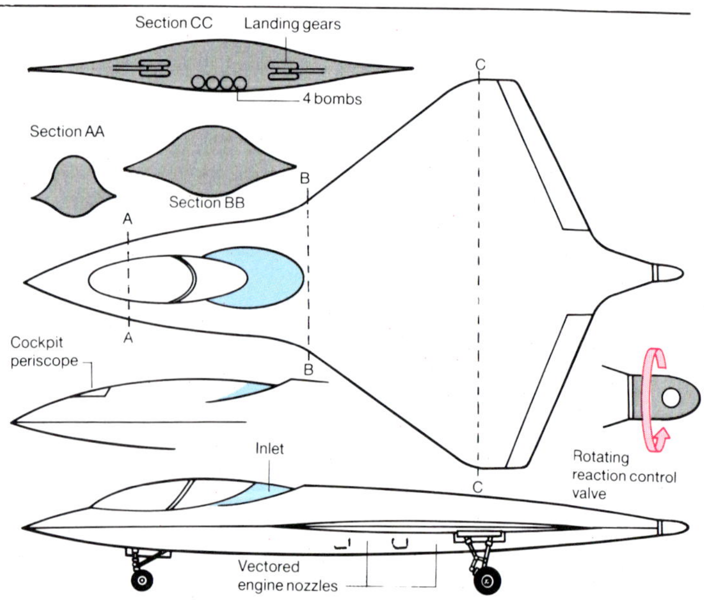


**Stealth CCV**
This suggestion for a future combat aircraft has a single engine with a flush dorsal inlet and a nozzle which can be vectored in any direction. There are ways of making it a VTOL, but as shown the nozzle is for enhanced inflight manoeuvrability. All weapons are accommodated internally, as shown in A-A, which also emphasises the blended wing/body shape.

**Stealth VTOL**
This is based on the well-proven "four poster" type of vectored-thrust turbofan, with nozzles disposed around the aircraft centre of gravity. The powerful RCV in the tail can point in any direction, and would counter pitch moments due to momentum effects in the dorsal inlet. The inset shows an alternative stealth cockpit with periscope.

looking like a length of pipe with pointed ends. This could be a VTOL if one could avoid the suckdown effect of the airflow around the pipe-like fuselage. It could then rapidly accelerate to supersonic speed and fly on a mixture of body lift and a slight downward component of engine thrust. Use of reaction control jets would eliminate the need for any aerodynamic fin, and the engine nozzles could even be vectored under computer control, not only about a transverse axis but in any direction, to affect necessary changes in course or flight trajectory. The nozzle angles would not have to be large, but the result would be a highly manoeuvrable aircraft which, like today's F-16/AFTI, would not need to roll into a bank before changing direction.

One of the commonly proposed stealth aircraft is the flying wing configuration, similar to the famous Northrop XB-35 and YB-49. Aircraft of this type are aerodynamically efficient, with a lift/drag ratio better than a tubular vectored-thrust machine, but they lose most of their stealth quality as soon as they make a banked turn. They are also highly visible when seen from above and below. In contrast, the tube form would look approximately the same from all aspects, and the large and diffuse lateral plumes from its four engine nozzles would not present an attractive target to an IR-homing missile.

Further advantages of the "flying tube" would be its suitability for low-level flight and its ideal shape for supersonic flight at all levels. The problem with the conventional aeroplane in low-level operation is excessive wing size. The latter is sized for takeoff and landing, and in high-speed flight – even with all high-lift devices retracted, and (with a VG wing) at maximum sweep angle – the gust response is very severe, causing frequent violent accelerations in the vertical plane which quickly degrade the efficiency of any human crew, quite apart from eating into structural fatigue life. With the near-wingless aeroplane there would be sufficient control of trajectory in the vertical plane to do NOE (nap of the Earth) flying, and thus make detection by the defences as hard as possible, but insufficient gust response to bother the crew.

**Above: In August 1985, after three years of talking, Britain, West Germany and Italy – Panavia partners – at last decided to build a new fighter. It might resemble the BAe EAP in this drawing.**

**Below: Very similar to the EAP, the Dassault-Breguet Rafale prototype is being built to a similar timescale, with American F404 engines. Unfortunately all these fighters need to land on airfields.**

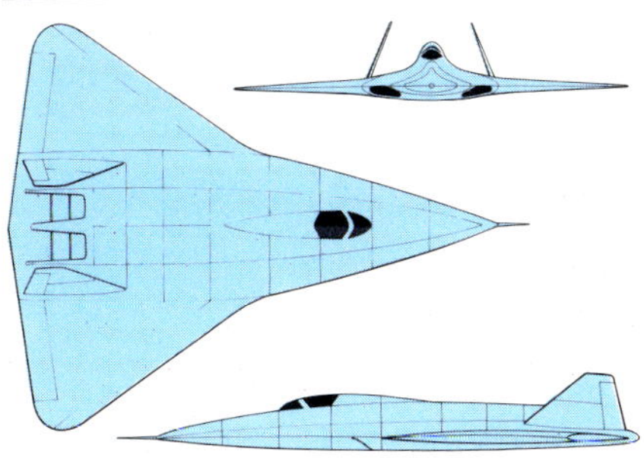

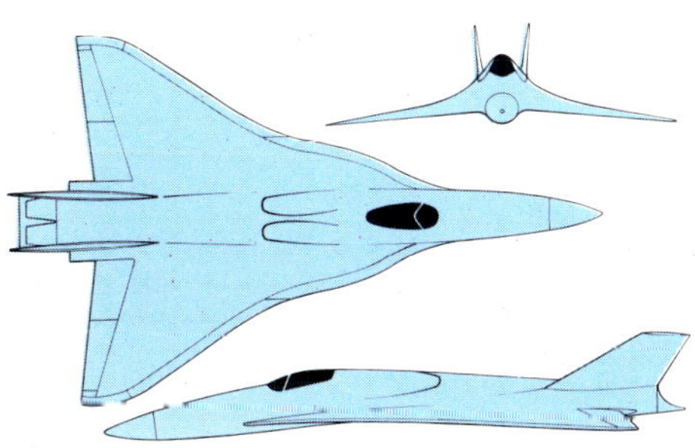

**The F-19**
At the top is a drawing purporting to show the Lockheed F-19 stealth fighter, based on one published by *International Defense Review*. Above is the author's guess. Of course, the actual F-19 need not look much like either.

Of course, an aircraft of this type would be wholly unsuited to low-speed loiter or any form of conventional landing. Loss of propulsion would mean ejection, but this is the procedure in a number of today's conventional aircraft. On the other hand, at supersonic speeds the lift/drag ratio – with suitable cautious shaping of the tube – could be no worse than a conventional ogival delta, in other words in the region of 9. There is little published information on the lift/drag performance of such near-wingless machines, and little point in thinking aloud about the possibilities. It is clear, however, that there is no law of nature forbidding the shaping of axial strakes which would not only enhance the lift but also attenuate incident radar waves. Looking a little further ahead still, it may be possible for a design team to create a lifting body with radar-attenuating strakes which can be pivoted through 90° in order to turn themselves into wings suitable for low-speed flight. The fact remains that, if we dare not go near airfields, such cleverness may be misplaced effort; the author believes future landings by combat aircraft will have to be vertical.

**Stealth/VTOL** This purely hypothetical tactical aircraft is the kind of thing the author believes could survive in a real war, which would destroy virtually all conventional aircraft on the opening day. The most obvious feature is its stealth shape, though it retains a conventional cockpit. It also needs no airfield, so it will be much harder for the enemy to destroy it on the ground except by chance in ones and twos, whereas known airfields would cease to exist. With internal weapon and sensor stowage it could be highly supersonic, but it is much more important to minimise emissions. The normal STOVL reaction control valves are used, rendering a radar-reflective vertical tail unnecessary. The author has not applied for a patent!

In a recent official submission the Director of DARPA (US Defense Advanced Research Projects Agency), Dr Robert S. Cooper, testified that, on his listing, the first three items out of 17 "key aeronautics technologies" were: 1, stealth; 2, supersonic cruise and manoeuvre; and 3, high-L/D hypersonic vehicles and boost/glide weapons. It is interesting that items 2 and 3 address themselves solely to flight at high supersonic speeds. Now however clever we may be, we can never flout the laws of physics. One of these laws states that the acceleration experienced by a body travelling in a curved path is proportional not just to the speed but to the square of the speed. Thus, if we go round a curve at 1,000mph (1,600km/h), we suffer four times the acceleration experienced at 500mph (805km/h). One does not have to think long about this to realise that at so-called hypersonic speeds, generally taken to mean speeds beyond Mach 3, bodies containing humans move in essentially straight lines. A straight-line military aircraft is a very specialized vehicle, and it has to operate well above NOE height, in full view of any hostile radars.

This is not to suggest that the hypersonic vehicle has no value, but for 30 years artists around the world have been disseminating beautiful pictures of hypersonic – or at least highly supersonic – fighter and attack aircraft followed by smoke trails or lines coming from the wingtips which are distinctly curved. So perhaps it needs to be repeated one more time: you can have speed, or you can have manoeuvrability, *but not both at the same time.* Dr Cooper, of course, knows this as well as anyone, and in his plea for supersonic manoeuvres he was doubtless thinking of unmanned vehicles able to accept accelerations of 12g and upwards.

There is no clearly defined upper limit for acceleration of an unmanned vehicle. Today's AAMs can pull several times more g than the best fighters, but because they tend to fly at highly supersonic speed their turn radius is likely to be much greater. This explains the apparent paradox that, provided it is not flying too fast, a fighter or attack aircraft can outmanoeuvre an oncoming missile. All its pilot needs to know is precisely when to start his turn. (The answer is to make missiles fly slower; then the target aircraft could not escape.)

Of course any mention of boost/glide vehicles automatically shifts the scenario to the fringes of space. Such vehicles are aerospace or transatmospheric vehicles (TAVs), probably shaped rather like the Shuttle Orbiter. Quick-turnaround TAVs would cover most of their mission in space but have the ability to re-enter the atmosphere and "skip" back into space by using aerodynamic lift and Mach numbers up to about 25. Such things have been around on paper for over 30 years, and in the late 1950s (Boeing having worked hard in the 1950s to get the USAF contract for X-20 Dyna-Soar which was eventually cancelled in 1963 "because NASA's manned spacecraft programme renders it unnecessary"), but such aerospace craft are probably still too nebulous for meaningful discussion here.

**Below: Britain's Hotol would take off by itself, but this proposed USAF TAV (transatmospheric vehicle) needs an auxiliary launch aircraft. Such a scheme would be unusable in wartime.**

## New rotary-wing aircraft

However, there are many other kinds of traditional aircraft which, like fighter and attack machines, are going through a major metamorphosis. Even the tactical attack aircraft, which traditionally was closely related to the fighter (and was often an obsolescent fighter no longer considered viable for front-line air combat use), is now diversifying into quite new forms.

One of the first to appear was the attack helicopter, as exemplified by the AH-1 Cobra and AH-64 Apache. These use traditional engine and rotor systems to fly new slender fuselages housing a crew of two, with plenty of armour, battlefield sensors and aiming systems for externally mounted weapons. Because of their agility and ability to hover, the attack helicopter is extremely effective against ill-equipped enemies, and as it is supposed to be well protected against hostile fire of up to 23mm calibre it can even have a fair chance of surviving against modern armies equipped in the style of the 1950s. Fairly obviously, it is not feasible to make a helicopter survive hits by any kind of SAM, and as modern armies are prolifically equipped with such weapons the helicopter is having to be rethought.

In a major programme aimed at creating the best tactical rotary-wing aircraft for the 1990s, the US Army has asked for proposals for an LHX (Light Helicopter Experimental). This is wanted in two forms, scout/attack (SCAT) and utility. These would differ mainly in fuselage and payload, rather in the way the AH-1 Cobra differs from the UH-1 "Huey". It is relatively straightforward for the US Army to fund advanced turbine engines, airframe structures and all-weather cockpits, but it is by no means certain yet what species of vehicle will be best for the LHX missions. As well as advanced conventional helicopters there are such contrasting systems as Sikorsky's X-Wing and ABC (Advancing Blade Concept) helicopters, a Hughes proposal with Notar (no tail rotor) and swept wings, and the Bell advanced tilt-rotor which in translational (cruising) flight behaves as an aeroplane.

Rotary-wing aircraft suffer from several fundamental drawbacks which increase their vulnerability. One is the dynamic system, where for reasons of high stress and weight it is impractical to provide redundancy or duplicate load paths. Thus, one hit can sever the drive and bring down the helicopter out of control. The rotors themselves are clearly vulnerable, compared with a typical wing, and loss of blade creates severe out-of-balance forces. Not least, the conventional helicopter is

**Below: Though fanciful in conception this Hughes helicopter shows the kind of vehicle the US Army is thinking of for LHX/SCAT. Hughes is researching a system in which the tail rotor is replaced by offset air jets.**

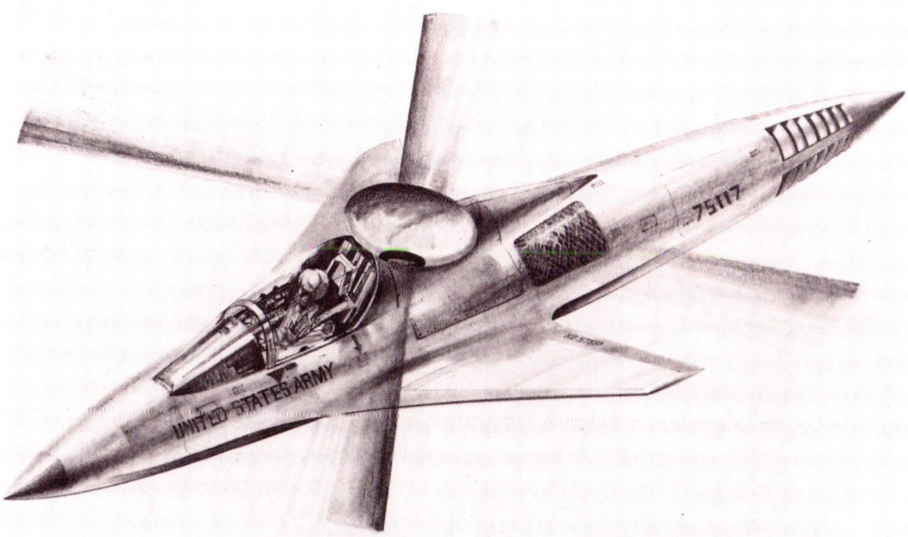

**Above: Here on US Army test, Sikorsky's XH-59A reached 303mph (487km/h) with its twin booster turbojets. It is lifted by a patented ABC (advancing blade concept) pair of counter-rotating rigid rotors.**

severely limited in flight speed, typically to below 200mph (322km/h), so defence systems tend to have plenty of time to engage the target.

The ABC helicopter, the winged jet-boosted helicopter, the stopped-rotor "X-wing" (exemplified by the Sikorsky S-72) and the tilt-rotor VTOL are all potentially capable of higher speeds in the 400mph (644km/h) region, which would reduce vulnerability. Such speeds do not approach those at which manoeuvrability becomes impaired, though they increase the necessity of having the very best instant-reacting sensors and other avionics for NOE flight control, all-weather navigation, target detection and identification and weapon aiming or delivery. In many ways the demands of the LHX-type rotary-wing machine are at least as severe as those of the much faster combat jet, in part because all-weather operation is needed very close to both the ground and to the enemy. This is amplified later.

## Developing transport aircraft

Transport aircraft have reaped the benefit of very powerful HBPR (high bypass ratio) turbofan engines, which have enabled aircraft size to be greatly increased. New technologies such as EBF (externally blown flaps), tested on the YC-15, and USB (upper-surface blowing), tested on the YC-14, have resulted in the capability of combining extremely large size and weight with jet cruising speeds and outstanding STOL (short takeoff and landing) performance.

All available technologies were combined in the McDonnell Douglas C-17, which for various reasons has been "put on a back-burner" while the USAF buys 50 new C-5B Galaxies. Development of the C-17 has continued under annual R&D and, in 1984, full-scale engineering funds, leading to a planned first flight in 1989. With the C-17 the main challenge was to achieve outstanding trucking capability with a totally new operational flexibility, such as all-weather use of short unpaved strips. This called for use of an advanced supercritical wing with very powerful flaps, and advanced directed-flow engine reversers. But the C-17 weighs as much as a DC-10 and has a wing of slightly less area, so its STOL capability is less than exciting. At maximum weight its takeoff ground roll on a paved runway is estimated at 7,600ft (2,320m), and the use of some form of powered lift has been considered.

Over 20 years ago a smaller aircraft in the same class was being built for the RAF, the Whitworth Gloster 681. This had vectored main engines and the option of 18 pure lift jets giving 198,000lb (89,800kg) direct vertical lift. It is typical of Western planning that this was cancelled in 1964, while 20 years later the USAF began to see the need for just such a vehicle.

Airfield requirements, or rather the lack of them, are clearly central to all future air force planning. There is no longer any doubt that airfields of the traditional kind are excellent in peacetime but cannot exist in a major war. Air forces, led strongly by the USAF, have for more than 20 years tried to dodge or ignore this, and even today the accepted official line is that air forces can go on much as before by augmenting airfield defences, providing HAS (hardened aircraft shelter) protection and buying aircraft that can "take off between the craters".

The notion that airfields would be the very first fixed-site assets to be wiped off the map, before any declaration of war, has been so unpalatable that it has been swept figuratively under the carpet. Much has been made of the economic problems, logistic supply difficulties and extra costs of trying to use STOVL (short takeoff/vertical landing) aircraft from remote dispersed sites, and all such arguments are still thought of in some quarters as a mere sales ploy on behalf of the Harrier! The Harrier was a pioneer aircraft which showed the way to go. Today's Harrier II is a greatly improved aircraft which combines enhanced capability with much reduced pilot workload and many other advances, but the NATO allies have not yet begun to build the third-generation STOVL aircraft which just might survive in sufficient numbers to be effective should there be a future war.

In the author's opinion NATO should already have prepared 30,000 or more operating locations for future STOVL aircraft, each offering natural cover and a takeoff run of at least 500ft (152m) ending in a natural "ski jump" ramp. Soviet satellites would soon identify most of these, but in any crisis only a small number would be actually operating STOVL aircraft, and it would be a considerable task trying to knock them all out. Today the Soviet missiles only need take out Gütersloh in Germany and Wittering in England to eliminate the STOVL force entirely!

Put another way, the known threats, listed previously, are so severe, so diverse and in some cases so unstoppable that only a fool would count on any airfield surviving more than minutes in major warfare. The only solution is to build off-airfield aircraft and disperse them throughout all the natural cover any region affords. The challenge lies in integrating the engines and airframe so that at takeoff the thrust is mainly upward instead of straight aft.

This urgent need to recognise the crudity of most existing propulsion systems, whose vectoring capability is an amazing zero, is additional to continuing need to improve the engines themselves. For over 40 years the pressure has been on to increase pressure ratios and turbine entry temperatures, cut down gas leakage paths, improve component efficiencies and reduce weight. Increased fuel prices accentuate the first four factors, and today there is very powerful additional pressure to improve reliability, increase overall engine life (and meantime between failure and overhaul – MTBF/TBO) and thus bring down what is today called the total cost of ownership. One has only to look at the history of the only vectored-thrust engine available today for STOVL aircraft, the Pegasus, to notice the way the emphasis which in 1960-70 increased thrust from 9,000 to 21,500lb (4,000 to 9,750kg) swung in 1970-85 to improving life, reliability and low cost, without bothering much about increasing power.

Of course, this engine and planned successors will probably, in due course, be boosted by PCB (plenum-chamber burn-

**MRASM attack**
Here an AGM-109 Medium-Range Air-to-Surface Missile (MRASM) lets go its 58 TAAM (tactical airfield attack munition) warheads, while a fighter tries to take off on an air-cushion transporter. In a real war nobody but an idiot would use cruise missiles or aircraft to attack an airfield, which would be the ideal targets for long-range ballistic missiles. The Soviet Union has thousands already targetted.

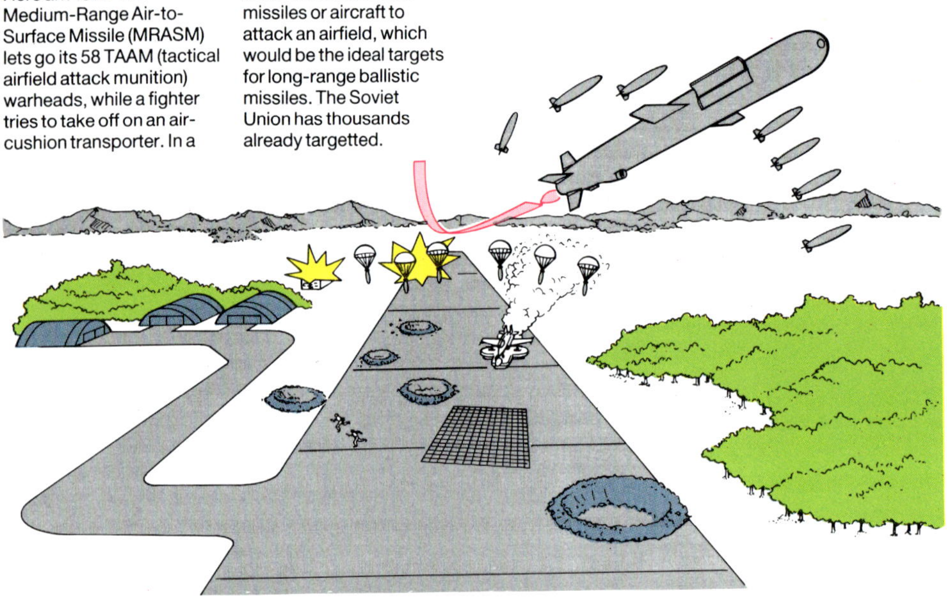

**Below: Hunting Engineering's JP.233 dispenses SG357 cratering munitions (one seen far left) and small HB876 area denial mines (foreground). More likely, airfields would be hit by missiles, and aircraft would be reserved for mobile targets.**

ing), which is essentially afterburning in the front nozzles which discharge fan air. Bearing in mind that thrust must be properly divided between the front and rear nozzles, the potential gains are very large because the reheat fuel is burnt in fresh air. Engines of Pegasus size can easily be PCB-augmented to ratings in excess of 30,000lb (13,600kg). This would obviously enable takeoffs to be made (rolling or vertical) at increased weights, but PCB is equally important in matching vectored "four-poster" engines to propulsion of supersonic aircraft. The drawback, of course, is that in the PCB mode not only is ground erosion in the hovering mode increased but the IR signature, on which enemy missiles can home, is multiplied anything up to ten times. Reducing this emission, by adding a surrounding shield or by whatever other method is practical, will probably provide a long-term challenge to engine designers.

Of course, integrating the engine(s) with the airframe so that the thrust axis can be vectored inevitably reduces the thrust and engine installed efficiency. It need not, however, exert a significant adverse influence on aircraft flight performance. Many geometric arrangements have been tried, including hinged "nutcracker" fuselages, pivoted fuselage engines and pivoting wingtip engine pods. The latter scheme, used on the German VJ 101C X1 and X2 supersonic prototypes of the late 1960s, has the advantage of extremely small penalties in structural weight or engine performance, but engine failure at takeoff results in severe asymmetric thrust forces which cannot be controlled. With any practical jet-lift aircraft, engine failure in the vertical flight mode means unpremeditated descent and probably ejection, but uncontrolled gyrations must be avoided at all costs. It follows that the immediate objective today, as it has been for the past 25 years, is to install one or more engines in a fighter in such a way that the thrust axis (or axes) passes through the aircraft c.g. or close to it under all flight regimes.

This is becoming much easier to achieve with the development of engines of shorter length. The J79 with afterburner has a length of 209in (5.3m), and even the more modern F100 has a length of 191in (4.86m), but the RB199 used in the Tornado has a length of only 127in (3.23m) complete with variable nozzle and reverser. An engine as short as this could, with or without its airframe inlet and inlet duct, be bodily pivoted in the fuselage to the optimum angle of about 55° for STOL rolling takeoff after a short run. Such an arrangement is obviously preferable to the current US fixation on two-dimensional (2-D) nozzles at the extreme tail. The latter impart a powerful nose-up or nose-down pitching moment when deflected, which has to be reacted by a canard foreplane, or by some other vertical thruster near the nose.

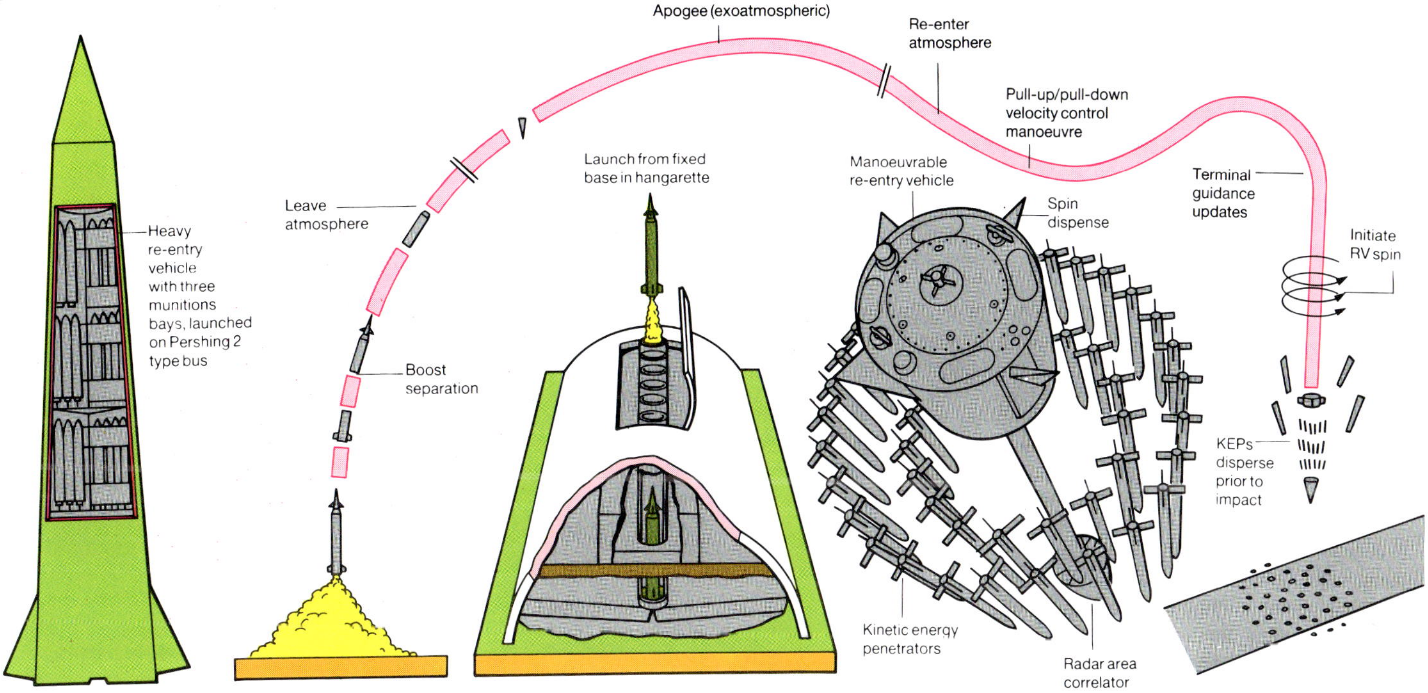


**CAM airfield attack**
Martin Marietta has proposed this conventional Counter Air Missile, based on its Pershing 2 missile. The only new hardware needed would be a payload bay which in the left-hand drawing has three tiers of KEPs (kinetic-energy penetrators). The next sketch shows a complete flight profile, which would obviously begin from a mobile launcher though Martin Marietta have suggested a fixed hangarette, whose locations would be known before the cement was set! Near the target the missile would come under pinpoint guidance of its radar area correlator; each missile could aim for a particular runway inter-section. KEPs are flung out by low-risk spin dispensing so that a large area is covered by each CAM. The most staggering thing about this urgently needed system is that not only does it not exist but it has not been ordered. NATO prefers a few manned aircraft!

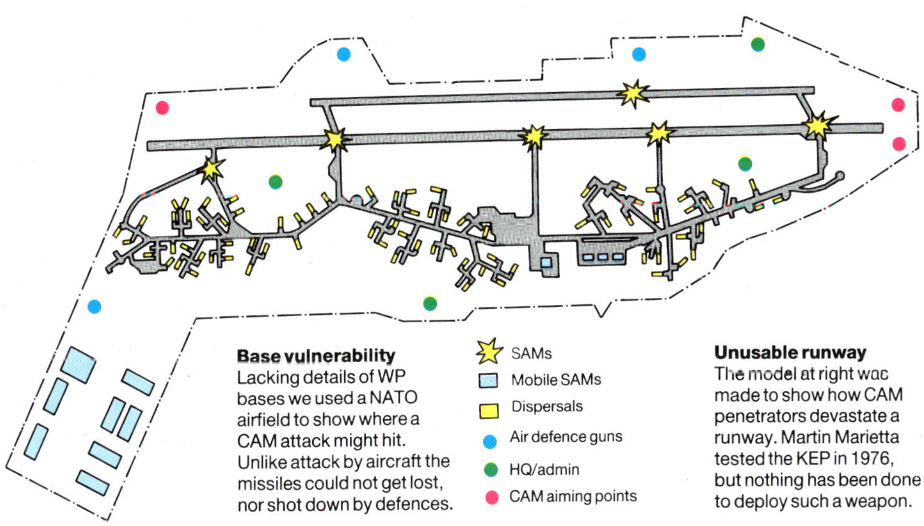


**Base vulnerability**
Lacking details of WP bases we used a NATO airfield to show where a CAM attack might hit. Unlike attack by aircraft the missiles could not get lost, nor shot down by defences.

**Unusable runway**
The model at right was made to show how CAM penetrators devastate a runway. Martin Marietta tested the KEP in 1976, but nothing has been done to deploy such a weapon.

This method of a vertical thruster near the nose leads to what may prove to be the preferred solution in the long term. If a short-duration rocket were to be mounted vertically under the nose of an otherwise conventional fighter, it could rotate the aircraft at the end of a very short horizontal run to a nose-high attitude approaching 90°, an angle high enough for the main engine(s) to lift the aircraft straight upwards. The rocket, which need not weigh more than a trivial 18lb (8kg) – less than the existing lower-thrust nose RCJ (reaction-control jet) – would jettison automatically at the end of a burn lasting about 0.8 sec. The main engine installation would need no major modification, and the performance of the inlet should not be impaired by the very high AOA (speed being very low ideed). The only adverse features of such a scheme are that the main gears or skids must prevent the nozzle(s) touching the ground, a reaction-jet control system would be needed (as it will for all combat aircraft) and airstrip erosion would be severe. None of these problems appears particularly difficult. After its steep initial climbout the aircraft is in all respects a normal fighter or attack platform.

At first sight such a scheme may appear laughable, dangerous and unworkable. But with modern computer control there is no evident problem. The rocket thruster to pitch the aircraft nose-up is a trivial addition whose performance could today be made precisely repeatable – tens of thousands of such thrusters have given repeatable brief impulses to payloads in space. The digital flight control system would impart the very small reaction-jet forces needed to hold the aircraft at the optimum climb-out angle.

Of course the basic objection to such a simplistic scheme is that recovery to a particular spot on the Earth's surface involves doing the procedure in reverse. Though with computer control the risks are precisely the same as those at takeoff, and

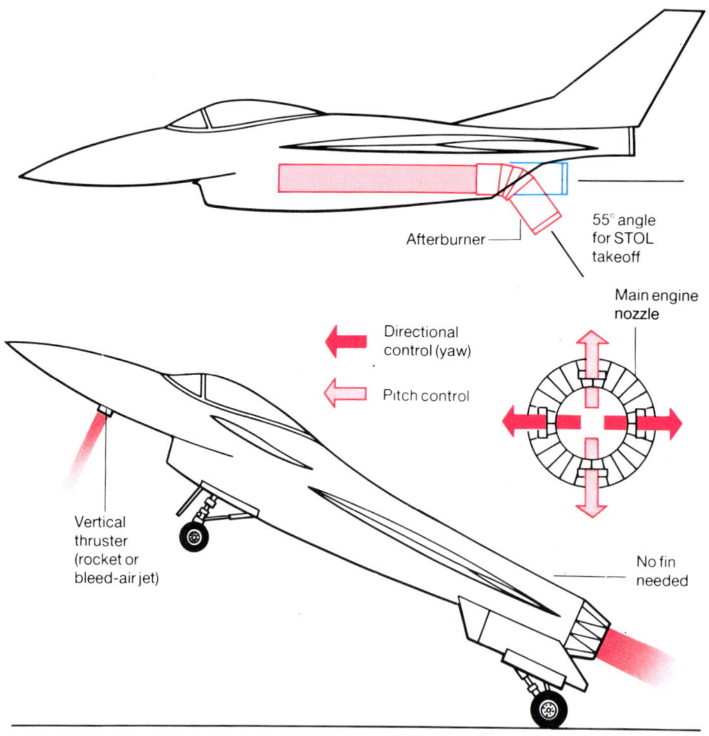


**2D nozzle**
In this fighter the main engine afterburner can pivot downwards to about 55° for a STOL takeoff. This 2D (two-dimensional) nozzle does shorten the run, but there must be no pitching moment imparted to the aircraft when vectoring thrust.

**Fixed nozzle**
In this fighter the engine nozzle has diverter flaps to control the aircraft in pitch and yaw. Otherwise a conventional unstable canard fighter, this aircraft would take off in a startling manner. After a forward run of only a few metres the nose would rear up almost to the vertical attitude, whereupon thrust would exceed weight and the aircraft would enter a steep accelerating climb. The problem is the precise computer control needed to reverse the process and make a safe and repeatable vertical spot landing.

**ACV takeoff**
Having the F-15, which has no off-runway capability, the USAF has suggested mounting aircraft of, say the 36th TFW at Bitburg on giant air-cushion vehicles, which could then hover across the craters and let the fighters take off, one of many ideas USAF is studying in its "sortie generation" program.

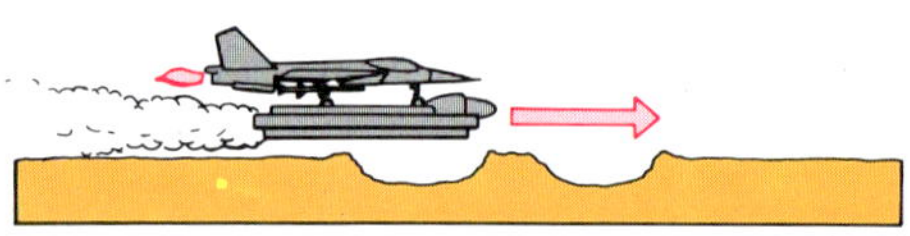

**Below: Impression of an F-15 lifting off an air-cushion transporter, which has deployed its braking parachutes after sliding over craters at a blasted airbase. How the F-15 gets back is unexplained.**

essentially similar to those in CTOL takeoffs from an long runway, the obvious impossibility of manual control poses a psychological stumbling-block. On the other hand the same situation applies in TF (terrain-following) flight through mountains in dense cloud, which has been standard operating procedure with some USAF units since 1967.

It is fair to question whether this form of ultra-STOL arrival, in which the pilot almost stops and then hopes to land, is any worse than CTOL recovery to a cratered runway in which he has to land and then hope to stop! For several years the USAF has devoted immense efforts to trying to solve the problem of how to stay in business once it has (reluctantly) accepted the idea that some day someone might actually crater its beautiful runways. (It does not yet appear to have addressed itself to the much bigger problem that there are many much more effective ways of knocking out an immovable airbase.)

These efforts have involved prolonged study of the aircraft, the runways and many other factors. Aircraft modifications might include low-pressure tyres (or addition of Flo-Trak pads around the wheels), adhesive sealing of fuel tanks, sealing of most other apertures against CB (chemical/ biological) attack, addition of a portable ground-based precision approach guidance system receiver (or possibly a self-contained on-board guidance sensor), 2-D engine nozzle(s) with some degree of vectoring including thrust-reverse, and long-stroke or other high energy absorbing landing-gear legs for running over rough surfaces (either unpaved or a quickly repaired runway). The main STOL objective is to take off or land in a ground distance of 1,500ft (460m), which has the drawback of requiring a series of major improvements to the (existing) aircraft, or introduction of a new type, while still ducking the issue of getting away from established airbases.

**Above: All USAF main operating bases for fighters have runway arrester systems for use in emergency. Here an F-16 engages a G&W Portarrest installation. Function is to save aircraft in peacetime.**

**Below: McDonnell Douglas has a contract to fly a STOL technology demonstrator version of the F-15 with canards and 2D vectoring nozzles. One feels such projects ignore thoughts of real warfare.**

Current USAF planning and discussion does everything except give up the idea of giant fixed airbases. Some form of special steep glideslope is seen as assisting "landing between the craters". Portable arresting systems, probably weighing many tons and costing many millions of dollars, are expected to be switched to and fro as particular bits of runway become apparently usable. Giant ACETs (air-cushion equipment transporters) are seen as solving the problems of ground mobility in the absence of paved taxiways. Each extra equipment item multiplies the logistic supply and maintenance problems, snowballing with extra men, rations and living quarters and ever-greater problems of concealment.

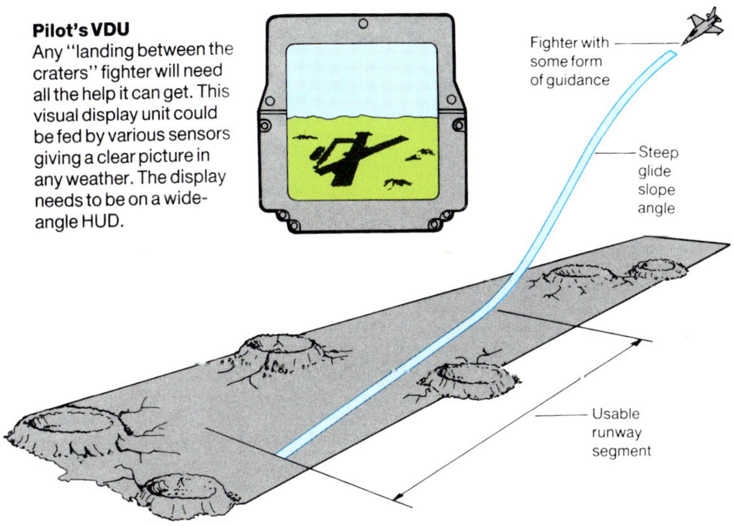


**Pilot's VDU**
Any "landing between the craters" fighter will need all the help it can get. This visual display unit could be fed by various sensors giving a clear picture in any weather. The display needs to be on a wide-angle HUD.

**"Between the craters"**
Lacking many other answers the USAF is engaged in big research programmes to see how it can keep flying in a post-attack situation. In the author's view the question is academic, but this sketch shows a "low-visibility precision touchdown system" which does not even exist on paper. The F-16 could be guided by an on-board sensor or some form of ground data-link. A steep glide slope would "minimise dispersion", so that the usable bit of runway could be fully utilised. Surely common sense tells us that in a real war NATO would not merely be confronted by a few craters in its runways!

If you start with 25-ton, 25-million dollar fighters with 25 tons of thrust blasting out at the back, you are then ill-placed to try to invent an air force that cannot be wiped off the map in a few seconds while the giant bases sleep. It seems blindingly obvious that it is more sensible to start with survivable STOVL airpower that needs no fixed airbase at all, and then go on from there to make such aircraft more lethal and even less observable.

When one stands back to take an overall view one can see that basic design of combat aircraft passes through quite distinct phases. Once jet engines had removed obvious limits on flight performance it was natural for designers in 1945-55 to spend much of their energy in increasing flight Mach number, from 0.8 successively up to Mach 3.2 (in the XF-103 of 1952-57), while simultaneously exploring new aerodynamic shapes suited to this new realm of performance. In 1955-75 the thrust upwards in flight performance evaporated, to be replaced by across-the-board improvements in engines, materials, systems (especially avionics) and weapons. Much excitement was generated by jet V/STOL, only to be dissipated by a strange failure to continue to believe in what had triggered this development – the crucial vulnerability of airbases. Further excitement was generated by the "swing wing" VG aircraft, only to be gradually eroded by a fixation in the 1970s on its penalties in place of its advantages. The final quarter of the century will see concentration on reliability, long life and the lowest possible total long-term costs, on which are superimposed strong renewed interest in close air combat and the gradual and seemingly reluctant acceptance of things that have been known but almost ignored since World War II.

One of the latter is what we now call stealth, which had been investigated by the British TRE (later RRE) at Malvern, in collaboration with RAE Farnborough in great detail during the war. Having almost wasted more than 30 years we are now at last taking the need for invisibility at all wavelengths very seriously indeed, though there will continue to be urgent debate on how far stealth should be allowed to compromise flight performance, payload and above all cost.

There seems little doubt that the trend during the rest of this century will be towards combat aircraft of reduced size, designed fully according to stealth principles, with every part of the mission controlled by VLSI (very large scale integration) microelectronics using VHSICs (very high speed integrated circuits) using digital data throughout and almost certainly optical fibre bundles as main data highways. Whereas in the past avionics system designers have packaged their aerials (antennae) and "black boxes" into or onto any convenient part of the aircraft, in future the aircraft and systems (especially avionics) will have to be designed together with total integration. Nearly 20 years ago the idea of conformal radars in which the antenna forms part of the airframe, opened up the possibility of system antennas stretching from wingtip to wingtip, or from nose to tail.

The point must be made that it is extremely tempting to credit ourselves and our allies with superior intelligence, cunning, industrial strength and all-round technical capability to that of our potential enemies. For many years it has been fashionable to rate the Soviet Union as distinctly inferior to the Western countries in across-the-board technology, with the notable exception of one or two specific areas such as superpower lasers. In a recent review of the US technological scene the combination of active and passive stealth was presented as a way to penetrate active defences almost with impunity, continuing, "With the ability to defeat an enemy's sensors we can negate large investments in systems. Stealth is required to meet the challenge of mission success, while staying away from high-attrition warfare."

Everyone wants to be better than the Bad Guys. But few countries even approach the Soviet Union in total dedication to sustained improvement in weapon systems and defence technology generally, and it is a fair bet that since 1980 the financial and manpower investment in stealth technology in the Soviet Union has been greater than the combined effort in the West. It would be amazing were this not the case. The obvious corollary is that by the 1990s Soviet combat aircraft will be fully stealth-designed and their anti-aircraft defence systems will be tailored entirely to the elimination of such aircraft, and not to the types in service in NATO today.

Already aircraft designers are being forced to take a keenly critical look at anything that forces a combat aircraft to emit any radiation. To provide a non-emitting propulsion system would appear almost to flout the laws of physics, though much can be done to ameliorate the problem. Where on-board sensors are carried to assist all-weather navigation, terrain-following, interception of moving targets, weapon delivery and recovery at base, much can be done to reduce emissions, and it is an outside possibility that emissions can be eliminated. After all, humans can move accurately under almost all conditions except total darkness without emitting radiation (other than body heat, which we cannot utilise), so in due course our vehicles ought to be able to do the same.

This is throwing much greater emphasis on optical and IR wavelengths, at the expense of radio and radar. On the very day this was written British Aerospace Dynamics announced the establishment of a diamond-tool machining centre of extraordinary precision, with lathe saddles riding on air bearings and positioned by using interferometric methods according to outstandingly subtle software programmes. The new facility can produce mirrors and lenses of extremely complex shapes in many materials including metals, germanium and zinc sulphide, with accuracy measured not in traditional "thous" but in microns (millionths of a metre). Such technological capability is already vitally important in enabling precision sensors to be created that are wholly passive in operation. These must in the very near term revolutionise air warfare.

The mind boggles in surveying the wealth of possibilities now open to the designer of sensors and weapons. Some of the possibilities becoming available with radars of both very long and very short (millimetric) wavelength are so unprecedented it is tantalising to remember that emitters are "out" – at least in the author's opinion. In their place we must rely on passive IR and optical systems. Traditional IR sights and FLIR (forward-looking IR) systems have relied on simple optics focussing IR (heat) rays on to a sensitive detector cooled (for example by liquid nitrogen) to about 70-80°K to get a good ratio of signal to noise. Usually the pilot can select a wide or narrow field of view, and the seeker head is normally stabilized through an aircraft roll angle of from 60-540°. The output is used to generate a raster (line-by-line) picture in a TV format, presented on a cockpit display. Usually the picture tones give grey tones between the coldest regions (black) and the hottest (white), but polarity can be reversed so that, for example, a ship can appear black against a white sea.

**Top left: Not part of a real aircraft (yet), these panels for a pilot and backseater are used in researching the cockpit of the next-generation ATF. Both crew have a synthetic picture of the terrain ahead; other displays differ.**

**Left: Collins' colour display is one of several nav/targeting systems to combine colour with stroke writing and raster (TV type) alphanumerics, overlain on various pictures called up by the pilot. Here a weapon reticle is centred on a bridge.**

**Right: Using a red laser at USAF Systems Command, Wright-Patterson AFB, in investigations into the design of helmet-mounted sight systems. Such sights have to be exactly positioned and focused.**

More important for the future is the SFPA (staring focal-plane array). The simplest way to give an idea of how focal-plane arrays work is to explain that they are almost a mechanical analogy of the eye, in that a lens focusses the incoming radiation on an array of numerous (millions) of very small sensitive receptors, all laying in the focal point of the lens. Such arrays are formed it a single (monolithic) slab of semiconductor material, and CAD (computer-aided design) is essential in order to achieve the unbelievable complexity on a microscopic scale. VHSIC processing is essential to keep track in real time (ie, as it happens) of what radiation is falling on the array, and special processors are needed, much faster than general-purpose microprocessors. An example is the GRID processor jointly being worked on in Britain by GEC and Marconi for array processing. To add two large arrays a normal microprocessor would need 1.5 sec, but GRID can do the task in 300 microseconds (0.0003 sec). This is because its architecture is tailored specifically to this one task. The next step is to combine one or more sensors, such as an SFPA, with their own specialised processors whose outputs then control a weapon flight guidance system.

**Top right: Research in progress at BAe Warton on future fighter cockpits, using both internal displays and a HUD presentation on an external scene. On the main panel are fuel state and attitude displays, while on the left are standby ASI, VSI and altimeters.**

**Right: Though the artist has presented a rather perfect situation, he has in fact shown a real cockpit. This is almost exactly the cockpit of the F-16C Fighting Falcon, now in wide service, dominated by the giant Marconi Avionics holographic HUD and with two multifunction displays in place of most traditional instruments.**

**Left: Two photographs showing a smart weapon landing on target. It is a GBU-15 (GBU = guided bomb unit), which has a warhead based on the Mk 84 bomb of 2,000lb (907kg). Guidance can be by TV or by an IIR (imaging infra-red) sensor in the nose.**

**Bottom left: Typical of modern smart bombs, this LGB (laser-guided bomb) is one of the French copies of the American Paveway series. Made by SAMP and Matra, most are matched to the Atlis II laser pod.**

The result is a brilliant weapon with CEP (circular error probability) of essentially zero. Weapons are today called dumb if they are unguided, smart if they incorporate guidance relying on outside help, and brilliant if they are self-guided. Most of the world's guided missiles today are in the smart category; they rely upon external radars or lasers to illuminate the target to provide radiation on which they can home, or they need direct human command guidance with the signals sent to them by radio or along trailing wires. Brilliant weapons are normally found only in the large strategic category, but by the 1990s brilliant weapons will exist in all sizes down to small anti-tank weapons. This is one of the chief ways in which SFPAs will revolutionise warfare, and especially air warfare.

Before going on to discuss weapons, the point must again be made that it is going to be difficult to give up radar. The chief drawback to IR arrays is that at such very short wavelengths the radiation is rapidly attenuated by atmospheric moisture, such as rain, snow and fog. In a winter blizzard it would be difficult to "see" a hostile tank at a range of a few hundred metres at IR wavelengths, whereas radars could see it clearly. A second factor is that, because nothing is sent out from the sensor, the round-trip time of flight cannot be measured, and there is no easy way of measuring range. Put another way, the IR radiation entering the sensor and being focussed on the array might come from one or 10,000 metres distant. On the other hand, apart from the vital and dominant advantage of being passive, the lack of range input is less of a handicap to a brilliant weapon. Such an FAF (fire and forget) missile simply goes on flying until it hits the target, and it is very unlikely that the distance to the target will be beyond the maximum effective range of the missile.

With the new species of brilliant weapon all that the carrier aircraft need do is bring them somewhere near their targets, or alternatively launch them from a distant stand-off position in such a direction that they can later acquire their target(s). Almost certainly the easiest kind of target is the surface ship. It is very large, emits IR and (almost certainly) various kinds of radio/radar radiation, is relatively slow moving and is "seen" against a generally uniform cool sea background. To give the anti-ship missile a better chance of approaching undetected it has become customary to design them to fly extremely low, in the sea-skimming mode; but most anti-ship missiles then go and throw away this advantage by pumping out their own radar signals in order to find and home on the target. Virtually all of today's anti-ship missiles use active radar homing, the only exception currently known being the Norwegian Penguin (now adopted by the US Navy and Marine Corps for use from F/A-18, A-6E and other aircraft), which uses passive IR homing.

Anti-ship missiles are a special case characterized by sea-skimming in a way impossible with any overland weapon. As explained in the final subsection of this chapter, Flying the Mission, the carrier aircraft itself needs to adopt the nearest thing possible to sea-skimming in terrain following, but so far no missile has been announced as doing the same. This is chiefly because in the past defensive systems have not had a significant capability against small missiles, which – it has been assumed – could fly to their targets well above ground level with very little chance of interception. Even today large sums are being spent on a wide variety of tactical and battlefield missile systems in which the missile flies at modest speeds in the neighbourhood of Mach 1 while in full view of enemy systems. The ability of

surface forces to destroy incoming missiles, bombs, cluster dispensers and other weapons is in future going to have to be reckoned with; there is no evident reason why with today's crop of ASMs (air-to-surface missiles) this capability should not soon reach near to 100 per cent.

Today's ASMs extend from wire-guided anti-tank missiles flying at 180mph (290km/h) over a range of a few hundred metres, guided from a helicopter, through numerous free-fall or glide bombs up to rocket or turbojet missiles which in the case of so-called cruise missiles can fly to a target nearly 2,000 miles (3,200km) distant from the point of release. Rather than review these in detail it is more informative to study the requirements and the design of the next-generation weapons for use later in the century.

It so happens that in the early 1980s much of the NATO effort on ASMs has been devoted to devices whose sole purpose is to disturb the surface of airfield runways, so that – while leaving the aircraft themselves serviceable – the enemy's ability to take off and land is supposedly interfered with. This line of thought began with the effort by France's Matra company to provide the Israelis with a concrete-dibber missile which was marketed under the name Durandal. This, being first in the field, has enjoyed a large commercial success, many thousands being sold even to the mighty USAF.

Eventually other companies decided to get into the act, important examples being MBB of West Germany and Hunting of the UK. MBB has the MW-1 system, which though directed mainly against armour is also useful in causing problems on enemy airfields. Hunting's JP.233 is specifically an anti-airfield system, and the aircraft dispenser discharges two types of munition, one being the SG357 for cratering paved runways and the other the much smaller HB876 area-denial munition. The latter, strewn in large numbers, poses a problem to the defenders in slowing up all measures aimed at clearing the region of munitions and repairing the runway.

**Above: Takeoff by an RAF Tornado GR.1 carrying two Hunting JP.233 dispensers under the fuselage. This can deliver 60 SG357 pavement-cratering weapons plus 430 small HB876 area-denial munitions. Together they can incapacitate an airfield.**

**Right: One of a Tornado's 430 area-denial munitions. This HB876 mine releases the ring of spring-steel legs which makes it stand upright. It is then exploded by touch, or by a randomly timed signal.**

**Below: A live trial of an MBB Kormoran anti-ship missile, from Marineflieger Tornadoes, against an old destroyer. With its rocket still burning, the missile enters via thin armour to explode amidships.**

In any general review it is obvious that certain objectives will continue to be highly desirable. One is to make missiles fly faster; this will improve their chances of surviving to the target and equally make it much more difficult for moving (or flying) targets to evade them. In some applications the increased kinetic energy of a faster missile is an asset in causing damage to the target. Another object is to minimise or eliminate emissions, as already explained in connection with aircraft. Another is to achieve guidance so accurate that it can be almost guaranteed that the missile will not only actually strike the target but strike it in the most vital spot. This enables proximity fuses to be replaced by direct impact fuses, which are much smaller and cheaper, and more importantly enables the size of warhead to be reduced considerably. One has only to compare the proximity-fused Roland SAM with the "hittile" (direct hitting) Rapier to see the advantages: at firing the former missile weighs 286lb (63kg), whereas Rapier weighs only 190lb (42kg), the difference being entirely accounted for by the fact Rapier's warhead explodes inside the target aircraft. Not least, tomorrow's weapon must be able to operate autonomously and have a considerable degree of inbuilt intelligence.

**Above: Bofors in Sweden test proximity-fuzed ammunition against targets, such as this air-launched missile, by firing at them whilst suspended in mid-air.**

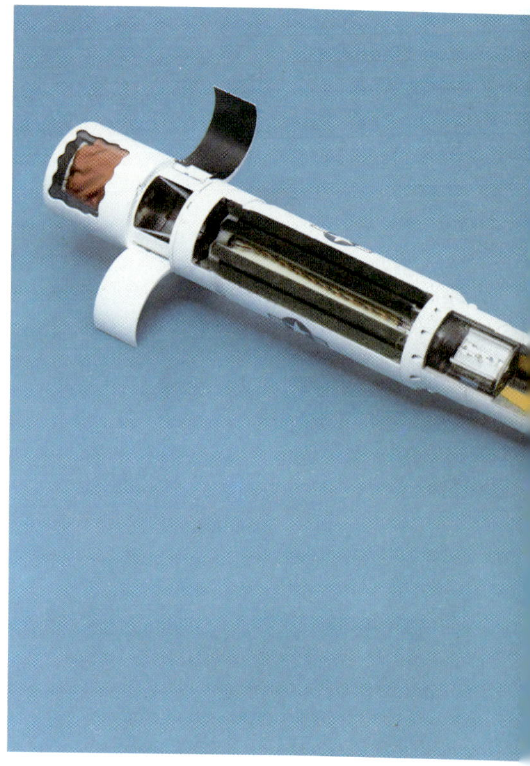

**Right: Avco's BKEP (boosted kinetic-energy penetrator), typical of numerous airfield-attack submunitions delivered in clusters to cause pavement heaving and cratering.**

Before elaborating on these objectives the need for higher flight speed may appear to conflict with an earlier comment that fighters can evade SAMs and AAMs by turning too tightly for them; the comment was made, "The answer is to make missiles fly slower." This presupposes a manoeuvring close air combat situation, something totally alien to tomorrow's crop of SAMs and AAMs. Certainly a missile travelling at, say, Mach 5 that misses its target by more than the lethal radius of a proximity-fused warhead is useless, because it has no chance whatever of turning round and coming back. Tomorrow's missiles must combine three things: precision guidance, intelligence to aim along the optimum interception trajectory (and not home on each successive "present position" of the target), and very high speed. With a flight time probably in the order of one second there is very little evasive action left to the target, which will probably be unaware it is being fired upon. Computer modelling indicates that, no matter what the target pilot does, a future high-speed SAM with precision guidance and intelligence is likely to score a direct hit at least 999 times out of 1,000. It would also, it appears, do almost as well against the smaller target of a typical ASM.

It is not proposed to describe individual weapon systems, but one now in advanced development is certainly a harbinger of the future. In 1984 initial development was completed of ADATS, the Air Defence Anti-Tank System, being developed by Oerlikon of Switzerland with Martin Marietta of the USA serving as chief partner under contract. ADATS is the first modern guided weapon to be developed to fulfil both roles. Although this is at present a land weapon system, it has been described because it is the first of a new family that will also be carried by aircraft.

For many obvious reasons – cost, hardware standardisation and operator training among them – it helps if weapon systems can be made as versatile as possible, provided performance in any role is not thereby degraded. There are many ways of shooting down air vehicles and about the same number of ways of killing tanks, and most could never be developed to do both tasks. The ADATS-type system is, however, an obvious bonus. The only degradation is caused by the missile having a direct impact fuse against armour and a laser proximity fuse for use against aircraft. The bulk and weight penalty is not serious. In any case ADATS was designed in the late 1970s and today could be done differently.

The existing system, development of which has in general been very encouraging, uses a missile 81in (2.05m) long and weighing at launch about 112lb (51kg), roughly midway between the weight of Roland and Rapier. Eight rounds are carried ready to fire in a slewing/elevating turret on an M113 vehicle, with eight more internally. The vehicle operates autonomously, with its own search radar. The turret contains a remarkable EO (electro-optical) sensor head comprising a FLIR (forward-looking infra-red), optical sight and TV for target acquisition, FLIR and TV for tracking and laser rangefinder. The TV and thermal-imaging FLIR are of course passive and immune to electronic countermeasures. Once within range the target is engaged by switching on a powerful carbon-dioxide laser which provides a guidance beam for the missile. The laser light is coded to match the missile, which ignores any other radiation on the same wavelength. The warhead is a shaped charge for use against armour, with surrounding steel fragments for lethality against aircraft.

What are the weaknesses of ADATS? One is its use of a surveillance radar, which invites an anti-radar missile (ARM); but radars (especially those on the ground) are rapidly becoming cleverer and more difficult as sources on which ARMs can home. More basic is the use of laser beam-riding guidance. This has important advantages over radar beam-riding, but results inevitably in a substantial missile with a proximity fuse.

Future ADATS-type systems will undoubtedly use a passive seeker with a focal-plane array. In combination with exceedingly rapid but complex digital data processing this would result in a self-contained FAF (fire and forget) missile so accurate that it would pick the right part of the target to hit. As previously noted, if such a missile were to impact at over Mach 5 the kinetic energy alone would probably be sufficient to destroy all but the largest aircraft, and only a small warhead would be needed. Against armour perhaps more needs to be learned about how best to combine a kinetic-energy penetrator with a shaped charge to achieve best results.

One of the more difficult missions to predict is helicopter anti-tank. In today's environment a specially designed helicopter such as the Apache has a good chance of surviving a number of missions, despite the fact that its sensors are low down in the nose and thus require exposure of the helicopter to enemy fire. Assuming that the helicopter can survive, what weapons can it use? Missiles can have semi-active homing at radar (centimetric), millimetric or laser wavelengths, or passive IR homing, or command guidance by trailing wires. Passive IR has all the advantages previously listed, while the seemingly primitive method of wire guidance is immune to countermeasures other than sudden obscuration of the target by smoke or other methods. Laser countermeasures have been studied for many years, and there are now several techniques for interfering with the beam, retransmitting on the same wavelength with slightly different coding or creating a reflective surface well away from the tank, beyond the effective range of a shaped-charge jet.

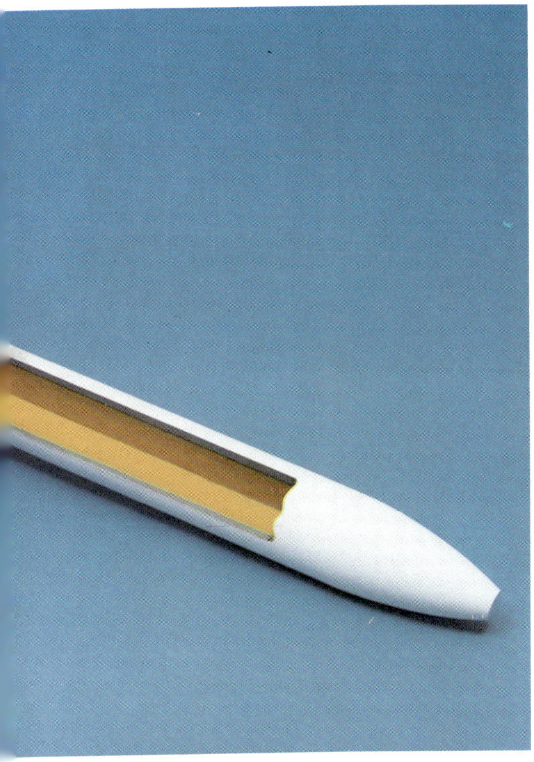

Above: One of the fastest atmospheric vehicles today, a Teledyne Ryan Firebolt is seen here cruising at Mach 4 at 100,000ft (30,500m). Firebolt is a reusable target; Mach 4 missiles could be similar.

Below: Test launching of an early HVM (hypervelocity missile). Probably the fastest air-launched weapon in the West, it has a sustained speed twice as high as the muzzle velocity of any aircraft gun.

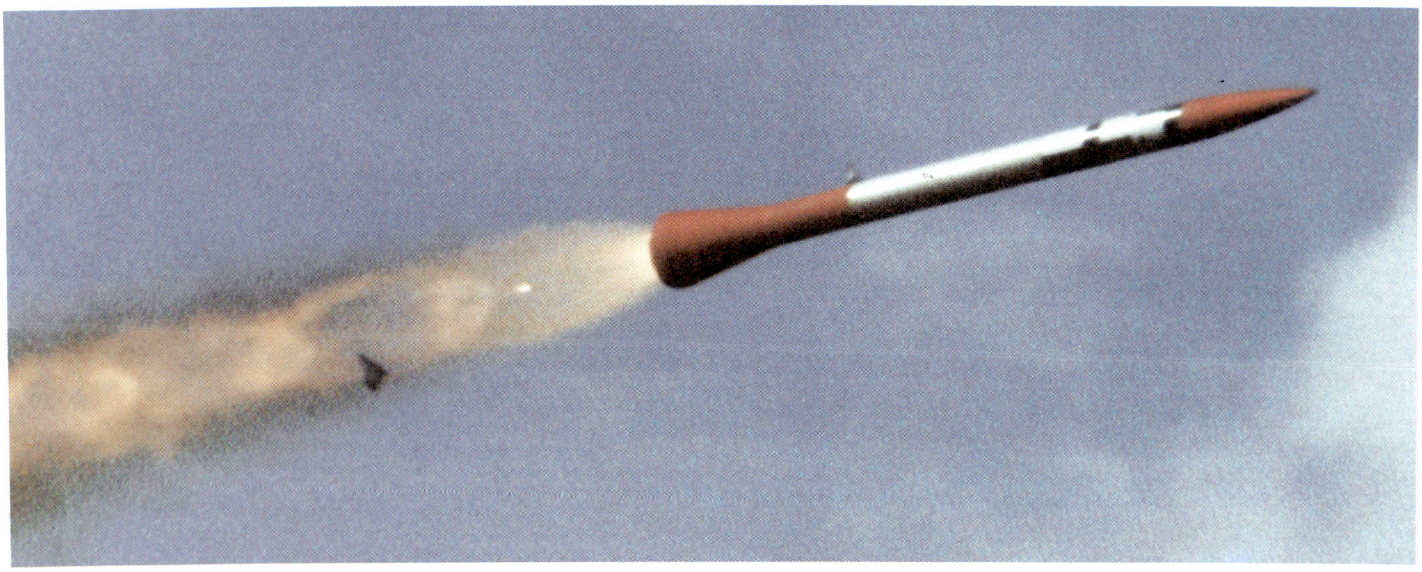

Yet laser guidance has been selected for the HVM (hypervelocity missile) now being developed by LTV as the next-generation anti-armour missile of the USAF (but not the only air-launched weapon, because there are others described later). HVM will be a simple tubular rocket weighing about 48lb (22kg), carried in 20-tube pods together with a carbon dioxide laser. The latter, looking far ahead, would in less than one second perform the functions of multiple target detection, classification and "prioritisation" (the official word) as well as 3-D ranging and doppler velocity measurement. The laser would have a coarse beam for missile capture and a fine beam for terminal guidance. Flight speed of HVM would be "over 3,355mph (5,400km/h), a speed high enough for an inert warhead of very dense material, such as depleted uranium, to punch straight through typical tank armour up to a range of at least 3.7 miles (6km). Development is proceeding as this is written in 1985, and it is hoped to achieve reliable hits on at least six tank targets simultaneously. The ability to receive laser guidance through the rocket plume was demonstrated in 1982, and fully guided flights have been made from ground and air launchers. Carrier aircraft will include the A-10 and F-16.

HVM is significant in combining multiple-target capability, precision guidance and hypervelocity flight. All are valued attributes in the anti-aircraft role, but the author believes in this case an explosive blast/frag warhead with proximity fuse may prove necessary. An alternative scheme would use a small charge to break up the missile into fragments shortly before reaching the target, relying upon their own kinetic energy for destructive effect.

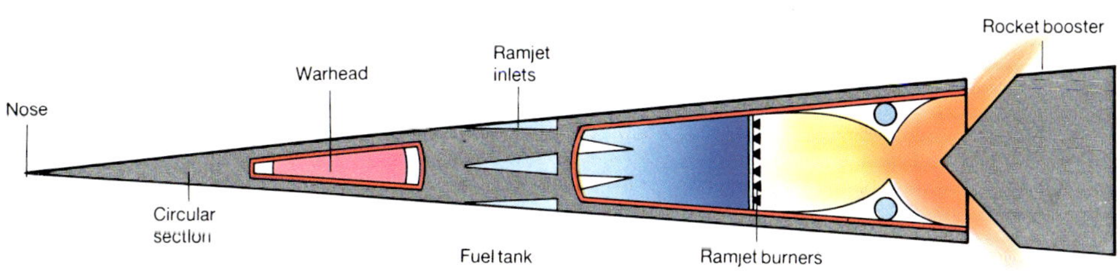


**Ramjet HVM**
British Aerospace has a contract to study a hypervelocity missile. It could look like this ramjet vehicle, depicted with its tandem boost rocket still attached. Crusing speed could be as high as 6,000mph (9,600km/h).

There are several anti-armour systems mentioned under development for the USAF's WAAM (Wide-Area Anti-Armor) programme. Two of particular interest are produced by Avco. The first was the ERAM (Extended-Range Anti-Armor Munition). This is a sophisticated mine packaged nine at a time into the standard SUU-65/B dispenser pod (which is then redesignated CBU-92/B). Groups of these could be carried by an F-16 and released over or near an enemy land force, the launch parameters extending down to low levels and at all available airspeeds. Low over the target the dispenser splits open and releases its ERAM submunitions, each of which parachutes to the ground. Inside each ERAM is a warhead called a Skeet, combining a sensitive IR seeker with a special warhead called either an SFFW (Self-Forging Fragment Warhead) or an EFPW (Explosively Forged Penetrator Warhead). Should the sensor of any Skeet detect a target on the way to the ground the warhead is ejected, it locks-on and is fired from a short distance above the target, the self-forging fragment jet piercing the top armour of the heaviest tank.

**Above: An impression of delivery by F-16 of Avco's SFW (sensor fuzed weapon), an ERAM which dispenses Skeet submunitions as described on this page. This and related weapons are featured under Land Warfare.**

Suppose the ERAM does not detect a target on the way down? This is no problem; in fact one of its operating modes is for hundreds to be strewn from the air along the expected route of an invading army. The mines parachute to the ground and lie there in wait, with three sensor aerials pointing diagonally upwards. Should a vehicle, especially a tank, pass within about 500ft (150m) the sensor triggers release of the Skeet submunition. This is fired by a small charge high into the air, its sensor scanning spirally over an area of about an acre (4,000m$^2$). Any vehicle within that region will put out warmth that will be detected by the Skeet, which will promptly knock it out with a blast from above.

Avco's other WAAM weapon is SFW (Sensor Fuzed Weapon), which is simply a direct attack mode using Skeets. These are packaged four at a time into a tubular submunition, which in turn is packaged in groups of 10 in the SUU-65/B TMD (Tactical Munition Dispenser) delivered by the attack aircraft. Thus, an F-16 could in a second or so let go about 480 Skeets to cover every vehicle in about a square mile of ground (2.6km$^2$). When the TMD is released it falls vertically but is swiftly braked by unfolding four fins which twist to impart not only drag but also rapid spin. The TMD splits open, the 10 submunitions are flung out and dispersed, and each then

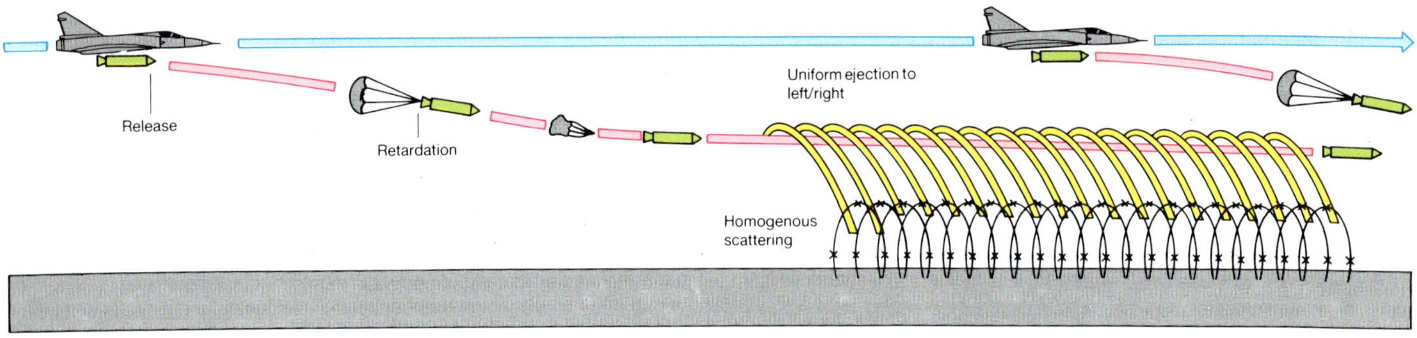


**Belouga dispenser**
This French dispenser, seen above carried to the target by a Mirage, is today in wide use with over 5,000 delivered. Quite a small store, with a drop weight of 639lb (290kg), Belouga is released as the parent aircraft overflies the target. After being braked it dispenses 151 bomblets, each weighing 2.65lb (1.2kg) in a controlled pattern selectable by the pilot. The bomblets are unguided but are themselves individually braked, so that each impacts almost vertically, for maximum effect. Obviously, such bomblets should be delivered by missile bus against airfields, and by aircraft against any kind of moving target.

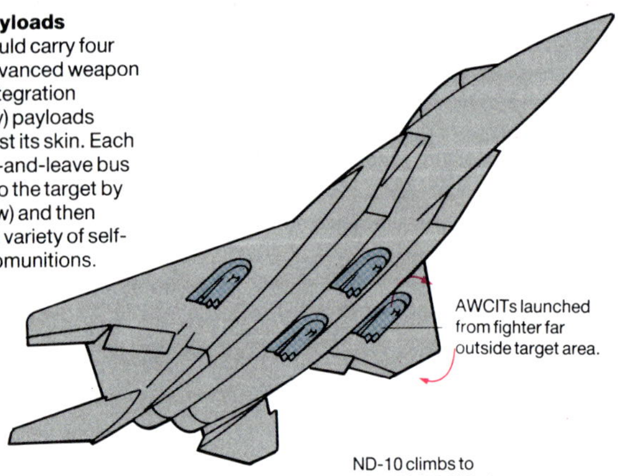


**AWCIT payloads**
An F-15 could carry four AWCIT (advanced weapon carriage integration technology) payloads flush against its skin. Each is a launch-and-leave bus able to fly to the target by itself (below) and then dispense a variety of self-guided submunitions.

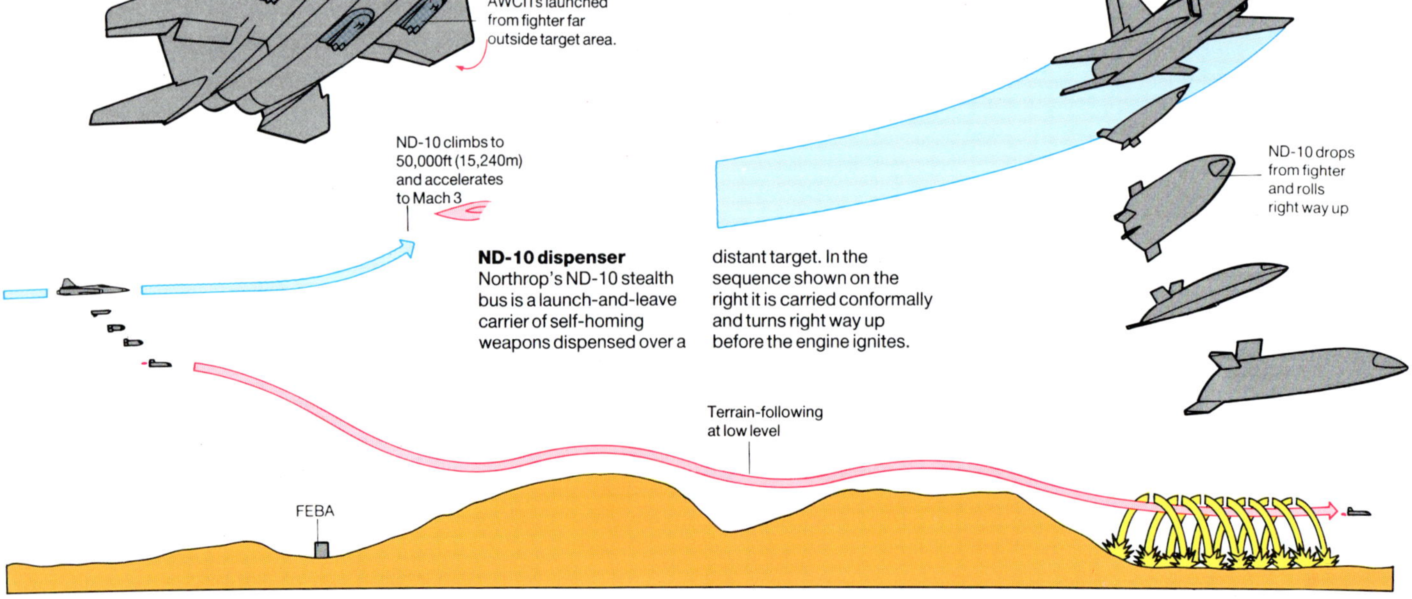


**ND-10 dispenser**
Northrop's ND-10 stealth bus is a launch-and-leave carrier of self-homing weapons dispensed over a distant target. In the sequence shown on the right it is carried conformally and turns right way up before the engine ignites.

Right: JP.233 payloads being dispensed by a Tornado. The two species are easily distinguished, SG357 being the larger.

Below right: SG357s in action. Why such munitions should be delivered by manned aircraft remains an unanswered question.

deploys a parachute which brakes it and sets it up in vertical fall over the target area. At a preset altitude the chute is jettisoned; what unexpectedly happens then is that a retro-rocket is fired which arrests the descent and then makes the submunition climb swiftly, while spinning it like a top. The four Skeet warheads in each submunition "articulate" outwards on levers and then are released to fly outwards to cover the local area. Each Skeet IR sensor looks for a target; if it finds one it automatically locks-on and, at the predetermined height fires the jet through the top armour. Should no target be present the Skeet simply explodes near the ground to cause airblast and fragmentation effect on soft targets that may be present.

The ERAM/Skeet system is the only one publicly known which can simultaneously take out a large number of hard targets by day or night, theoretically in any weather. The limiting factor in the last instance is the aircraft, not the system, though clearly in very heavy snow (for example) the detection range and reliability would be degraded. Attacks can be made along or across the line of a linear target, such as an army on a highway, and the system is equally effective against array (randomly dispersed) targets.

## Aircraft guns and air-to-air missiles

In air-to-air operations there is at present no obvious alternative to the traditional gun and guided missile, but both are continually being improved. Where guns are concerned the Soviets are probably ahead, because of their heritage of world-beating aircraft guns and the long time since a major advance appeared. In the West we have got into a morass of unstandardised ammunition, even the tiny RAF having guns of 7.62, 7.7, 20, 25, 27 and 30mm calibre! Apart from the GEC Minigun carried by (non-British) Lynx helicopters and M61 Vulcan pod used by RAF Phantoms, all these guns are single-barrel. In the USA, however, the standard fighter gun has been the six-barrel M61 for almost 30 years (except for the M39 fitted to the Tigershark). In the late 1960s a new 25mm gun was being developed to arm the F-15, but this was cancelled after severe development problems.

The next-generation guns in all countries are likely to have much shorter (so-called "telescoped") square-section propellant charges, which make full use of the available space unlike a round case, though a square gun breech is structurally more difficult. Another feature is likely to be the elimination of the cartridge case, or at least its replacement by a case consumed when the charge is fired. This at a stroke offers numerous advantages, including reduction in cost (and probably ammunition bulk and weight), elimination of case-induced stoppages and elimination of the problem of either getting the case to separate cleanly from the aircraft or else storing it on board.

For the rather longer term there is the obvious possibility of using guided projectiles. Today there are only a few weapon systems that combine firing from a rifled barrel with terminal guidance, and most are of 5in (127mm) or 6.1in (155mm) calibre. For air combat use much smaller calibres appear desirable, because though large-calibre guns up to 105mm have been carried by aircraft none has had the high muzzle velocity needed for air combat engagements. The question thus resolves itself into matching the pros and cons of today's calibres of 20 to 30mm, with large numbers of unguided projectiles, against on the one hand the prospects for reliable firing accuracy and on the other the alternative of much smaller numbers of guided projectiles, probably of larger individual size and mass. The problems with attempting to discuss the prospects here lies in the very early state of the art, its generally high security classification and, above all, in the wealth of possible methods that could be used both for guidance and for trajectory control.

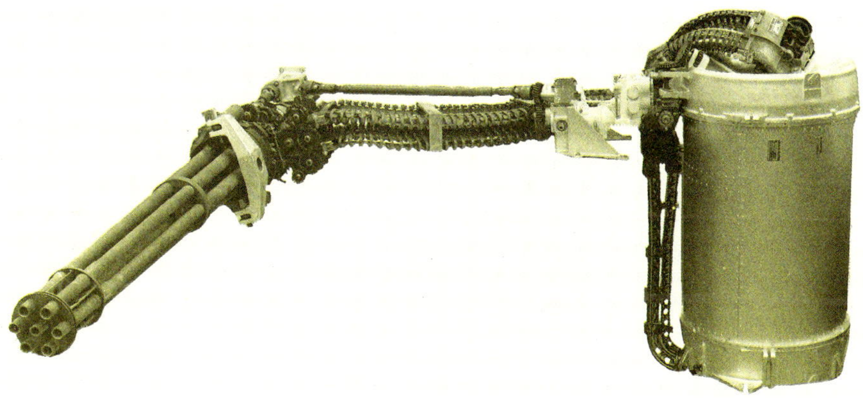

Above: Though over 35 years old in design, General Electric's M61 "Gatling" remains the standard gun in all US fighters except the AV-8B, which has a 25mm version.

At present it is probably more rewarding to put development effort into aiming systems which ensure that every round fired scores a hit, even with today's primitive guns and ammunition. For many years the USAF has been getting results close to the ideal with the IF/FCS (integrated fire and flight control system). In live combat between an IF/FCS F-15 and a PQM-102 target (rebuilt F-102 Delta Dagger) virtually 100 per cent hits have been scored over ranges in the order of 2,000ft (600m) not only in easy situations but also in the most difficult conditions with both aircraft pulling g and the target having a high crossing velocity. Under such circumstances an unaided pilot would be very unlikely to pull the trigger; still less would he expect to score a hit. But if the fighter's trajectory is automatically guided by the fire-control system, and the trigger circuit automatically closed at precisely the right moment, it is possible to make every round strike home, At present this seems a better way to go than developing costly guided projectiles for bigger guns.

Such dramatic advances in the accuracy of guns inevitably tend to reduce the demand for AAMs. In the 1940s such missiles were widely expected to replace guns in very short order, and nobody had the foresight to see that guns would continue into the 21st century. The AAM has in the past looked great on paper but performed abysmally in actual warfare, with a few notable exceptions such as the NATO AIM-9L advanced Sidewinder in the Falklands war in 1982.

The tendency has been towards two distinct classes of AAM, the short-range IR-homing (and thus "fire and forget") type, such as Sidewinder, and the medium-range SARH (semi-active radar homing) family which steers towards the source of radar signals sent out by the fighter and reflected from the target, as exemplified by the AIM-7 Sparrow. The two families have been taken a long way by such US companies as Ford Aerospace and Raytheon, but these very competent design teams know better than anyone the limitations of the available hardware and are trying not to lose control of the market.

There are many drawbacks to existing AAMs, not least of which is a disappoint-

**Future AAM**
This 1990s fighter is armed with 16 small AAMs carried in an internal bay.

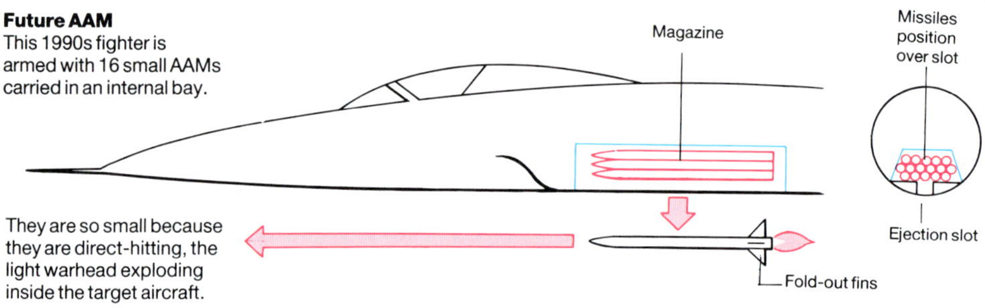


They are so small because they are direct-hitting, the light warhead exploding inside the target aircraft.

**Above: The backseater's view as an F-15B fires an AIM-7 Sparrow against a target flying at higher altitude. This could hardly be simpler, especially as the target is contrailing at close range.**

**Below: Western air forces desperately need AIM-120A AMRAAM, which will at last bring fire-and-forget capability to long radar-guided ranges. This was the first test firing in March 1981, from an F-16.**

ing performance in warfare. The IR-homing dogfight AAMs are a fine basis on which to build, but sadly the British Aerospace Taildog (later SRAAM), which could have been the starting point for the NATO Asraam (advanced short-range AAM), was judged too advanced and replaced by something like an improved Sidewinder. In the medium-range field the position was far worse, in that the target aircraft had to be illuminated by the radar of the interceptor right up to the moment of impact. This could prove deadly, and the American AIM-120A Amraam (advanced medium-range AAM) has for five years promised to rectify the situation. Sadly AIM-120A has run into severe problems, both of a technical and cost nature, and even though it is one of NATO's most urgently needed weapons it will arrive much later and cost more – according to the media, a cool million dollars per copy.

On the other hand, where AAMs are concerned there is no question but that most of today's species are overlarge and expensive, and got that way because of their inability to strike the target reliably. From the early days 30 years ago with Falcon, Sparrow and Blue Sky (Firestreak) the objective was to bring a large warhead somewhere within lethal distance of the target aircraft, using a sophisticated proximity fuse to detonate it at the point of closest approach. As a result AAMs generally have been large and heavy, and even with the benefits of later electronics and other miniaturisation medium-range AAMs today are never found below the 425lb (193kg) of BAe Sky Flash and are usually well over 500lb (227kg). The warhead of AIM-7F Sparrow weighs 88lb (40kg) and that of AIM-54C Phoenix no less than 133 lb (60kg). Soviet warheads are almost certainly heavier still.

One has only to look at the Rapier SAM to see what could be done if one could make an AAM that hit its target. Rapier's warhead mass is classified, but it can fairly be described as "grapefruit sized". It probably weighs less than one-tenth as much as that of Sparrow; indeed the entire Rapier missile weighs less than the Sparrow warhead! Despite this the Rapier warhead is semi-armour-piercing, designed to pass through thick skin of modern supersonic aircraft even at narrow grazing angles. The fuse, of crush type, is behind the warhead and thus detonates the head after it has penetrated inside the target aircraft. Using hollow-charge and other forms of multiple directed energy even a small warhead detonating inside an aircraft can prove instantly lethal.

The corollary is self-evident. If we could invent a direct-hitting AAM we could make it about the same size as Rapier and still destroy our targets. Such AAMs could be carried internally, as well as hung out in the airstream, and a fighter could easily carry a dozen at a time. This would give what is called high combat persistence: the capability to engage numerous successive targets.

The point must be made that the weight and drag of even Sparrow and similar AAMS is considerable, and exerts a significant adverse effect on the carrier aircraft.

**Matra Super 530D**
Matra's R 530 AAM had an extremely poor record, but today's Super 530D is a much better weapon, though for some reason the French appear to expect to meet targets flying at Mach 2.5 at 75,000ft (22,860m), as in the upper diagram, or Mach 1.5 at 40,000ft (12,190m), as in the lower. As these are just the kinds of target modern AAMs like it is hardly likely that a sensible enemy would be so obliging.

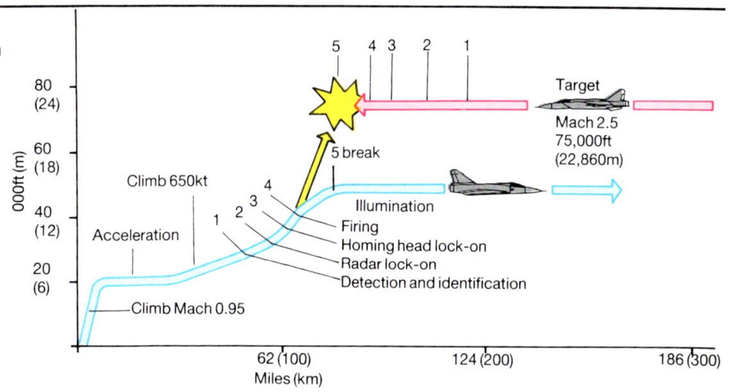


**The fast AAM**
In the diagram below the blue arrow shows the distance covered by a Sparrow AAM at Mach 3 (ignoring the acceleration from rest). The red arrow shows that in the same time of 3.8 seconds an AAM flying at Mach 10 could intercept a target 8 miles (12.9km) distant. Unfortunately Mach 10 also means the missile cannot manoeuvre.

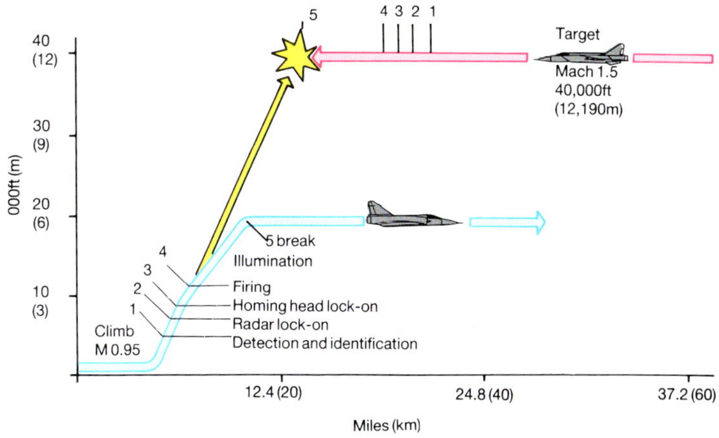


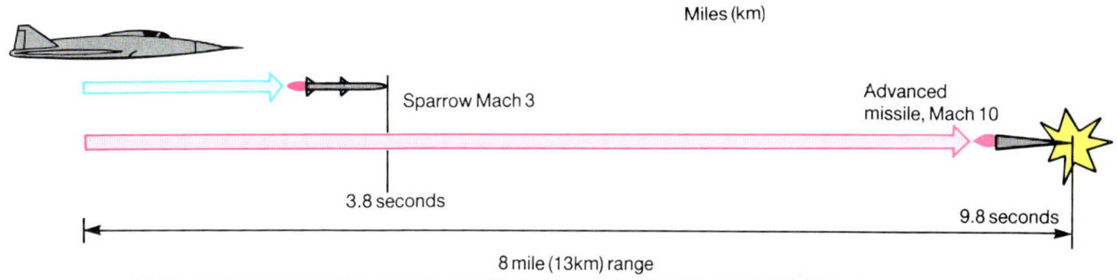


If we could slash AAM weight to one-tenth of its present level, without significantly reducing motor burn time, we could have stand-off interceptions by really small and agile fighters which automatically have such major advantages as greatly reduced RCS (radar cross-section) and significantly reduced IR emissions. With VHSICs, optical data buses and many other new technologies now becoming available there is no reason why tomorrow's fighter should not be in the 7-ton class or less. Unfortunately for the West, such fighters seem much less exciting than ATF ideas with six engines and a length of over 100ft (30m).

**Above: The air inlets on this study for an AAM of the 1990s feed a ducted rocket, probably the ideal long-range AAM propulsion system.**

**Below: The large AAMs on this Swedish JA37 Viggen interceptor are RB71s, or BAe Dynamics Sky Flashes. These are the most advanced radar-guided AAMs in service.**

Considering as the baseline for this discussion the multirole tactical fighter/attack aircraft involved in a land war, the obvious features of any mission are that it cannot start from a known and pre-targeted place and it must do its utmost to avoid making emissions en route on any wavelength. It also has to obey the laws of physics, so over undulating terrain the NOE profile has to be flown relatively slowly. There is some evidence that with full stealth design a better profile is that which imposes the limiting vertical accelerations in a "roller coaster" ride while holding maximum low-level airspeed. This is going to be a less-undulating flight, at a greater average height above the ground, than could be flown by a slower aircraft, while being entirely below the non-undulating trajectory that would be followed by a highly supersonic aircraft.

No existing combat aircraft can fly at Mach 2 or more at low level; structural stresses and fuel consumption would be colossal. The fastest (clean) is probably the Tornado, with demonstrated speed in excess of 800kt (912mph, 1,468km/h, about Mach 1.2). With a heavy external load of bombs, tanks and ECM jammer pods the speed would be much less, and terrain following would be possible in most regions at below 200ft (91m) AGL (above ground level).

This problem of relating speed, height AGL and terrain quality is one that will be with us for as long as men indulge in armed conflict. There is no obvious way to escape from the need to attack at the lowest possible level, and with a human crew on board there are quite low limits to acceptable vertical acceleration, either in atmospheric gusts or because of the undulations of terrain following. Even a pilotless cruise missile has to fly a fairly easygoing profile at present, because to stress the airframe for a harsh NOE ride at high speed would greatly increase airframe weight and fuel consumption and make it difficult to achieve the design range.

This matter of mission range, or radius for a vehicle flying a two-way round trip, exerts a central influence on the design of the aircraft. In any case where the radius is difficult to achieve it is possible to cheat with inflight refuelling. The campaign in the South Atlantic in spring 1982 would have been impossible without this technique, nor could the Falklands have been routinely flown to by the RAF subsequently.

The RAF is short on tankers, but the tanker strength of the USAF is enormous and serves as a force-multiplier on a global basis. For any future combat aircraft to be designed without a probe or a boom receptacle will surely be most unusual except in the special cases of small countries absolutely convinced they will never need to fly

**Above: In a gentle turn an F-15B gradually comes in under a KC-135 so that the boomer can fire the boom into the fighter's left wing. The USAF is the only service to use the rigid boom method of air refuelling, which puts the responsibility on to a crewman (the boomer) in the tanker.**

**Tornado attack mission**
The large diagram gives a very simplified profile of a typical mission by a Tornado all-weather interdiction aircraft. After takeoff, possibly with a load of ordnance of some 20,000lb (9072kg), the aircraft cleans up and, still with wings spread, climbs to a high cruise altitude where the amazing economy of the RB.199 engines really pays off. Nearing hostile airspace the wings begin to sweep as speed increases in a dive down to low level, which even in peacetime training means 200ft (61m). Various ride levels can be chosen depending on circumstances. Nearing the FLOT (forward line of troops), crossing into enemy territory, a hard ride at very low level would be selected. The height would be extremely low but would depend on airspeed (see upper diagram opposite). In this automatic terrain-following mode the TFR (terrain-following radar) and computer generate an imaginary ski-toe shape which races ahead with the aircraft (green in main diagram for position 1). At position 1 the TFR "E-scope" (top), HUD (blue, centre) and cockpit flight director all command a pull-up to avoid whatever has penetrated the ski-toe (it is a steep hill). The pilot can do TF flying manually, but normally control would be automatic, the flight controls being automatically commanded to follow the minimum safe altitude (blue flight profile). At point 2 the aircraft has responded, satisfying the pull-up command, the ski-toe being seen on the E-scope to be just clearing the hill ahead. At position 3 the ground return on the TFR has fallen below the ski-toe, resulting in a command to push over back towards the ground. At position 4 the diving Tornado's TFR detects the next hill, suddenly penetrating the ski-toe, and at once commands the exact pull-up command that will just clear the hazard. This procedure continues, manually or automatically, all the way to the target. At night or in bad weather TF flight would always be fully automatic, and of course it needs nerves of steel on the part of the crew. At position 5 the crew have noticed a target of opportunity (not previously planned for attack) and the pilot switches the HUD to the CCIL (continuously computed impact line) mode. As shown in the enlarged HUD screen at 5 (far right) the pilot flies the Tornado so that the CCIL (the slightly sloping line across the centre) is centred on the target, release of weapons being manual. Alternatively either crewman can mark the target, enter the position into the MC (main computer) and make a precise automated attack. No other aircraft offers its crew so many alternative forms of navigation, target designation and ranging and ordnance delivery method. Moreover, as the scale of nautical miles along the bottom shows, the radius of action is considerable; 850nm is 979 miles (1,576km).

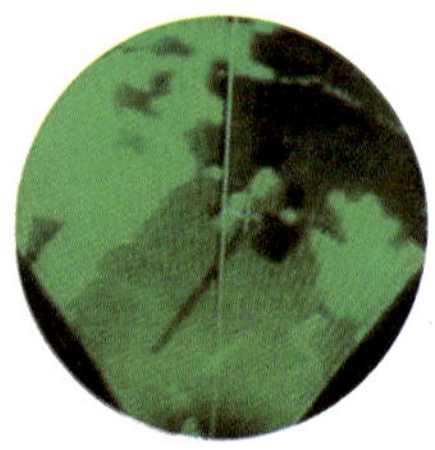

**Displays**
A planned target runway can be seen coming up on the radar several miles ahead. Closer (lower radar picture) the details are clear. The HUD (below) shows the symbology for a preplanned visual attack in one pass, from an offset initial point.

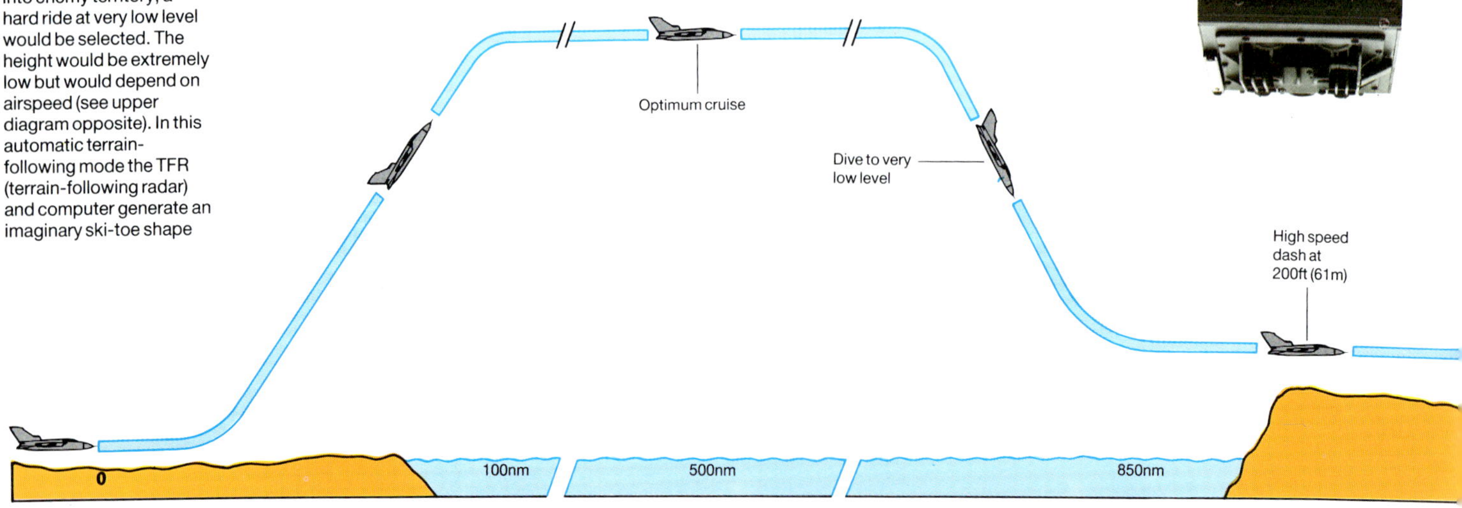

**Above: Though it is all happening at once the crew of this Luftwaffe Tornado are relaxed, and probably enjoying the smooth flight possible with this remarkable aircraft. No other Western combat aircraft (but possibly the Su-24) can attack at such high speed at low level in comfort.**

a mission extended either in distance or in time. There is even no technical objection to air-refuelling of cruise missiles, which could thus be made smaller for a given task and be more difficult to detect. A computer-controlled missile could clunk into the drogue basket even more rapidly and positively than a manually flown aircraft.

In the case of manned aircraft the air refuelling can be programmed soon after takeoff, or near the target or on the return journey; or it can be provided, not necessarily as an emergency measure, to an aircraft caught low on fuel by unforeseen winds, combat damage, bad weather at base or for many other reasons. In any case, it is a colossal plus on the side that has it available, and except in a highly localized conflict it makes many missions possible, with adequate reserves for contingencies, that could not otherwise be flown at all.

Navigation has to be precise throughout the flight, without using emitting navaids other than receiving emissions from ground-based aids such as VOR or Omega. Tacan and doppler appear to be undesirable on this count, though perhaps a vertically directed beam from a radar altimeter can be tolerated. INS (inertial navigation system) techniques are wholly acceptable, and we have come a long way since early INS boxes burdened the F-104G in 1960 and caused severe problems in warm-up and alignment time, reliability and even progressive inaccuracy with time from takeoff. Today's laser gyros can result in INS installations that are close to being ideal for even small fighters, with alignment problems almost eliminated (even on a carrier). It is also to a considerable degree acceptable to use satellite navigation sys-

**How low?**
These highly stylised mission profiles (right) show that flying a World War 1 S.E.5a at 100mph (160km/h) it is possible almost to keep the wheels on the ground. An F-84 at 500mph (800km/h) has to keep well clear of even fairly gentle terrain. As vertical accelerations depend on the square of the speed, the Tornado flying at 900mph (1,450km/h) cannot fly a violently undulating profile but is forced to stay a little higher above the ground. It can still fly lower than any other aircraft at any given speed.

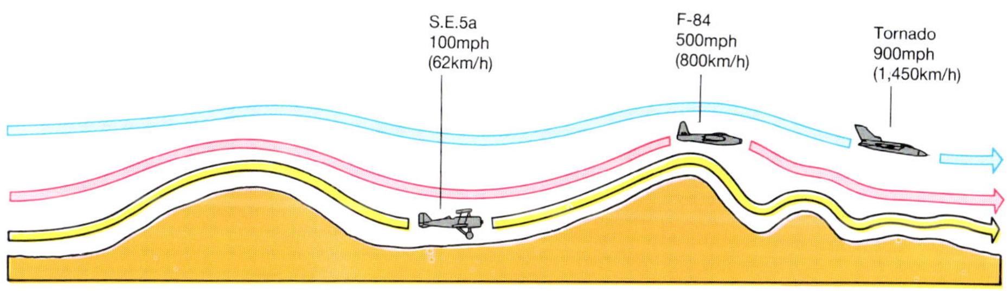


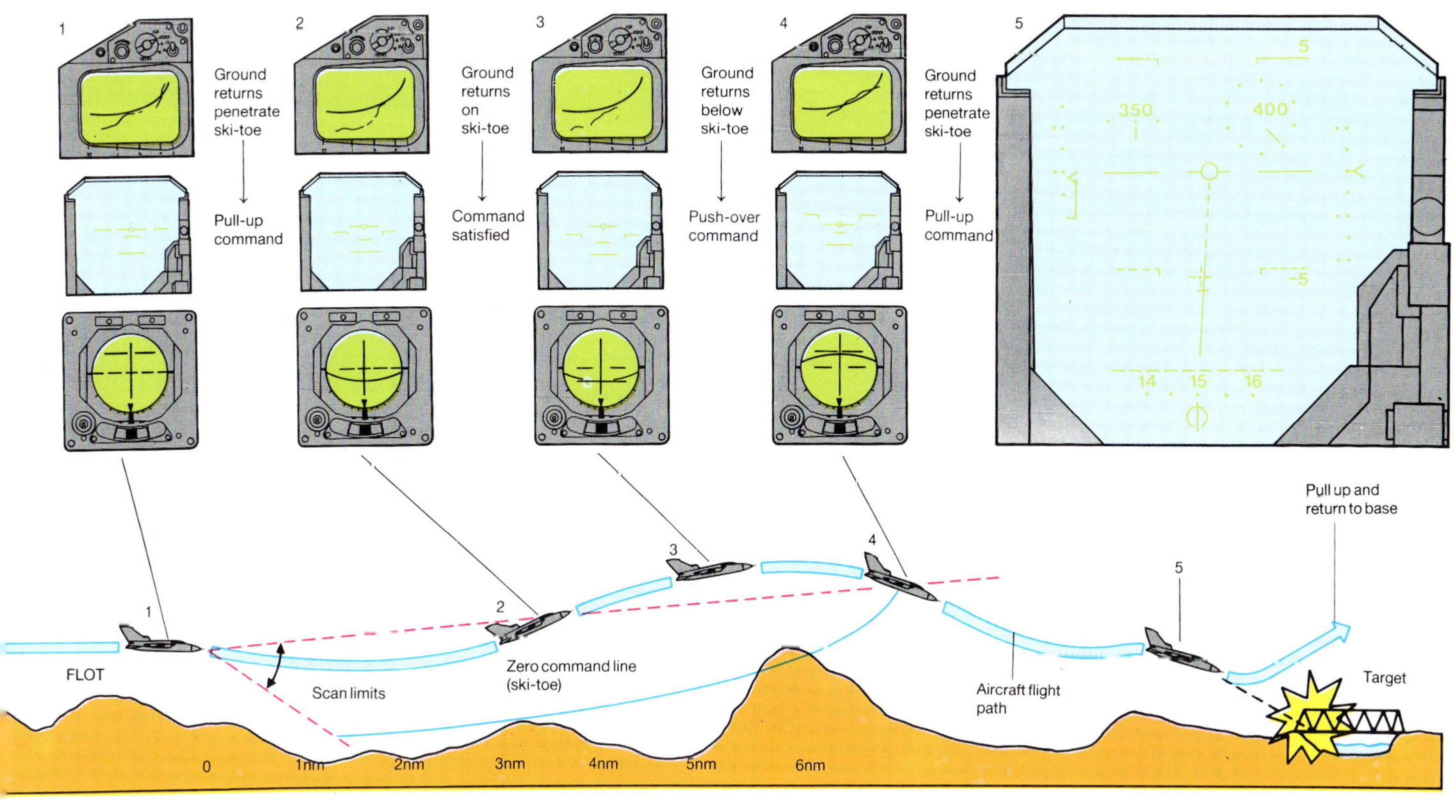

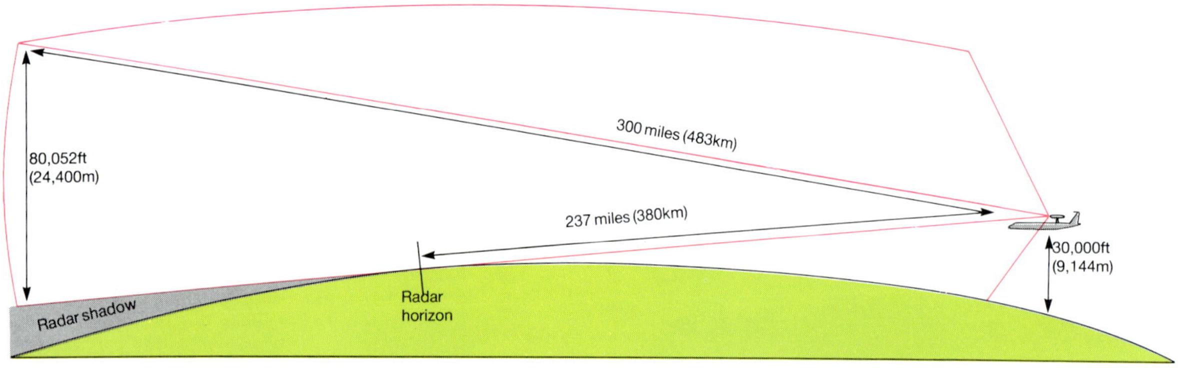


**AWACS ranges**
Basic geometry of a high-flying surveillance radar is the same for all platforms, though some radars are clever in being able to see OTH (over the horizon). Here some effective ranges are given for a Boeing E-3 Sentry at operating height against targets at great height and near sea level. The diagram also shows upper and lower elevation limits. (The Earth's curvature is exaggerated.)

**Above: Now fully operational in support of USAF tactical aircraft in England and the USA, the Grumman EF-111A follows the US Navy's lead with the Prowler in being a dedicated EW aircraft able to fly with the aircraft it is protecting. Though costly, such platforms are invaluable.**

**Below: This splendid picture was taken during a US/Egypt Bright Star exercise in 1984 and shows that, provided it has a nice five-mile runway, the USAF can emulate the "bad guys" and live under canvas. Aircraft are a Core E-3A, B-52Gs and a regular C-130 (MC-130s participated).**

tems, and even guidance from a friendly AWACS which can also pass on details of many other things unknown to the combat pilot.

It is impossible to over-emphasise the value of supporting aircraft such as AWACS for control and direction and a dedicated EW platform for providing sophisticated jamming of hostile systems with speed of response, system management and total radiated power far beyond the capability of a cost/effective tactical fighter. What is still very far from clear is how such supporting aircraft can survive even for seconds in any airspace within 100 miles (161km) of the enemy.

One of the major problems for future pilots penetrating hostile airspace will be the automatic control of aircraft trajectory in NOE flight at high airspeeds. There is no way this task can be performed manually, even for a short period; conversely it is an ideal duty for an automatic electronically controlled system. The difficulty lies in the need to have an exact input, in real time, of the physical profile of the terrain and obstructions built or sailing thereon, within a sector of about 2° left/right ahead of the aircraft.

The pilot can get a perfectly adequate "picture" of what lies ahead with a FLIR or LLTV (low-light TV), both of which operate at night and make no emissions. But these cannot duplicate today's TFR (terrain-following radar) in providing the essential

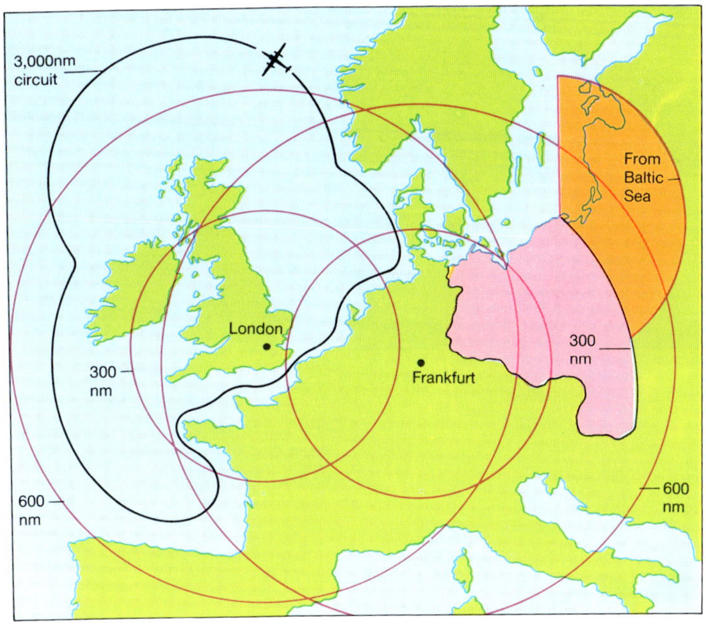


**TR-1 coverage**
Able to fly 12-hour missions at 70,000ft (21,340m), and with a ceiling (without payload) of 90,000ft (27,430m), Lockheed's TR-1A is a remarkable reconnaissance platform which carries many classified sensors including systems for the precise location of hostile emitters such as surface radars. What is not yet explained is how this very large and slow aircraft can avoid being shot down. Though probably painted with the same dull black "iron ball" radar-reflective coating as the much faster SR-71A, there seems to be no way that the TR-1's presence and exact location could be unknown to air-defence forces across the WP frontier. It has to fly high to see as far as possible eastwards.

geometric information that must be fed in to any terrain-following system. In the case of an obstruction, such as a radio mast, it may be more efficient to fly around rather than over, and without accurate inputs of obstruction height and range, and range-rate (closing speed), a computer lacks the information necessary for it to take the vital decisions. It may be that purely passive methods can be made clever enough to measure ranges, perhaps by some form of computer controlled binocular scene-matching and triangulation (in a more accurate extension of the human judgement of distance), but the author is unaware of such a method.

Thus it could well be that for at least the foreseeble future we shall have the ludicrous situation in which almost undetectable stealth aircraft will be unable to penetrate hostile airspace except with a working TFR. This is as sensible as a blackened-face member of special forces advancing on his target while lighting the way ahead with a torch.

Techniques of air/air combat were exhaustively discussed in another Salamander book *Modern Air Combat*. Not enough is yet known to hazard guesses – and they must be guesses, because the subject is classified – on just how the use of stealth type aircraft will alter these techniques. Certainly there will be renewed emphasis on detecting IR (heat) and RF (radio frequency) emissions from all target aircraft, and in making AAMs home on such emissions, even if they are brief and intermittent. This means even greater fidelity in missile guidance systems, which in future will tend to be tuned to the expected emissions of particular types of hostile aircraft. Indeed, with increasing miniaturisation, a next-generation capability is the dual-wavelength missile.

For many years the Soviet Union has developed most of its families of AAMs in two versions, one group with SARH (semi-active radar homing) and the other with IR guidance. In the IA-PVO (manned interceptor force), at least, it has been usual to carry equal numbers of each species, selected by the pilot according to the weather and other conditions. The dual-wavelength missile would simply have two homing heads, side-by-side or in tandem (but obviously with the receivers for both having a good view ahead), with microcomputer switching while the missile is in flight. Should one target emission cease or become confused (by countermeasures, for example), the alternative wavelength would be used full-time. Dual homing systems would be equally valuable in attacks on naval surface targets.

## Conclusion

To sum up, therefore, future air warfare is going to depend even more than it has in the past on electromagnetic radiation at all wavelengths. People who can manipulate EM radiation to conceal, confuse and mislead are going to have a good chance of surviving even in contact with the enemy. Since 1938 ECM has been an obvious key to survival in perilous airspace, and future historians may be hard-put to explain why the subject was largely ignored in the West from 1945 until about 20 years later.

Today we can be very clever in playing tricks at RF (radio frequency) wavelengths, but the problem of defeating the passive IR (heat-homing) missile remains. Short of using man-power or elastic there is no easy way to eliminate IR sources in our aircraft on which hostile missiles can home. The amazing thing is how little has been done, except in helicopters, even to reduce or screen these sources. Clearly we prefer a 100 per cent fighter that gets shot down in the first few minutes to a 95 per cent one that might last hours or days. Pulsed IR radiation emitters help, but modern SAM and AAM systems are specifically designed to defeat them.

Inevitably the discussion of future advanced-technology warfare is limited by considerations of security. This chapter was written without information from official Western defence sources, because such material is highly classified. What does appear to be a cause for concern is the extent of the gap between the kind of war that any reasonably informed observer can dream up in a few minutes and the "old hat" kind of war for which our current hardware is suitable.

In the design of combat aircraft even the next generation – in the West, at least – appears to address itself to the requirements of the proverbial "last war" instead of to the one that matters, which happens to be any "next war". To spend billions on fragile aircraft that sit on precisely targeted airbases appears to be faulty procurement. To fool around with rapid filling of craters, hovercraft transporters and slightly steeper landing approaches in order to get down "between the craters" appears to show a certain lack of imagination.

**Below: The biggest – though not necessarily the best – AAMs in the world are those carried by the MiG-25 known to NATO as AA-6 Acrid. This Libyan aircraft has one IR-homing (left wing) and one using semi-active radar homing.**

SECTION FIVE

# LAND WARFARE

"We should expect the battlefield of the 21st Century to be dense with sophisticated combat systems whose range, lethality and employment capabilities surpass anything known in contemporary warfare. The airspace over the battlefield will be saturated with aerial and space surveillance, reconnaissance, and target acquisition systems. Air defence weapons will exist to deny the use of these aerial platforms. The conflict will be intense and devastating, particularly at any point of decisive battle, thus making it extremely difficult to determine the exact situation. In such an atmosphere of confusion, command and control will be exceedingly difficult. It appears that no single weapons system can be fielded to cope with the total battle requirements. The battle will be waged with integrated systems of all arms and services. Battlefield mobility will be an absolute essential for success. One other aspect of the future battle is drawn from the growing proliferation of nuclear, chemical and biological weapons, coupled with the enemy's apparent permissive attitude regarding employment of these weapons. It is imperative that forces plan from the outset to fight dispersed on this 'conventional-nuclear-chemical-biological-electronic' battlefield."

"The Battlefield of the Future",
*Airland Battle 2000*,
U.S. Army Study (1982)

"Dispersion is not the only means for protecting the troops. In this regard of enormous significance is, primarily, a rapid closing with the enemy and the making of powerful strikes against him, the careful camouflaging of the troops, the construction of fieldworks and the use of the protective properties of the terrain, as well as the able and prompt use of mobility and manoeuvrability of the motorised rifle and tank subunits."

"The Revolution in Military Affairs",
General N. A. Lomov, *Revolution in Military Affairs*, Moscow, 1973.

Since the beginning of the nuclear era in 1945 there have occurred many significant changes in the conduct of armed hostilities. Military scholars have searched diligently for some common denominator of change, but the search has been frustrating and the goal elusive. Although nuclear weapons have cast a heavy shadow over post-war military planning and operations, they have not been employed in any conflicts and to identify any single clear denominator has not been possible.

Military leaders since the earliest times have always tried to improve their tactics and weapons, but the introduction of new weapons or tactics has accomplished genuinely revolutionary effects in only a few instances. Land warfare has been revolutionised three times since the turn of the 20th century. The first occurred in the 1914-18 war with the *denial of mobility* to ground forces by machine guns and artillery; the second saw the *restoration of mobility* by tanks and aircraft; and the third has been seen in the *defeat of technically modern forces* by guerrilla warfare used as a strategic weapon, rather than for peripheral harassment. Since most modern land forces are designed, equipped and organised around the use of tanks and other armoured fighting vehicles (AFVs), there is always the possibility that some new kind of defensive weapon could make tanks and AFVs obsolete and thus constitute the fourth revolution in land warfare since 1900.

The forty years since 1945 has been a period during which rapid changes in the technology of war fighting have occurred. Although the new military capabilities have been employed on a comparatively small scale in peripheral conflicts and have not been decisive, they nevertheless, have been demonstrated in combat and field-tested extensively and are widely recognized everywhere today.

A brief historical reference can put considerations of advanced technology

**Right: A Soviet soldier in full NBC clothing runs through a cloud of chemical smoke in a training exercise. The Soviet Army has vast stocks of chemical weapons and shows every intention of using them in any future conflict in Central Europe.**

**The land battle is rapidly becoming vastly more complex and technologically oriented, using weapons systems of ever greater sophistication. Revolutionary changes are approaching, and as a result the form and shape of a future battlefield is becoming difficult to predict.**

into a better perspective. In 1815, Napoleon Bonaparte was defeated in a decisive battle at Waterloo and during the century which followed the Western world experienced tremendous advances in technology. This great surge was brought about by a number of factors including political, economic and sociological ones. The Industrial Revolution which took place at the same time relieved considerable manpower from the needs of agricultural production and contributed to the development and expansion of defence industries as well as the creation of mass armies. New and expanded railways, accompanied by rapid electrical communications made the movement and control of the larger military forces practical. There were some changes in tactics brought on by newer weapons, but these were largely either by way of modernization or evolutionary in nature.

The Battle of Waterloo involved around 70,000 men on both sides in an area only about 6×13 miles (6×8km). The battle took place over about 36 hours, with the major and final action occurring in a single day. By contrast, in 1944, following the 6 June Allied invasion in Normandy, the Allied armies continued to build up their strength to around 1.5 million men over a period of about 45 days, fighting German forces which eventually numbered about 1 million, before being able to break out of the Normandy peninsula and advance toward the German frontier.

During the 1973 Middle East War, the Arab and Israeli forces found themselves dispersed over an immense battlefield: on average, one man per 430,600sq ft (40,000m$^2$) and with losses (casualties and prisoners) exceeding the entire number of participants in the Battle of Waterloo – and this during less than 20 days of combat.

Probably the most important task facing military leaders in peacetime is to anticipate how a future war might have to be fought, in order to prepare their forces for an outbreak of hostilities. Advances in military technology must naturally be a major feature in such an assessment. Past results, however, indicate that advance perceptions or conclusions have frequently been highly inaccurate, often with disastrous results, which have either led to defeat or have taken several years to rectify.

Due to the uncertainty of knowing precisely what the next war would be like in reality, we must try to comprehend the elemental features of warfare which makes combat operations of all sorts resemble one another and, at the same time, examine how new weapons developments have altered the operations, doctrine and tactics of armies that are using them now and how they might be employed in the future.

**Below: Current MBTs, such as these US Army M1s, possess unparalleled firepower, armoured protection and mobility, but the shape and size of the next generation of MBTs may be totally different.**

The tank was a British innovation that was introduced late during World War I in response to the machine-gun and improved artillery, and because the relatively static trench warfare demanded some means to advance across uneven terrain under heavy fire. The early tanks were awkward and were employed in small numbers without any doctrinal or operational concepts for their use. As a result, their early employment was seldom very effective nor decisive.

Between the two World Wars, tanks and more appropriate concepts for their use were developed although the development was both intermittent and rather slow. The newly established armoured forces had inherited the role, organizational structure and traditions of the obsolete horse cavalry and, early during the campaigns of World War II, armoured leaders recognized that they needed to develop new concepts.

The new tactics quickly demonstrated that the tank could restore fluidity, mobility and manoeuvre to the battlefield. The mobile characteristics of the tank enabled it to travel either on roads or cross-country. The cannon mounted on the tank could produce greater accuracy than artillery in direct (flat-trajectory) fire, and tanks could penetrate quickly into weakly defended areas, beyond local or frontal defence areas and could then threaten enemy logistics, supplies, reserves and lines of communication. As tanks improved in design, construction and concepts for their employment, it became obvious that the armoured forces, in order to be truly effective, had to be employed together with infantry and artillery. Thus, the foot soldier had to be provided with mobility equal to the tank as well as some protection from hostile fire. The supporting artillery had to be able to keep up with both the tanks and the infantry.

**Above: US MBT designers always seek to optimise firepower, armour, mobility, fire control and crew comfort, usually at the expense of maintainability and complexity, as on the latest M1 Abrams.**

During World War II, the Soviet infantry rode into battle on top of their T-34 tanks, but at the war's end, the victorious Red Army still had over a million horses in its ranks. German, British and American infantry were transported in trucks, tracked or half-tracked (but mostly unarmoured) vehicles. When nuclear weapons had become a reality, military planners were forced to recognize that while tanks and their crews had some measure of protection from the radiation and fall-out side effects of nuclear explosions, the infantryman, on foot or in an open vehicle, needed some protection in order to remain effective. Because the modern mobile army, with tanks as the central system, fights together with its weapons and personnel supporting one another, the newer weapons systems had to be designed and developed with this in mind as well as all other aspects of the nuclear battlefield.

The main battle tanks (MBTs) of today's and tomorrow's armies reflect application of the most modern technology. While all tanks resemble one another, there are still some significant differences. Tanks that are deployed with armies today as well as those likely to find their way into armoured forces in the near future all represent mature weapons systems. This means simply that the fundamental design criteria are well-established and the operational product will represent the adjustment or compromise in the emphasis placed upon one or another of the fundamentals

The three basic or fundamental design criteria for tanks are: *firepower, mobility* and *protection*. In the final analysis, General Staffs, who must make the determining judgement, cannot optimize *all* of the criteria; their selection will reflect the emphasis one army or another places on each item – "trading-off" between the factors to arrive at their selection. Thus, for example, one army may place firepower in first order of priority while another will place protection or mobility first. Doctrine and strategy and, as we shall see, even political considerations, play a major role in determining emphasis and selection.

There have been many changes in the MBTs between those of today and which are now being developed for future use and those which saw action in World War II.

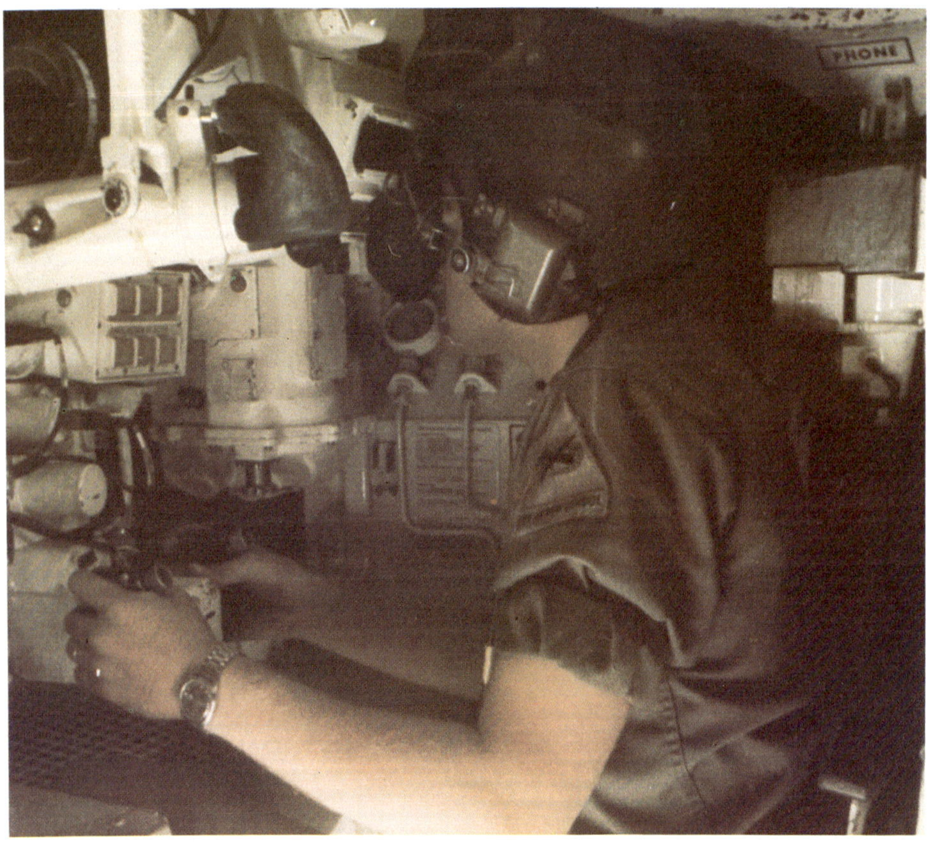


**Left: The gunlayer's position in an M1 shows how sophisticated modern MBTs have become, as technology is used in the endless search for tactical and technical superiority over any potential enemy.**

The calibre of the main guns, for example, has doubled from 70-80mm to between 120-125mm. The thickness of the frontal armour has doubled and the firing range of the guns has also increased from 875-1,750 yards to about 2,200-4,400 yards (800-1,600m to about 2,010-4,020m) against infantry targets.

The newer tanks have night sights, thermal sights and electronic fire computers which enable them to begin fire on targets within seconds after sighting with a high kill probability. The night sights now in use allow accurate firing up to about 1,640 yards (1,500m).

It is evident that present and future tank designs are considerably more complex and heavier than the World War II models (with a few exceptions, such as the German Tiger II). For example, the German Leopard II and the American Abrams tanks both weigh about 55 tons. This is about the same weight as the British Chieftain whose replacement, Challenger, equipped with compound armour, like the German Leopard and American Abrams, reportedly weighs in around 60 tons. Increased weight means that more powerful engines must also be provided if the same degree of mobility is to be retained. The Soviet Union, on the other hand, kept the weight of their tanks to 40 tons or less for a lengthy period, though there are reports now that the new Soviet T-80, equipped with compound armour will weigh between 45 and 50 tons, or more. Weight also has a direct effect upon speed, mobility, size and endurance, so these factors must also be addressed.

**Below: The US Army's HIMAG test-rig has an external 75mm ARES gun with an automatic loading device, thus deleting the human loader. This leads to a lower, lighter, but much more complex, vehicle.**

Tank designers, considering weight and size factors, must also bear in mind bridge and road bearing restrictions and the ability of the tank to manoeuvre through built-up areas and under bridges, as well as to be able to be carried by rail cars and tank-carriers. Until quite recently, the average tank (of about 50 tons) with its main gun, propelled by an engine of around 1,000hp, required a crew of four; a commander, driver, gunner and ammunition loader. During the 1970s designers concluded that if the size of the crew could be reduced then the size and weight of the tank could be made smaller. It followed that if the tank gun could be loaded automatically, the loader position could be eliminated from the crew. This was of particular importance, since the loader, whose function required that he stand in the turret of the tank, made it necessary that enough room be provided for him to stand and move in the body of the tank but be able to reach the gun and load with internally carried ammunition.

The interior of even the larger tanks is quite crowded, with the loader, the gunner and commander having to fit into the turret while the driver is squeezed into the body. (France and the Soviet Union select tank crew members on the basis of size, about 5ft 5in/167cm, being a maximum height for selection.) Loading the tank gun automatically would permit a lowering of the height of the tank, thus reducing weight and silhouette.

One result of deleting the loader is to be able to lower the tank's height, while maintaining the same concept for turret design, with the commander and gunner sitting inside the rotating mass. A more radical solution is to move these two down into the hull, thus enabling the turret to be made very much smaller, as on the Swedish EDES-19 design.

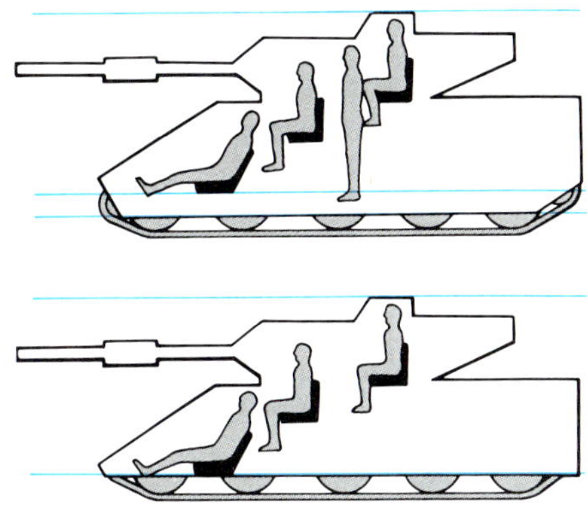

**Automatic loader**
A critical factor in a tank's height is the loader, who must stand upright to be able to feed rounds into the breech. Automatic loaders allow the tank height to be lowered, a major tactical advance (bottom).

**Tank test bed**
US Army Tank Test Bed (TTB) vehicle is based on M1 automotive components. It is being used to test automatic loaders, a variety of crew configurations, different surveillance devices, and other tank components.

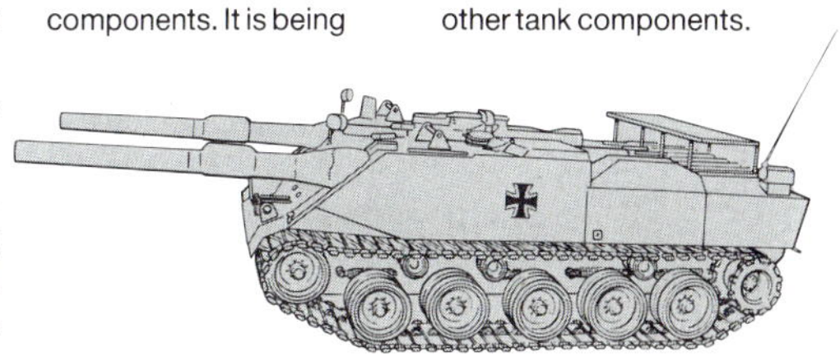

**Turretless tank**
A concept on test in the FRG is a turretless tank with two 120mm smoothbore main guns. This increases the rate of fire and results in a low profile, but may have the same tactical limitations as previous tank destroyers.

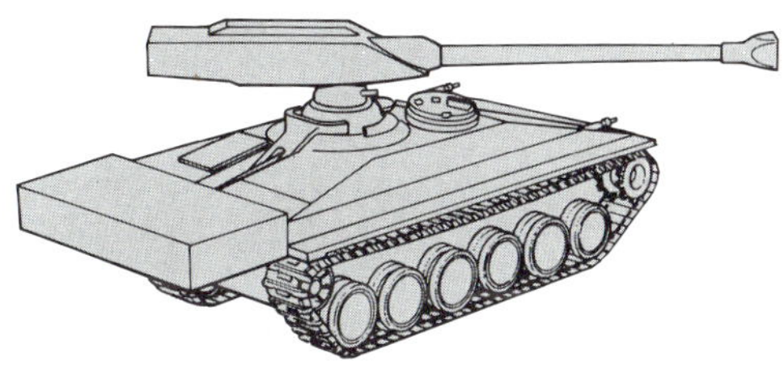

**Swedish EDES-19**
EDES-19 is a Swedish study for a next generation tank with an automatic loader and external gun. An articulated loading arm reloads the gun within 2 seconds, regardless of barrel alignment.

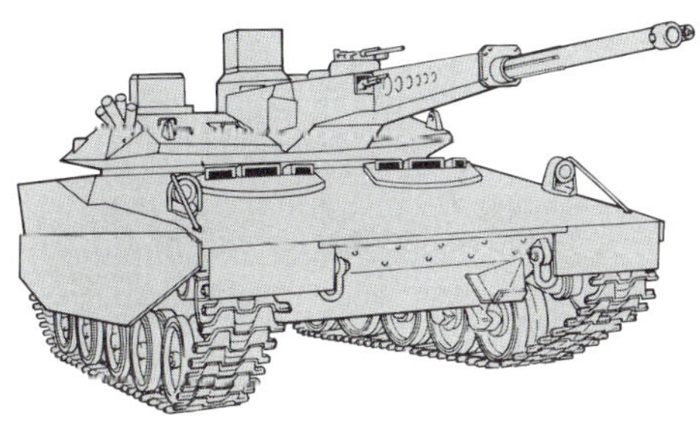

**US RDF tank**
Turretted version of the proposed US RDF tank has a 75mm automatic cannon, which is capable of engaging both anti-tank and anti-aircraft targets. The fully airportable tank weighs just 14.8 tons (15,037kg).

Another important consideration in reducing tank crews from four to three men is the saving in overall manpower. All armies now have manpower and training problems and to reduce the trained tanker requirement by 25 per cent is quite an advantage. Conversely, for the same manpower armies could man 25 per cent more tanks without increasing overall manpower levels.

Designers also postulated that with all of the technical aids now available to the tanker, the commander and gunner positions could be combined, Thus, the tank crew could be reduced to three, and possibly two members. Some designers have gone farther to propose that the crew could be reduced to one man who would drive the tank and operate the guns (autoloaded), much as the single-seater fighter pilot does in his aircraft.

## Battlefield robots

Radical proposals are even being made for remote-controlled, automated tanks – ie, robots. Some small robot tanks were produced in limited numbers during World War II, but were used principally for mine-clearing operations. However, these new proposals are for fighting robots, capable of attacking enemy tanks and other targets. This is an attractive concept which, following the recent developments in robotic technology appears highly feasible. A US Army study in 1983 pointed out: "Because of high manpower costs and declining birthrates, robots will be needed, particularly on certain high-risk missions. Robot 'point men' could detect booby traps or ambushes and warn human soldiers following behind. Robotic tanks could roam the battlefield locating or planting mines. They could operate uninhibited in nuclear, biological or chemical environments."

The US Army has begun development of a number of robot units. ROBAT (Robatic Obstacle Breaching Tank), which will be produced in the near future, permits a combined arms force attacking an objective to breach minefields in stride and under fire in daylight or darkness. ROBAT is a first step and is not totally automated. However, the US are also developing an "autonomous land vehicle" – a tanklike device capable of manoeuvring around an open battlefield using its own computerized "intelligence" – and Robot Defense Systems of Thornton, Colorado, has already created a prototype machine called PROWLER (Programmable Robot Observer With Logical Enemy Response), which is a mobile robot sentry resembling a miniature tank. The current prototype has two M60 machine-guns and a grenade launcher. The final form will be equipped with microcomputers, artificial intelligence software and distance-ranging sensors, allowing it to patrol the perimeters of areas such as airfields, military bases and storage depots and to identify intruders. The sensors and weaponry can be specified for the task.

Another robot "soldier" – ODEX I, developed by Odetics, Inc., of Anaheim, California, has six legs and can climb, crawl and simulate most human movements. The current version is initially programmed to investigate suspected minefields, lay smokescreens, make reconnaissance patrols and other simpler battlefield tasks. Future versions are expected to be able to stalk enemy tanks, fire rockets, throw hand grenades and even act as a battlefield medical aid man to help evacuate wounded under fire.

The most severe problem with all these proposals for fighting robots or automated sentries/guards would appear to be not so much with the operation of the original unit but to lie in the command and control of numerous robots, and confidence in them.

**Above: Programmable Robot Observer With Logical Enemy Response (Prowler) vehicle, an unmanned vehicle for surveillance tasks, with on-board sensors and weapons, made by the US Robot Defense company.**

**Other military applications of robots could be mine emplacement and clearance, vehicle recovery, ammunition (and other) handling, expendible grenade carriage, besides fighting sentry duties (below).**

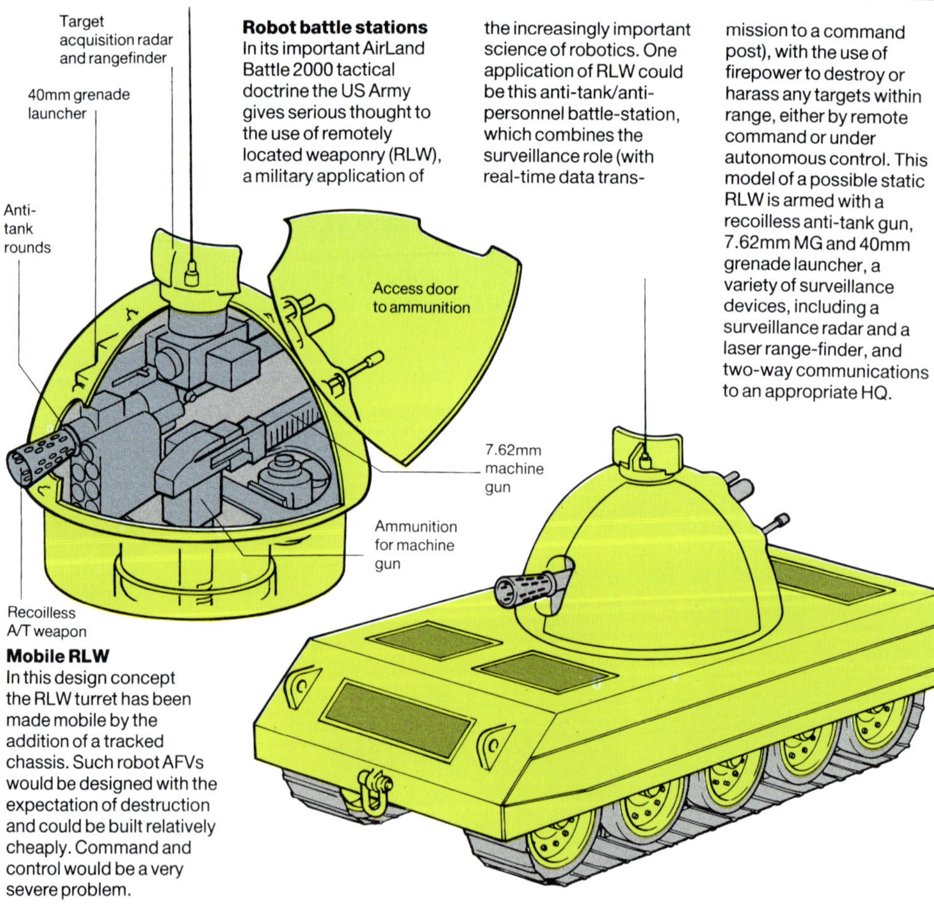


**Robot battle stations**
In its important AirLand Battle 2000 tactical doctrine the US Army gives serious thought to the use of remotely located weaponry (RLW), a military application of the increasingly important science of robotics. One application of RLW could be this anti-tank/anti-personnel battle-station, which combines the surveillance role (with real-time data transmission to a command post), with the use of firepower to destroy or harass any targets within range, either by remote command or under autonomous control. This model of a possible static RLW is armed with a recoilless anti-tank gun, 7.62mm MG and 40mm grenade launcher, a variety of surveillance devices, including a surveillance radar and a laser range-finder, and two-way communications to an appropriate HQ.

**Mobile RLW**
In this design concept the RLW turret has been made mobile by the addition of a tracked chassis. Such robot AFVs would be designed with the expectation of destruction and could be built relatively cheaply. Command and control would be a very severe problem.

## Problems with automation

The most recently fielded series of Soviet tanks, the T-64 (apparently a transitional model), the T-72 and T-80, all use autoloaders and have three man crews. The Western armies have shown reluctance to move to autoloaders and their potential to reduce crew sizes due, in part, to problems and malfunctions experienced with cur-

**Above: The sleek lines of the Soviet Army T-80 are largely due to the fitting of an automatic loader and the deletion of the fourth crew member, a step which Western armies are strangely reluctant to take.**

**Below: The British Army, as a result of its World War II armoured warfare experiences, always gives priority to firepower and armoured protection, as epitomised by these late-model Chieftains.**

rent autoloading mechanisms and also because the modern tanks have been proven to require about 25 per cent more maintenance than the older, simpler versions – more maintenance with fewer people to perform the work usually means sub-standard maintenance. On the battlefield, a poorly maintained tank or one with any malfunction is a certain recipe for disaster. The four-man crews of the current generation of tanks normally perform about eight man-hours of maintenance per day under routine operational conditions. Accordingly, crew size is not related to tank size and weight, but to crew endurance and maintenance as well.

The tank designers who have been seeking to place the main gun entirely outside the tank must rely entirely on autoloading and other automated procedures to aim, fire and reload the gun. In theory, design concepts for these procedures are sound and, in practice, for a tank designed fundamentally for defense, adequate. Sweden, for example, produced a "topless" turret approach in its S-tank; but other Western nations have continued to believe that current levels of reliability for equipment militate against such designs – at least for the moment.

The Soviets, consistent with their general approach that "more of everything is better", have sought to have a large number of relatively less complicated, more expendable tanks with effective guns but less armour protection. Soviet tanks have been designed with the principal armour protection in the front, a low silhouette (wide, with a sloped shape to increase the effectiveness of the armour) and capability for high speed. Israeli tankers who captured and operated Soviet-built tanks after the 1973 Middle East War expressed surprise that the Soviet tanks seemed to "drive like a sports car".

American tank designers have always seemed to try to optimise all of the factors and have featured heavy armour, heavy firepower, advanced weapons fire control and a high order of comfort for the crew. Generally, the US tanks have performed well in combat, although they require a considerable amount of maintenance to do so. British tank designs emphasize high firepower and armour protection, and trade off mobility to achieve these factors. The German tanks emphasize high firepower and mobility, while accepting less protection and weight. French tanks are designed for very low weight but enhanced mobility. The Israelis, on the other hand, after long experience in using tanks designed by others, have come up with their own tank, the Merkava (Chariot), which features a good degree of armour protection and unusual ammunition and storage features to accommodate a high rate of ammunition use. The Israelis have emphasized armour protection in their design to produce a tank which can be used mainly in a defensive role. The Merkava is also unusual in that it is designed to carry up to six wounded soldiers in its compartment or can carry troops and supplies such as AFVs do. Merkava is also structured to permit rapid refuelling in the field.

As indicated earlier, the firepower (gun calibre) of tank main guns has increased by two times since World War II. During the 1960s, the NATO nations developed and deployed a new generation of MBTs, reckoned to be the third generation since the introduction of tanks in 1918. The more complex and sophisticated of this newer generation, the US M60, the German Leopard I and the French AMX-30, outclassed and outgunned the Soviet T-54/55s, their counterparts. Some ten years later, the NATO systems were matched technologically by the Soviet T-62 and a new fourth generation of Soviet tanks, now in the field – the T-64, T-72 and T-80. Fourth generation is generally taken to mean a tank with new features including a large calibre, smooth-bore gun of high velocity which can fire the Armour Piercing, Fin-Stabilised, Discarding Sabot – APFSDS – projectile; a laser rangefinder and automatic loader; a high power-to-weight ratio; one which incorporates special armour which is designed to defeat High Explosive Anti-Tank (HEAT) ammunition (the shaped charge), and the High Explosive Squash Head (HESH) round as well. (An exception is the British Challenger, which is a fourth-generation tank but mounts a rifled as opposed to smooth-bore gun.) The T-72 incorporates all of these features, and the more advanced T-80 is reported also to have advanced armour with a newer smooth-bore gun which fires ammunition equipped with depleted uranium penetrators.

The NATO nations face the Soviet/Warsaw Pact improved armour with the German Leopard II, the US M1 Abrams (and M1E1), a new French AMX, and the British Chieftain and Challenger.

The increasing lethality of the guns has been exemplified by the T-54/55 with its 100mm gun; the T-62's 115mm smooth-bore gun which could fire four rounds per minute (compared to the three round capability of the T-54/55); and the T-64 which features a 125mm smoothbore with a 4 to 6rpm capability. The NATO nations went from the US M48A3, with its 90mm gun firing HEAT rounds, and the British Chieftain, with a 120mm gun firing APDS rounds, to the new Challenger's 120mm gun which fires the modern APFSDS ammunition as does the German Leopard II. Up to 1984 the US M1 Abrams and the French AMX-30 mount 105mm guns (firing APFSDS), but since then M1E1 Abrams (and earlier tanks, in a retrofit programme) are mounting a 120mm gun.

There are undoubtedly advantages and disadvantages over the main guns which ordnance experts can argue; however, in the case of the NATO shift in gun calibre, the decision seemed to have been influenced by politics to some degree. As NATO nations sought to increase the standardisation of their arms and as the various members of the alliance endeavoured to obtain a share of the arms production for their industries, political pressures came into play. The United States and the Federal Republic of Germany undertook an agreement in which the Germans would acquire the American AWACS aircraft and the Americans would adopt the German Leopard II tank. As the details of the agreement were worked out, however, it was understood that the US M1 tank would be equipped with the German 120mm gun under an executive agreement on tank commonality. The Leopard II would use the American gas turbine engine in its propulsion system and both tanks would share fire-control systems. After debate in the American Congress, some changes were mandated and the US adopted a tank design that would be compatible with the ultimate objective of standardisation.

In the meantime, the long-standing dispute concerning the relative merits of

**Above: Early versions of the US Army's M1 Abrams have the British rifled 105mm L7 main gun, but later production models will have the West German-developed smoothbore 120mm gun, making this a very powerful and battleworthy MBT indeed.**

**Below: British Chieftain tank scores a direct hit during a live-firing night exercise at the Suffield training area in Canada. Great technological effort is put into maximizing the chance of a first-round hit for modern MBTs.**

the British rifled-bore gun and the German gun came into the forefront and both of these guns then had to be re-evaluated against the British-designed 105mm gun using a newer and more powerful round. Finally, in 1978, the US decided to purchase the German gun. While this US decision resolved the particular dispute, it did not solve the logistic decision which standardisation was supposed to eliminate, so that the US at least may have to handle two sizes of ammunition for these guns for the time they are in the inventory. Other NATO members may or may not decide to employ 120mm guns but, in any event, they also will have their older tanks for some time and will have to supply required ammunition in varying sizes. The selection of a particular item of equipment is seldom as simple and straightforward as designers would like it; the relative technical merits of a particular system or subsystem may not be the deciding factors for the General Staffs.

## Tank mobility

After firepower, mobility is a fundamental criterion for tank design. The basic contributors to the tank's mobility are the tracked system upon which its movement is based, the suspension and the automotive power. Like its civilian counterparts, the tractor and bulldozer, the tracked system of the tank enables it to move over rough or uneven terrain and across country. Naturally, it can move at higher speed on roads and the maximum speeds of the current generation of tanks (about 28mph, 45km/h for the Israeli Merkava to the higher 42mph, 68km/h for the Leopard II) are achieved on firm surfaces rather than rough terrain. On cross-country travel, the maximum speed of the tank is also limited by the weight of the tank, and its suspension system. Speed and agility are also related to protection in that they enable a threatened tank to move quickly into a position of lesser danger.

What is most important to understand is that tank commanders cannot move their vehicles for more than a few hundred kilometers without expecting to lose most of their command to mechanical breakdown. Tanks are neither designed nor built for long distance moves. With their 40 to 60 ton weight and moving on tracks, they are optimised for a cross-country movement of some 9 to 18mph (15 to 30km/h), not high-speed forced marches on roads. The normal rule of thumb for tank commanders is that a division with 300 tanks on a three-hour road march (about 60 miles/100km) could expect to have around one-third of the tanks break down. Accordingly, tank units prefer to make their longer moves by rail, or to use tank transporter trucks, and to reserve their ground movements to circumstances in which their unique qualities of mobility can be employed against the enemy. Even so, a few hours of bouncing across rough terrain, with an excellent suspension system on a modern tank, can still take a toll on crew endurance and efficiency.

**Above: Leopard 2, current Bundeswehr MBT, epitomises the German philosophy which gives priority to mobility and firepower. The use of Chobham armour also ensures good protection, but combat weight has increased to 54.3 tons (55,150kg).**

## Anti-tank weapons

When discussing the subject of protection, the natural questions are, "Protection against what? What are the major threats against the tank?" As has been pointed out, the tank has come a long way from the time it was designed to be a protected artillery weapon which could be moved across a broken field under fire to support the infantryman. In World War II, the tank restored the war of movement in which the pace of the land battle was determined largely by the tank and the foot-soldier was often subordinated to the armoured column. In recent times, particularly in the 1973 Middle East War, it was dramatically established that the infantryman can now play a much greater role on the battlefield since a single soldier with a guided missile can stop a costly, heavy and sophisticated tank. This did not necessarily mean the end of tanks as useful weapons, as some commentators believed it would; major military forces hastily adjusted their priorities and tried to compensate.

Tank designers have arrived at various shapes for the actual tanks, and of individual (especially vulnerable) parts of tanks, as a basic protection against anti-tank weapons. For example, if a tank's armour is applied at 60 degrees of slope it doubles the distance, or "thickness" that a projectile must penetrate. The Soviet T-72,

**Below: Despite 30 years difference in design dates the M1 (right) is no different conceptually from the M60 (left); it is larger, more effective and better armed, but also heavier and considerably more expensive.**

the American Abrams and Sweden's S tank all have a high slope on their hulls; the British Chieftain has a high slope on the turret. This, then, presents one of many problems for designers of anti-tank weapons; how they have tried to overcome them is described later.

Since many tanks, particularly the Soviet models, are designed with heavier protection on the front against anti-tank guided missiles (ATGM), a new Swedish-designed ATGM is one of many systems being built to attack top armour. Tank designers must therefore cope with the problem that taking armour weight from the front to be placed elsewhere might well weaken the frontal protection.

Re-shaping the overall structure of tanks can help, but only a little; the S-tank, for instance, with its smaller, lighter construction, three-man crew and sloped armour is calculated to have 60 per cent less chance of being hit than "conventional" tanks, because of its configuration – and therefore it is thinly armoured.

There have been anti-tank weapons almost as long as there have been tanks. Today, there are two classes of anti-tank weapons in use: *passive* and *active*. Mines are passive weapons; the active weapons are missiles, rockets and guns. Mines are a long-time anti-tank weapon and mine technology has advanced considerably in recent years, although the basic principle remains the same. Tank crews are always very anxious about reports of mines simply

**Above: Latest improvement to MBT protection is "reactive armour" which is bolted on in small sections as shown on this Israeli M48. This adds less than 1 ton (1,016kg) to the MBT's weight.**

because, in most instances, they are unable to discover mines until their tank is stopped by one – and the most recent mine technology increases the difficulties.

The latest "track-buster" mines can be deployed by helicopter, artillery or special distribution vehicles. A ton of modern mines (about 1,200) can cover an area about 110 by 1,100 yards (100 by 1,000m) and a tank entering the mined area has more than 70 per cent possibility of hitting a mine and losing a track. An immobile or otherwise incapacitated tank is a certain target for destruction by other armoured vehicles or infantry.

All modern forces have a wide variety of mines and dispensing systems. At present, while the Soviet forces have large numbers, their systems are not believed to be as effective as the new American-developed system being introduced for NATO units, or the current British Barmine system.

The US system, called GEMSS (Ground-Emplaced Mine Scattering System), is a trailer-mounted system, the M128, which can be towed by either tracked or wheeled vehicles and dispense both anti-tank and anti-personnel mines. The anti-tank mine has a magnetic influence fuse which can attack the full width of the targeted vehicle, while the anti-personnel mine (directed against the infantry accompanying the tanks) is a ground blast fragmentation mine and is activated by automatically deployed tripline sensors. Both mines have a built-in self destruct mechanism which is set by the dispenser operator to provide a choice of a long or short time interval. Once the self-destruct time has elapsed, the minefield is neutralised, permitting unhindered movement of friendly forces. A trained team of combat engineers can place 800 mines in just 15 minutes.

The anti-tank guided missile (ATGM) can now be added to the list of mature weapons. In the 1973 Middle East War,the Egyptian forces successfully used Soviet ATGMs against Israeli tanks, and this success caused the Soviets (who have always had a heavy emphasis and investment in their armoured forces) to initiate a detailed study of their own vulnerabilities to these weapons. The Soviet publication, *The Armed Forces of the Soviet State* (1975) concluded: "The battle between armour and anti-tank missiles has now shifted to the scientific research laboratories, the proving grounds and industry . . . obviously the traditional method of improving the survivability of tanks – by increasing the thickness of the armour – is far from being the only solution and probably not the best one to the problem."

The earliest anti-tank projectiles, the armour-piercing (AP) round relied on brute

**HEAT rounds**
In a High-Explosive Anti-Tank round, a jet of molten metal pierces armour up to five times the warhead diameter. This effect does not depend on the round's velocity, and is used in many anti-tank weapons.

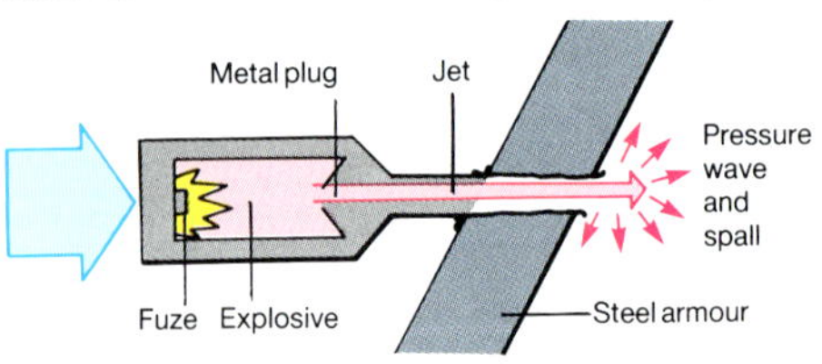


**HESH rounds**
High-Explosive Squash-Head rounds crumple on impact. The explosive is then detonated and shock-waves do great damage, including detaching scabs of armour which hurtle around the interior of the fighting vehicle.

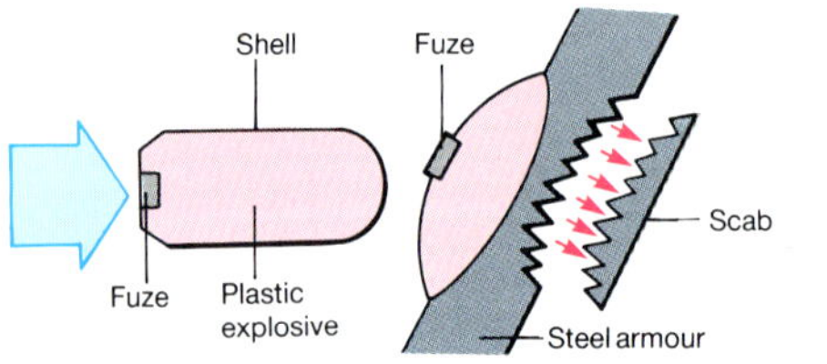


**APDS rounds**
Armour-Piercing Discarding Sabot rounds have tungsten carbide cores and a sabot which is discarded as the round leaves the barrel. The US plans to use depleted uranium in place of tungsten carbide.

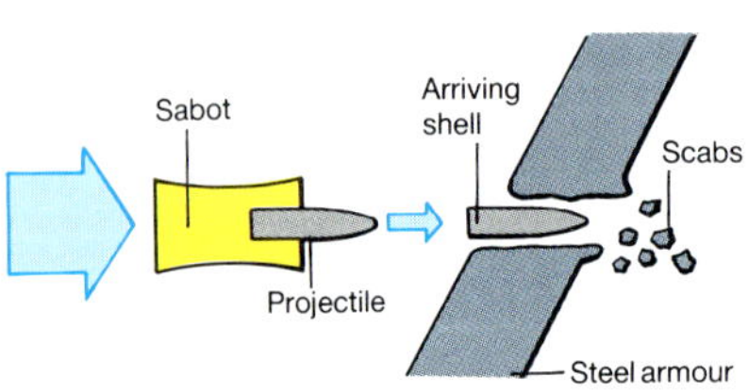


**Chobham armour**
The British-invented Chobham armour is a material of secret composition. It is claimed to provide good protection, as shown here, against HEAT, HESH and APDS rounds, and is widely used in NATO.

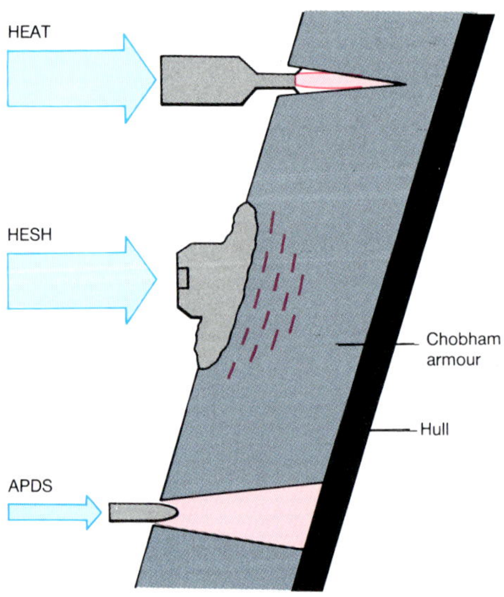


**Reactive armour**
Reactive armour consists of bolt-on sections which explode when hit by an anti-tank round. This dissipates the jet of a HEAT round, reducing its effect, and may deflect an APDS round enough to prevent penetration.

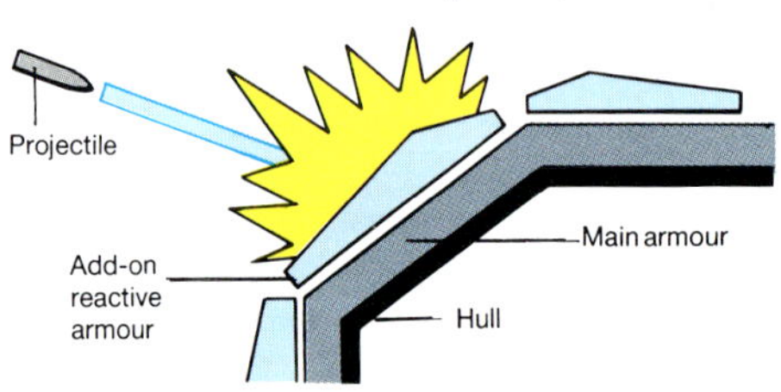

force to penetrate tank armour. During World War II, the *shaped charge* was developed as the principal threat to the tank. The shaped charge relies upon chemical energy rather than brute force for its destructive power. When the inverse-cone of the shaped charge projectile strikes the armour, the released energy creates a very high temperature which melts through the metal armour and permits the burning explosive to penetrate into the tank. It was found that the shaped charge could be employed in a variety of weapons and, since it did not need high velocity or force, could be used by individual soldiers or mounted in light fixed-wing aircraft and helicopters.

Later, the High-Explosive Squash Head (HESH) round was developed, using plastic explosive which spreads on the armour plate on contact and is then detonated by a fuse causing pieces of armour to "flake-off" inside the tank. The HESH round can be defeated by spaced or laminated armour.

In response to the shaped charge, British scientists developed a composite, laminate armour, called Chobham armour (after the location of the laboratory at Chobham). This recent innovation employs layers of metals, ceramics and plastics which absorb and break up the high velocity and shaped charge rounds, enabling the Chobham-equipped tank to survive otherwise disabling or fatal hits. The American M1, German Leopard II and the Soviet T-80 all employ this type of armour protection.

**Above: Sense-and-Destroy Armour (SADARM), a new anti-armour system, integrates a microwave sensor with a self-forging warhead, to produce a true artillery-delivered, autonomous weapon system.**

As the composite armour diminished the tank's vulnerability, at least to the smaller shaped-charge rounds, a new type of anti-tank round was developed – the long-rod penetrator or "shoot" round, mentioned earlier. This projectile resembles a long dart projecting from a plug. The dart is made of a tungsten alloy or depleted uranium compound and literally "drills" its way into armour with great energy.

As a response to the penetrating rod, there has been introduced a new type of armour – "reactive" or "active", using an explosive charge in its outer surface to interdict the path of the rod or dart. Several types of active armour have been developed, but the concept is not yet widely employed. At this time, only the Israelis are reported as having used active armour on their M48 and Merkava tanks (during their Lebanon campaign).

The effectiveness and efficiency of ATGMs result from their design, manoeuvrability, rate of fire, guidance system and penetrating power. The first generation of ATGMs (developed by the French in the 1950s) were slow, manual command line-of-sight (MCLOS) weapons which required the operator to track the target continuously and to guide the weapon through a trailing wire, meanwhile remaining exposed to enemy fire. The second generation system, the semi-automatic-command, line-of-sight (SACLOS), requires only that the operator keeps his target in the command system's aiming mark while

**SADARM**
The 155mm SADARM is the latest US attempt to produce an effective artillery-launched, autonomous anti-tank weapon system. A standard M483A1 cargo round contains four sub-munitions: a sensor, an explosively formed penetrator (EFP) warhead, and an orientation and stabilization device. The latter is a novel single-wing similar to a maple-seed. The sub-munitions are base-ejected over the target and then descend, spinning to set up an automatic scan. On target detection the sub-munition fires its EFP warhead. This form of top attack is very promising, but tank designers are taking urgent steps to add armour protection to this previously invulnerable area.

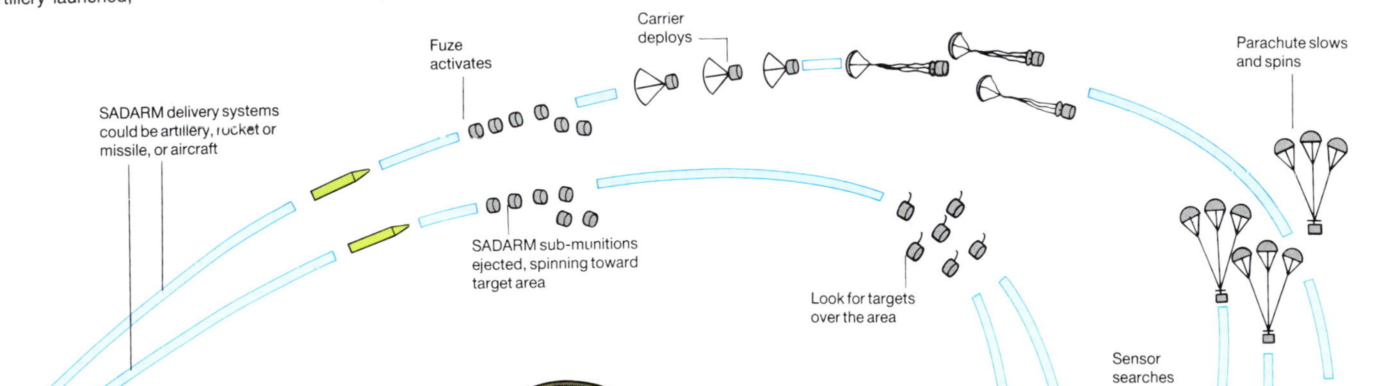


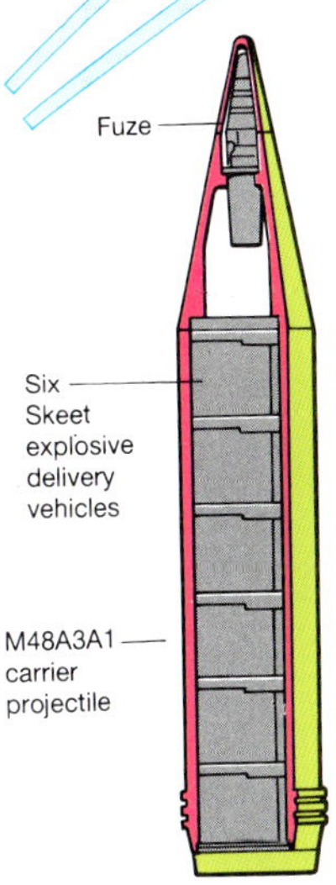


**Cargo shell**
The M483A1 155mm cargo shell is used to carry different sub-munitions in the SADARM system. The shell contains an M577 fuze and a transportation section which carries either three parachute rounds, or up to six of the "maple leaf" rounds, which are also known as "Skeets". The Explosively Formed Penetrators (EFP) can attack tanks from the top or, if they do not detect targets, implant themselves in the ground to act as conventional anti-tank mines. The system is claimed to be effective against all armour threats.

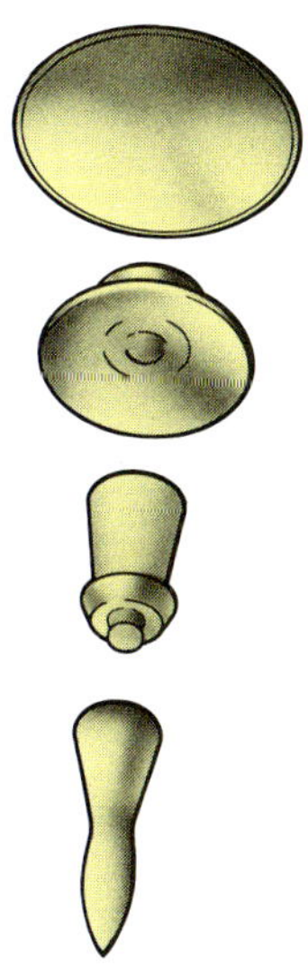

**Explosively Formed Projectiles (EFP)**
The business end of the Skeet is a metal disc, curved to form a lens. On target detection a charge is fired and an explosive pressure wave forges the disc into an armour-penetrating projectile (whose shape is decided by the purity, contour, thickness, diameter, and curvature of the lens, as well as the shape of the pressure wave), which is fired at high velocity at the target below. These sophisticated techniques have been developed using computer modelling, a good example of the possibilities opened up by modern technology.

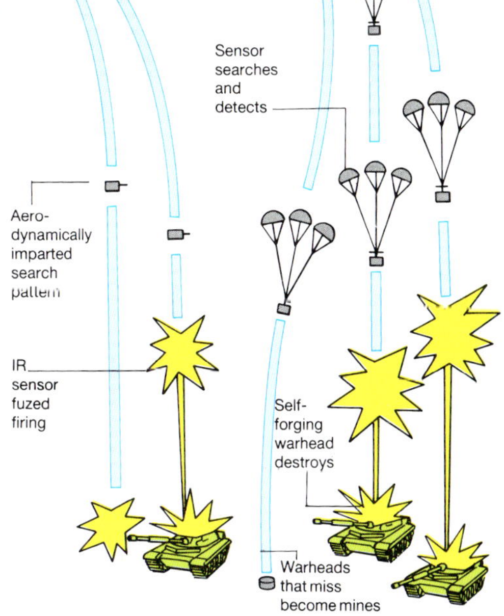

electronic guidance systems direct the projectile to the target. The more recent laser-seeking missiles, such as the US Hellfire or the Soviet "Spiral" system need a ground or airborne laser designator to mark the target.

Each of the current guidance systems has some disadvantages. For example, lasers have difficulty in penetrating target areas obscured by smoke or fog, and the laser device itself can be detected by special viewing devices and attacked. Systems using infra-red seekers can be decoyed by burning vehicles or flares.

Now in the final stage of research and development are micro-processors which can discriminate between real and false targets as well as devices which can recognise images and detect targets, and these can be expected to be fielded in the near future. Obviously, ATGMs which use such sophisticated seekers for guidance that they can be "fired and forgotten" are more effective in that a sweaty, frightened operator (or especially a helicopter platform) does not have to remain on target for a protracted period, and enable the heavily engaged operator to fire at one target and then immediately turn to engage another. ATGMs such as Hellfire and AT-6 home in on targets without guidance; however, these have not yet been widely deployed.

For the time being, a defending force constructed around both ATGMs and tanks has changed the battlefield relationships to one which favours the defender – just as they have changed the factors of time and space in modern combat. ATGMs can be used decisively against armoured vehicles and committed in small numbers, thus making it possible for lightly armed troops, militia or guerrillas to fight against armoured groups (as the Soviets have been discovering in Afghanistan).

Future tanks will have more powerful and more efficient propulsion systems, greater manoeuvrability and agility, and more effective armour and guns. Today's ATGMs can be expected to be less effective against them. But the technology of ATGMs, while mature, is still only in its second (soon third) generation. One can conclude, therefore, that the struggle for dominance of the battlefield between the two technologies is going to continue with the result that higher and more rapid rates of attrition can be foreseen in the armoured battle.

## Armoured personnel carriers

Current armoured vehicles (other than tanks) are also provided with increasingly powerful weapons, such as 25mm guns and anti-tank guided missiles. There has been a general increase in armour protection for other battlefield vehicles and this is now featured on rocket launchers, reconnaissance vehicles, anti-tank carriers, engineer equipment (such as bridge-layers) and other special purpose vehicles such as communications, command, control and intelligence vans.

While the tank is the "centrepiece" of the modern army, we must recall that all of the arms support one another and cannot operate in isolation. With the possible exception of nuclear systems, no one arms system can be decisive. The increased mobility provided by the tank demanded that the infantry keep up with the armour. On a potentially nuclear battlefield, the infantry soldier requires some protection from radiation effects and blast of nuclear weapons as well as from small arms and artillery. Thus, the first generation of armoured personnel carriers – regarded as "battle taxis" for the soldiers – were

**Folgore infantry anti-tank weapon**
After all other systems have tried to destroy advancing tanks, the infantry need anti-tank systems of their own. Most of these use the recoilless principle, firing rocket-propelled projectiles with hollow-charge warheads. Typical of the latest in this field is the Breda 80mm Folgore, a recoilless cannon, which can be fired from the shoulder, a bipod or a tripod, with a maximum effective range, using the tripod and a range-finder, of 1,094yds (1,000m). The diagrams below show the flight trajectories and times of the Folgore projectile. The speed of such projectiles is critical, and the Folgore can reach a target at 1,000m in just 3 seconds (an average of 746mph, 1,200km/h), during which time the tank target travelling at 30mph (48km/h) could have moved 44yds (40.25m).

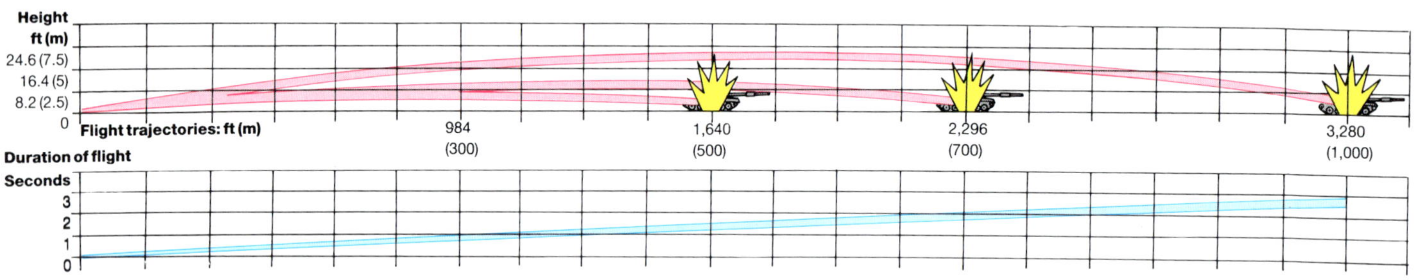

Below: One of the most advanced and successful of contemporary anti-tank systems is the US Army's TOW, seen here being tested against an M47 target.

Below: The same TOW missile impacts on the M47 causing instant and demonstrably disastrous damage. An M47 crew would not survive such an attack.

Below: Swedish FFV AT4 anti-tank weapon. One of many disposable weapons now available, a new warhead is claimed to give much enhanced behind-armour effects.

developed. The earlier APCs were thought of as re-establishing the dragoon – an infantryman who rode into battle, but fought on foot.

During World War II, the major combatants (except for the Soviets) devoted about 30 per cent of their armoured vehicle production to infantry transport. When, in 1955, the Soviets began their full-scale infantry mechanisation, their BTR-152 (a wheeled vehicle) and the BTR-50 (a tracked vehicle) appeared; they were neither unusual in design nor comparable to the contemporary vehicles used by the NATO nations.

When the Soviet BMP first appeared, however, it was a radical innovation, combining the capability for the troops to fight while mounted and having significant firepower to support its section. The Soviet adoption of the BMP was originally directly related to the emphasis stated by former Soviet leader Nikita Khrushchev that future wars would most likely be fought with nuclear weapons – the BMP was designed to be the vehicle for a nuclear war. The BMP (*Boyeva Mashina Pekhoty* – infantry combat vehicle) was first deployed with Soviet Motor Rifle regiments during 1966 and made its first public appearance in the November 1967 Moscow military parade.

The BMP had a number of shortcomings and an improved model was introduced in 1970 called BMP-1 and, at the time of its introduction was clearly the best production model AFV in service with its 73mm smoothbore gun and the "Sagger" ATGM. It retained its supremacy until the German Marder was introduced in 1971, of which the current model mounts a 20mm cannon (smoothbore) in its turret as well as the MILAN ATGM. The Marder is better protected than BMP although, weighing some 29 tons against the 16 tons for BMP, it is not amphibious. But in Central Europe, where the rivers have steep banks which are also inclined to be soft, making entry and exit difficult, the swimming capability of BMP as an advantage may be more apparent than real.

**LAV-25 wheeled light armoured vehicle**
Successful in the US competition for a light armoured fire support and troop-carrying vehicle, the LAV-25 was developed by General Motors of Canada from the Swiss Piranha. The GM diesel engine, with the 8×8 drive, gives a 60mph (98km/h) road speed and a cross-country average speed of 17-22mph (27-35km/h). It has a Hughes 25mm Bushmaster gun and 7.62mm coaxial MG (both stabilized), and later versions will have TOW II ATGW. It is air-portable by C-5, C-141, C-130 and CH-53E helicopter (underslung).

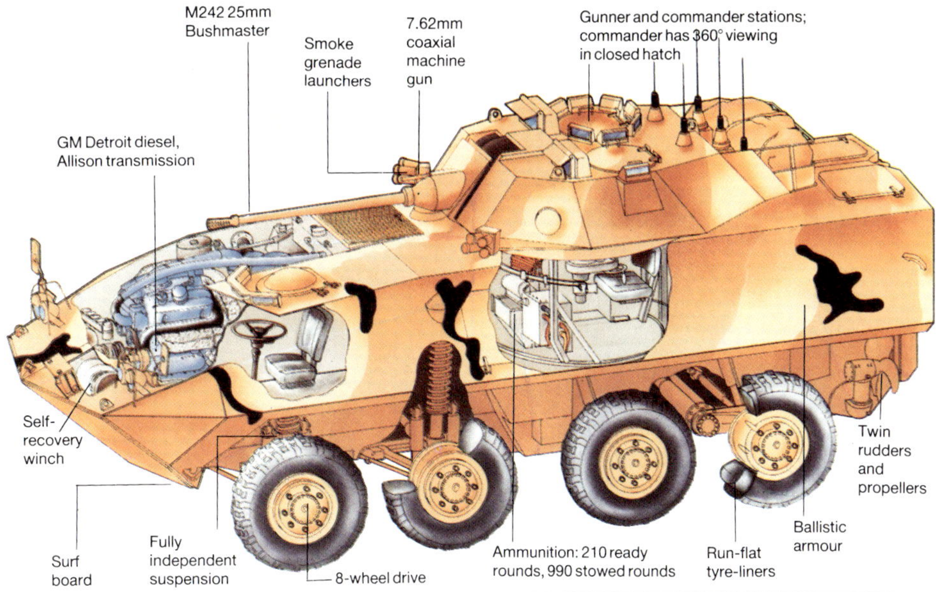


The newest AFV, the US Bradley Fighting Vehicle System (BFVS), is produced in an infantry version as well as a cavalry model. The Infantry Fighting Vehicle, which weighs about 50,000lb (22,700kg) can carry an infantry squad whose members can fire either mounted or dismounted. It has a two-man turret mounting a 25mm stabilized cannon, its primary weapon, supported by the TOW ATGM and the 7.62 coaxial machine-gun. The six 5.56mm firing port weapons positions (which mean infantry do not have to poke their M16s through the firing ports) are along the side and rear of the vehicle. The overall mobility of this IFV is comparable to the M1 Abrams tank, and its "space laminate" armour is designed to protect against small calibre weapons, provide overhead crew protection against artillery airbursts and hull protection against anti-personnel mines. The commander and gunner are provided with day and night thermal sight capability and the driver has an image intensification night vision system as well. The Bradley IFV, today, is considered superior to the current production model BMP in mobility, armour protection and the range and accuracy of its main armament.

The AFV is not universally accepted by military leaders. During the 1973 Yom Kippur War in the Middle East, numerous battlefield photographs were published showing Soviet BMPs whose turrets had been literally "blown off", and one leading Israeli officer, Major General Israel Tal, stated that the fighting vehicle concept was "a complete fallacy". Since 1973, more than fifty "analytic" articles have appeared in Soviet military journals directing their attention to the BMP as "the weakest link in the armed forces".

Although ackowledging the many outstanding features of the new US Bradley Fighting Vehicle, its critics have pointed out the vulnerability of its aluminium armour to such shaped-charge weapons as even the portable Soviet RPG-7 because of the fact that aluminium vapourises when struck by a HEAT round. Other critics argue that newer AFVs tend to approach tanks in size, weight, fuel consumption and cost but lag far behind in combat effectiveness – these critics suggest that nations would be better advised to follow the Israeli example and develop their tanks to carry some soldiers rather than to expend further resources on AFVs. If these latter critics prove to be correct, the current crop of AFVs may be the last, or at least to be only transitional vehicles.

**Below: The US Army's M2 Bradley is the latest tracked infantry combat vehicle to enter service in a major army, and is the culmination of many years of development and dispute. It is a sophisticated and well-equipped vehicle with a heavy armament and is specifically intended to fight in company with the M1 Abrams MBT.**

Helicopters are playing an increasing role in armoured warfare, being employed in a number of ways, such as in a "gunship" role as airborne artillery, as an anti-tank weapons system, and to insert troops in the enemy's rear areas for disruption or move them quickly to counter enemy armoured advances or exploit advantageous situations. It is clear that when the helicopter is used as a troop carrier, it replaces and supplements and, at times, fulfils the function of the MICV or APC.

Between April and June, 1982, British forces engaged in combat in the South Atlantic, in the Falkland Islands, and their helicopters were employed in a variety of tasks such as assault, transport, use as flying cranes to move heavy equipment and casualty evacuation. Some newer roles, such as airborne early warning and special communications were also identified. On 21 May, a detachment of Sea King HC4s from 846 Naval Air Squadron lifted some 912,000lb (413,680kg) of supplies and transported 520 troops, a remarkable feat in which Sea King was employed at 10 times the normal rate in terrible weather and most difficult flying conditions. British commanders found that their logistics demands often had to take first priority, sometimes over tactical requirements.

During the same period, in June, Israeli forces invaded Lebanon and greatly expanded their previous use of helicopters. Like the British, the Israelis began to use their helos for airborne early warning and also began to use them as gunships – a use which was rare in the past.

Helicopter designers have explained that they believe the craft now in service, or coming on line, will likely remain in

**Above: Helicopters give vital support in land operations. In the South Atlantic War an RN Sea King detachment, used at 10 times normal rates, moved 912,000lb (413,680kg) of supplies and 520 troops.**

**Below: There are two types of attack helicopter. One is purpose-built (eg, the US AH-64), the other is adapted from another type, such as the AH-1 (Cobra), based on the ubiquitous UH-1 (Huey).**

production through the 1990s, since their elements of design are fundamentally sound. Heavy fighter aircraft can protect themselves against ground and air fire with their high speed and on-board weaponry and because of their relative stand-off distances from hostile weapons. The helicopter, on the other hand, is relatively slow and its armament is essentially inferior to that of fighters; however, it can make better use of the terrain because of its flight characteristics.

Designers are seeking to provide additional armour protection for future helos and to retrofit current equipment. Designs now under way are making use of space-age plastics for protection and considerable attention is being paid to providing increased survivability for the craft and its crew against crashes from low altitudes. Designers regard survival as being equivalent to mobility.

Future helicopters will also have improved all-weather capability, long-range reconnaissance equipment, weapons with long-range accuracy and small silhouettes with low infra-red signatures. Evolutionary models of the US AH-64 and the Soviet Mi-24 "Hind" are examples of development in helicopter design-to-production techniques employing advanced technology. A major factor in which helicopter designers encounter difficulty is in design of rotor assemblies – a historic weak point.

Brought about by the increased lethality of anti-helicopter weaponry, tactics employing map-of-the-earth flying and "pop-up" manoeuvres demand better visibility. Thus, the new Bell 206 and Hughes 500, as well as the Westland Lynx make use of periscopes in mast- or roof-mounted sights, and other special devices, which will enable them to "skulk" behind hills and other terrain features and select targets and make their reconnaissance, or direct weapons fire on the enemy.

The special role in anti-tank warfare foreseen for the Soviet Mi-24 "Hind", the US AH-IS Cobra (with TOW) and the British Lynx (also with TOW) indicate that dedicated anti-helo vehicles may be under development that will accompany forces into the field in the future.

The US AH-64 Apache, now being deployed in the field, is a good example of a quick-reacting airborne anti-tank weapon system. Equipped with a 30mm chain gun and 2.75in rockets, and capable of speeds up to 146 knots (270km/h), this attack helicopter can use its Target Acquisition Designation Sight and Pilot Night Vision Sensor (TADS/PNVS) to navigate and attack under adverse weather conditions and at night.

**Attack helicopter anti-tank operations**
The attack helicopter has become an invaluable element of the anti-tank defence, especially in its ability to redeploy rapidly to meet attacks from unexpected directions. In this scenario an AH-64A is attacking advancing Soviet tanks with its Hellfire ATGW. The tanks are being marked by laser designators from another helicopter (OH-58D) on the left, and by a ground-based designator on the right. Both designators could be used for any other laser-guided ordnance (eg, tube-launched artillery).

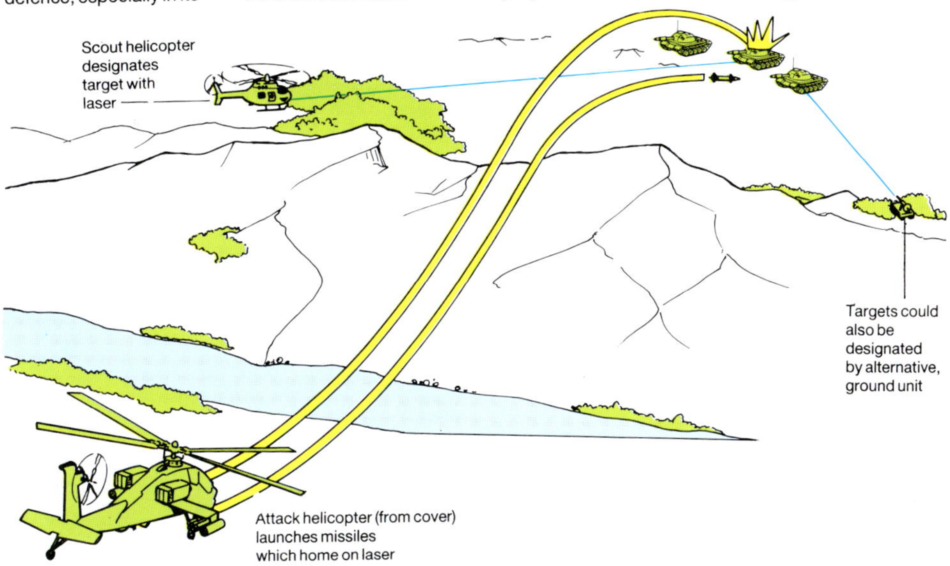

**Left: US Army's AH-64, a sophisticated attack helicopter. It seems inevitable that such helicopters will be used as "fighters" in air-to-air combat to establish local air-superiority.**

**Below: Bell OH-58D advanced scout helicopter with mast-mounted sight and two twin Stinger missile pods. Its main mission is to find and designate targets for the AH-64's Hellfire missiles.**

The role of the infantry soldier has changed very little during recent centuries. His fundamental mission is to seize and hold ground. An American military journal stated this quite simply in a recent heading that proclaimed: *"Nuclear weapons destroy, naval forces can blockade, but only ground forces can retain or establish political power."* Modern armies use their tanks to break through defences supported by artillery and airborne weapons systems but it is their infantry which attacks and holds territory in battle.

What has changed is the battlefield environment of the soldier as it is altered by advances in technology and weapons development, and in the greatly increased mobility of the soldier. These factors have initiated changes in doctrine, tactics and deployment of the ground forces as well as their equipment and supply. The most significant change in tactics and equipment has already been mentioned concerning the development and use of the APC/IFV and the use of helicopters to transport and support the infantry soldier.

Infantry weapons themselves primarily consist of small arms – rifles, pistols, bayonets, grenades, and crew-served weapons such as machine-guns, cannon, and ATGMs as well as the larger recoilless rifles (rocket/missiles). Basically, small arms are anti-personnel weapons, although they can be used against helicopters and other aircraft, AFVs and lightly armoured vehicles. At present, it is not possible to predict what the effect of improved small arms may be in some future conflict. However, with the increase in the quantity and quality of tank guns and anti-tank weapons against armoured vehicles, it is likely that armoured attacks will be interdicted more easily. When this occurs, the role of the infantry will become mainly to advance and clear the way for a counterattack. During this type of action it can be anticipated that high rates of ammunition expenditure will occur in violent firefights. Standard infantry tactics determine that

**Above: US Army squad weapons for the late 1980s: M16A2 rifle/M203 grenade launcher (left rear); M16A2 rifle (right rear); M249 squad automatic weapon (front). All use NATO 5.56mm standard ammunition.**

**Below: The British Army's L70 Individual Weapon (IW) also uses the NATO 5.56mm with the Belgian SS109 round. The buttless "Bull-pup" design results in a neat, very accurate and effective rifle.**

**Remaining velocities**
The diagram below shows the remaining velocities of three rounds: the NATO standard SS77 7.62mm, the US M193 5.56mm and the new NATO standard 5.56mm SS109. SS77's superiority at ranges out to 437 yards (400m) is clear, but its remaining *kinetic energy* is always greater than that of the US M193, due to its greater mass.

**The NATO competition**
NATO agreed that the 7.62mm round (left) was too large and led to rifles that were heavy and difficult to control. In the 1960s the US had switched to the Armalite 0.223in (5.56mm) round (centre) but NATO decided that, although 5.56mm calibre was best, the actual round should be the Belgian SS109 (right).

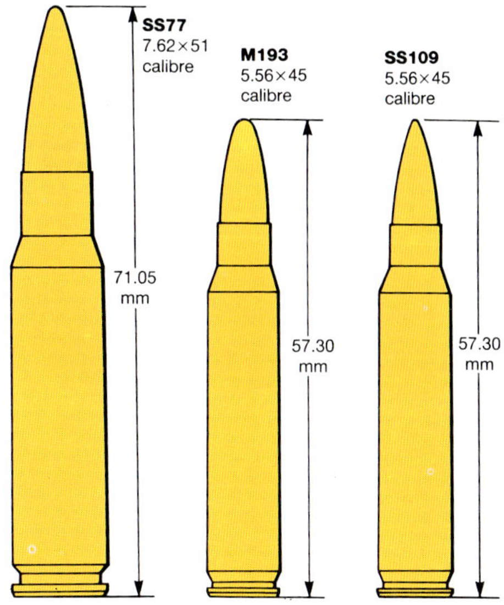


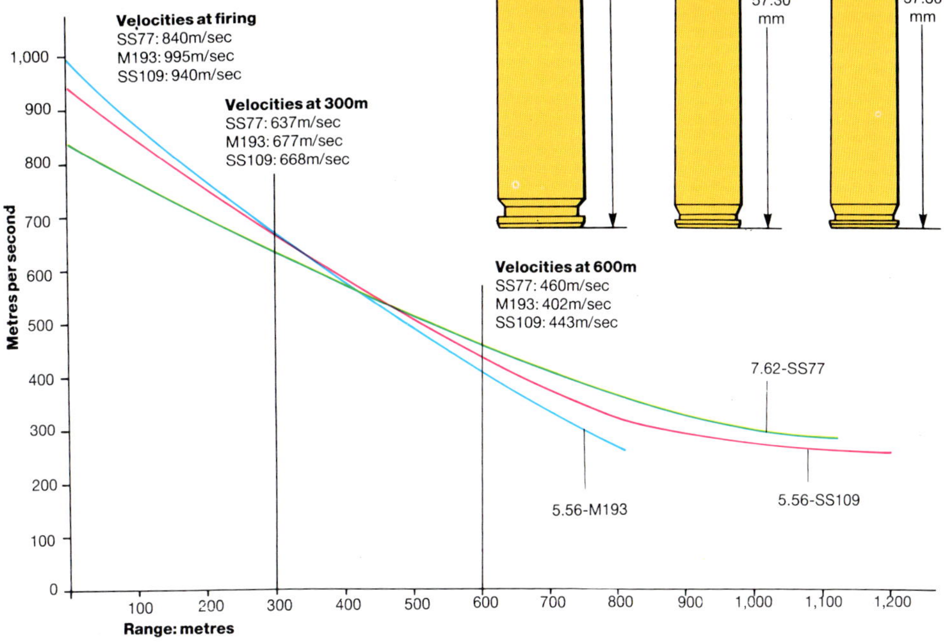


**PERFORMANCES OF SS109 BULLET**

| Target | Range of penetration |
|---|---|
| 8 soft steel plates, thickness 1mm located at 25mm one from another | 50m |
| 1 soft steel plate, thickness 10mm | 240m |
| 10 pine boards, thickness 254mm located at 25mm one from another | 50mm |
| Armour constituted by a plate, thickness 6mm hardness 450 HB | 25m |
| Armour constituted by a plate, thickness 5mm hardness 450 HB + 2 soft steel plates thickness 3.5mm | 50m |
| Lethal range of the ammunition: 1650m | |

**New NATO Standard: Belgian 5.56mm SS109**
This is dimensionally identical to the US M193, but the difference lies in the design and weight of the projectile. The brass jacket surrounds a lead core but, having a steel nose which increases the perforating power and, due to its higher mass, SS109 has a higher kinetic energy at all ranges than M193.

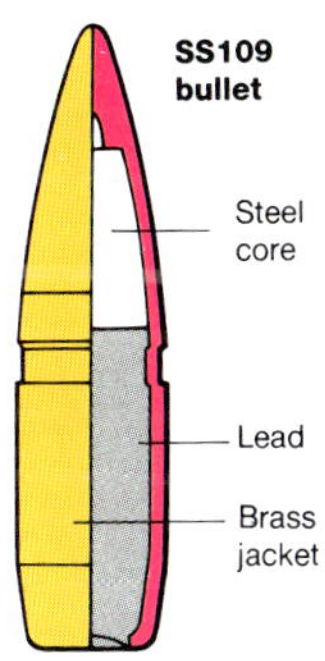


**COMPARATIVE DATA**

| Cartridge | 7.62-SS77 | 5.56-M193 | 5.56-SS109 |
|---|---|---|---|
| Calibre×case length (mm) | 7.62×51 | 5.56×45 | 5.56×45 |
| Mass of cartridge (g) | 24 | 12 | 12.3 |
| Mass of the powder W (g) | 2.9 | 1.65 | 1.63 |
| Bullet length | 28.8 | 18.8 | 23 |
| Mass of the bullet C (g) | 9.3 | 3.565 | 4 |
| Length of the test barrel mm | 558.8 | 508 | 508 |
| Muzzle velocity Vo (m/s) | 835 | 995 | 947.5 |
| **Time of flight** | | | |
| To 300m (s) | 0.4 | 0.366 | 0.376 |
| To 600m (s) | 0.97 | 0.94 | 0.933 |
| To 1000m (s) | 2.09 | | 2.16 |

most small arms rounds are fired in a suppression or enemy neutralisation role, and to give cover for the movement of individual soldiers and to prevail in any firefight.

The newer military assault rifles are all highly effective at the ranges expected to be encountered in combat, namely from 330 to 440 yards (300 to 400m). These ranges were determined after detailed studies of combat since World War II. The NATO nations have generally concluded that the 5.56mm round has sufficient range and accuracy to recommend its adoption over the 7.62mm round, since it decreases the weight of the ammunition, the weapon, and the training time for the soldier. Some proponents of the older 7.62mm round argue that the lethal power of the heavier round is greater, but these arguments are similar to those which took place when the British Army went to the 7.62mm from the .303 round and when the US Army shifted from the .30 calibre round. Most soldiers accept the 5.56 and are content with a rifle that weighs less than half that of the older models, uses a magazine with 20 to 30 rounds and, on full-automatic, can fire 30 rounds in less than 3 seconds. The Soviet forces have also opted for a lighter round, going from 7.62mm in the AK-47 to 5.45mm in the AK-74/AKS-74.

There has been a most interesting application of advanced technology in sniper rifles. Since the new night vision equipment and improved optics have become available, the sniper can work at night as well as in daylight, a factor which caused the British land forces some problems during the campaign on the Falklands in 1982, as Argentinian snipers were particularly effective at night. Both NATO and the Warsaw Pact forces train and employ snipers. The US have equipped their snipers with a 7.62mm rifle which is an adaptation of the M14, accurised with a telescopic sight and three ranges of ammunition: long range, medium range and subsonic, which is used with a suppressor. The British forces use their 7.62mm L42A1 rifle and the Soviets used a semi-automatic rifle (the only nation to do so), the 7.62mm Dragunov (SVD). It is likely that improvements in sighting equipment and special ammunition will

**Below: One of a number of weapons systems designed for the 5.56mm SS109 round is the Italian Beretta 70/.223. The AR70 assault rifle, shown here, has only 85 parts, which is at least 20 per cent fewer than in any similar model.**

**Above: Turkish soldier firing a Heckler & Koch-designed G3 7.62mm rifle, manufactured locally. Many countries are now able to undertake their own small arms and ammunition production.**

represent the next techological advance in sniper rifles.

A most interesting technical innovation is the prospect of caseless ammunition for the rifle. During the middle of the 19th century as breechloading rifles became common, the cartridge case was adopted because it enabled the breech of the weapon to be well-sealed at the time the rifle was fired. The ejected cartridge case also removed unwanted heat from the breech. A return to caseless ammunition would save not only the cost of the case, but lighten the weight the soldier must carry. Caseless rounds for modern small arms have been slow to evolve because of the difficulties in obtaining a good seal, protecting the propellant from moisture and an undesired ignition of the ammunition, or "cook-off" when the round is chambered.

Recently, the West German manufacturer, Heckler & Koch, have developed a new rifle, their G-11, which fires a 4.75mm caseless round. This round is about half the weight and almost half the size of the 5.56mm round, so it is not difficult to see that caseless ammunition offers some interesting possibilities for the infantryman and is expected to appear in production quantities sometime in the near future.

The bayonet should be mentioned briefly, although it may seem odd to include this weapon in a discussion of advanced technology – it is not considered to be a serious weapon by most modern military forces but was used to good effect in close-quarter fighting in both Vietnam and the Falklands. But recent changes to this traditional weapon have been made by the Soviets, whose bayonet is designed to be used together with its sheath as a wirecutter, for digging and for sawing. Its features enable the soldier to use it as a more or less all-purpose tool (which is what soldiers have been using bayonets for since World War I).

Thus the bayonet is mentioned, not as an example of advanced technology, but rather to illustrate *adaptive* technology from the user's viewpoint. It is a similar principle to that which inspired the slot in the new and very modern Israeli Galil rifle. The Israelis had observed that their UZI submachine-guns were frequently damaged by soldiers who were using the breech as a bottle-opener for crown-type bottles. When the Galil was designed, a "bottle-opener" was built in!

Another individual weapon of a traditional sort is the hand grenade. Often referred to as "personal artillery" by infantrymen, the grenade performs well at the cinema, but less well in combat. Only about half the individuals in a target area, a radius of about 6ft (2m) from the exploding grenade are likely to be injured, and less than 10 per cent of the wounds inflicted are likely to be fatal. With a normal thrown range of 130 to 150ft (40 to 45m), the weight and hazards of employment do not make the grenade a particularly useful weapon, except where it may be anticipated that several of the enemy are grouped close together in a room, trench, cave or similar locale.

However, the grenade has been improved by a modest technological development. The US M26 and British L2A1 and similar grenades have been improved by using notched wire coiled around the inside of a fragmenting thin casing. The explosive blows the notched wire into 1,250 fragments, making it a reasonably effective anti-personnel weapon.

The other items of the infantryman's arsenal – pistols, mortars, machine-guns and submachine-guns – have mainly changed in response to improved manufacturing techniques, or changes made possible by adoption of improved ammunition. Application of modern technology has mainly given the ground soldier an improved anti-tank capability, improved mobility and lighter, more portable weapons.

However, advanced technology has provided major improvements in items of personal equipment for soldiers today and tomorrow. The fathers and grandfathers of the modern infantryman, veterans of World War II, Korea, Malaya, Algeria, Vietnam and the Congo, all used pretty much the same kinds of equipment, little

**Right: Advanced Combat Rifle (ACR) is AAI's contender for the next US Army rifle. It fires two types of caseless round (one 5.56mm and the other a sabot 5.56/4.32mm design) in single shot, three-round burst or automatic modes.**

**Above: Infantry weapons, especially small arms and equipment, must be able to function with total reliability in all types of climate, from the humid jungle and dry heat of the desert to the Arctic cold.**

changed from World War I. The boots, rainwear, packs, ammunition belts and other personal kit looked quite similar and any veteran could describe the best way to polish boots, clean webbed items, shine brass and maintain kit. The young soldier of today would find it difficult to understand all of his Dad's descriptions. Modern technology has given many of today's foot soldiers new boots that are very water resistant, proof against penetration from spikes and other objects, and with metal hook eyelets for quick removal, replacing the traditional lacing and eyelets. The new nylon and Cordura materials for packs and pouches are tougher and lighter than the old canvas webbing and the old brass, steel or gunmetal buckles and snaps have been replaced by light alloys, plastic or patented Velcro fasteners which are lighter, cheaper and do not corrode or require any polishing and – most important – can be opened or removed while wearing protective clothing or gloves.

The newer technologies are now providing improved, lightweight body armour to protect the individual soldier, new microclimate cooling systems for wear in hot climates, and new helmets which contain communication systems and improved clothing for cold weather. Today's load-carrying equipment is more fundamentally functional, more adaptable and lighter than it has ever been. The latest types of Gore-tex raincoats and capes allow condensation from body heat to escape, but keep out rain (something that generations of military raincoats seemed unable to do). The newest helmets (like the recently adopted US Kevlar models) are extremely lightweight and comfortable to wear, give excellent protection – but are less convenient for use as a cooking pot or washbasin.

The new night vision devices, which are discussed in more detail elsewhere, permit soldiers to function at equivalent levels during day and night operations, and are sensitive enough to see through fog, haze and current ranges of artificial smoke, although recent research suggests that it may be possible to produce a smoke which such devices might not be able to penetrate, a counter-countermeasure!

Finally, modern technology has contributed greatly to improvements in the soldier's nourishment. Until quite recently, soldiers in the field had to eat from tinned rations, and make do with various types of preserved foods, poorly prepared field rations or local foraging. Generally, the Western armies have been more concerned with feeding their troops than those in the Warsaw Pact. For example, most Western armies have field kitchens at the company level; the East European forces (most particularly the Soviets) have them only at the regimental level. Whereas preserved foods are still on the menu, in the West at least the most modern "freeze-dried" and irradiated foods make it possible to feed the soldiers nourishing and palatable meals while in combat, and the new rations are light in weight and have exellent keeping properties.

**Left: US infantryman in the latest combat uniform and equipment. The application of technology has led to greater efficiency and effectiveness, but nothing is conceptually different from that worn or carried by infantrymen in World War I.**

Mention of airborne forces tends to conjure up images of the films of World War II depicting skies darkened by troop-carrying aircraft dropping thousands of paratroopers on to the battlefield. Improvements in electronic warfare, air-defense capabilities and the new command, control and communications systems have all eroded the ability of the parachute troops to perform their earlier missions of major shock force applied at strategic or critical points by a surprise vertical envelopment. Some military scholars have gone so far as to question whether any future role for the airborne soldier exists at all, and to suggest that the air-landed and heliborne troops have only a limited and special role as they function closely together with their comrades on the ground. The scholars include in their reasoning such recent cases as the British campaign in the Falkland Islands and the American action in Grenada, and also the possibility of power projection expeditions in the Third World.

Nonetheless, advanced technology has contributed to improvements in airborne forces in the same ways it has done for the conventional ground soldiers: in improved communications, weapons and means of transport. The new communications and position location equipment have helped to overcome the paratrooper's long-standing problem of becoming isolated from the main forces and from other airborne elements. There have been some notable improvements in airlift equipment, aircraft, and in the parachutes used for personnel and equipment, particularly with regard to more accurate delivery, packing and extraction. Parachute equipment improvements have mainly tended to benefit the special forces such as the British Special Air Service, US Special Forces, and Soviet Spetsnaz troops. The highly advanced steerable parachutes and other exotic equipment are very useful for the selected and highly trained special units, but parachute battalion commanders of the more traditional units do not particularly like the idea of mass drops involving a thousand or more soldiers steering themselves around the sky and possibly off the selected drop zone.

What kind of future roles will there be for the traditional parachute soldier? Most military leaders believe there will continue to be a need to have elite, highly trained and capable soldiers who can be para-dropped and used in situations similar to those in which they have been successful in the past, as well as in situations which might be brought about by the possible use of tactical nuclear weapons. The US still keeps its 82nd Airborne Division, the British, French and Israelis their Parachute Regiments as do other nations. The Soviets maintain seven Airborne Divisions with about 50,000 men. While wartime employment of airborne forces may be more limited than in the past, it is unlikely that these elite formations are kept on hand only for parades.

**Left: A mass of US paratroops. Airborne troops are independently minded and highly trained infantry. Considered the elite in most armies, there are still doubts on their role in general war.**

**Above right: US Army UH-60 helicopter lands air-mobile troopers. Helicopters have given great mobility to infantry, without needing the special training and equipment as for parachute operations.**

**Below right: BMD airborne fire support vehicle, one of a range of fighting vehicles specially developed for the very large, well equipped and efficient Soviet parachute forces.**

**Below: Canadian paratroops landing in the Arctic. There are many situations in limited war where parachute insertion may be the quickest and most effective way of getting forces directly into battle zones.**

When the day of the electronic battlefield truly arrived, it can be said that the artillery were leading the way with the use of improved communications, calculators, observation means, survey capabilities and so on which were provided by the increased use of electronics. In fact it was the employment of artillery that extended the lethality and the destructiveness of the battle, from the 18th century onwards. During World War II, almost 60 per cent of all battle casualties were caused by artillery fire. The big guns caused from 50 per cent of casualties in urban and forested areas, and up to 75 per cent in open and desert terrain. Any combat veteran will rank artillery as the enemy weapon he fears the most.

Major technological advances in mobility and munitions have occurred in recent years. For example, artillery is increasingly self-propelled in order to keep up with the tanks and motorised infantry. Even helicopter gunships deployed with modern armies could be considered as improved non-divisional artillery.

Today's artilleryman has a wide range of nuclear and conventional munitions. In addition to improvements in the standard High Explosive (HE) rounds and smoke and other chemical shells, artillery can fire the new Beehive rounds (a shell resembling a shotgun shell with thousands of small darts) and an increasing number of improved conventional munitions (ICMs), some of which have rocket boosters that increase ranges by more than 50 per cent.

## Improved munitions

New guidance systems exist which home in on point targets such as enemy fixed positions, tanks and personnel carriers. The US Copperhead and similar systems are designed to be used against high-value targets. These systems use the reflection of a laser beam projected on to the target by a forward observer. The projectile is fired into the general vicinity of the target and, when it nears the target area, searches for the laser reflection. An extension of this type of system enables other weapons to use laser designators; the US developed GLDD (Ground Laser Locater Designator) can be used with systems such as Copperhead, Hellfire and Maverick.

Newer systems, some of which are now fielded and others which are in development, employ independent terminal guidance procedures such as Radar Area Correlation which, as it nears the area of the target, enables the projectile to compare its live radar reflection from the target area with reference scenes stored prior to launch. The projectile then makes course adjustments based on the comparisons, producing virtually pinpoint accuracy, and allowing use of smaller nuclear warheads or conventional munitions. Pershing II, for example, uses this type of guidance system. The newer Precision Guided Munitions (PGMs) will allow for very accurate placement of sub-munitions which can project over a wide area mines and other incendiaries and a whole new inventory of poison gases and illuminators. They also eliminate the dependency upon observers and laser reflections which can be obscured by weather, terrain, smoke or chemicals. With the increased range and improved munitions, modern artillery has been made significantly more effective because of the contribution of modern electronics, in particular communications and computers.

Artillery employment is classified as *direct* or *indirect*. Direct fire is simply when the gunner fires at a target he can see, while indirect fire is upon targets which are concealed from the direct observation of the gunner. Since the early days of World War I, most artillery fire has been indirect.

Indirect fire is customarily over ranges up to 25 or 30 miles (40 or 50km), while the range of direct fire of artillery, tank guns, anti-tank guns and smaller cannon is usually no more than 2,180 to 3,280 yards (2,000 to 3,000m). Current indirect fire weapons are expected to hit a target area of about 330ft (100m) in diameter from 15

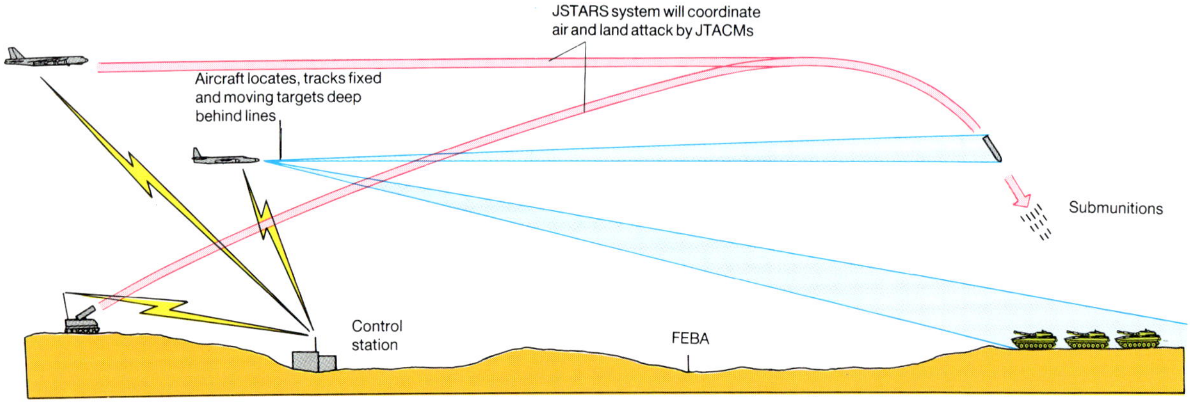


**Joint tactical system**
The US Army/USAF Joint Tactical Missile System (JTACMS) tried to tie all elements in the "deep battle" together; it collapsed in mid-1983, but is shown to emphasise how some form of integration is essential. TR-1s will be used to obtain deep surveillance information and will pass this in real-time to battle headquarters, supporting B-52s and ground missile units.

**Left: US M198 is one of many 155mm towed howitzers now in production. Technology can still produce greater range and more effective shells for lighter guns.**

**Above: US Ground/Vehicular Laser Locator Designator (GVLLD) is one of many devices now available to increase the effectiveness of artillery and air support.**

**Above right: West Germany, Italy and UK cooperated in the very successful FH70 155mm towed howitzer programme, which has resulted in an outstanding weapon.**

**Right: The US Multiple-Launch Rocket System (MLRS) is a free-flight rocket weapon offering rapid-response, heavy volume, long-range fire support.**

miles (25km). In order to site and aim the guns, accurate surveys and mathematical calculations are necessary. These actions which traditionally were accomplished manually are now almost exclusively done by electronic computers. The essential element of modern artillery systems employment is communications. Without effective communications and modern electronics the gunner might just as well go back to the direct fire of the Battle of Waterloo. (Artillery in direct fire is normally reckoned to be at least twice or more times as effective as indirect fire, but the loss rate to the artillery is also increased by about ten times!) Western armies are increasingly employing the latest in electronic technology and improved munitions; the Soviets have used about 50 per cent of their artillery in direct fire missions and although they may have to have some concern about the losses, they are obviously less worried about the problem of vulnerability of associated communications to electronic warfare.

Rockets have a special place in modern artillery. The Warsaw Pact forces rely considerably upon their rocket weapons, the BM-21 and derivatives in particular. It is a truck-mounted multiple launcher with 40 122mm tubes. A battery of BM-21s can lay 30,000 rounds of high explosive on a target in 20 seconds, a feat which would require 320 155mm artillery rounds.

The new multiple-launched rocket system (MLRS) recently developed together by the US, UK, Federal Republic of Germany and France will be deployed in the immediate future. This free-flight, area fire system has a primary mission for counterfire and enemy air defence suppression and carries improved conventional submunitions as well as a scatterable mine warhead. This system has the potential to add a terminal guidance warhead (to defeat armour) and a binary warhead for chemical weapons.

The main disadvantages of rockets are their relatively poor accuracy when compared to the tube guns, as well as rapid range adjustment. New technology is overcoming both disadvantages and it is likely that, with the improved precision guided munitions (PGMs) becoming more available, rockets may replace most of the heavy guns now in use, although logistic support will probably remain a problem for some time.

The next stage of development for rocket-artillery weapons will be the Cannon Launched Guided Projectiles (CLGP), already in their early production stage. The current version, the US designed Copperhead, is launched from a 155mm tube and then guided on to its target by an observer who uses a laser designator. While future versions of CLGP will probably use micro-electronic circuits to operate individual seeker heads, the Copperhead must remain in sight of its laser indicator in terminal mode – obscurants can defeat this capability. Its proponents point out how successful it has been in the Middle East (except perhaps in sandstorms), but fog and rain are rare in the desert!

Still in research and development is another US concept, the SADARM (Seek and Destroy Armor) which has submunitions being ejected from their carrier warheads over the target area in which they seek and destroy their own individual targets. When this concept reaches the production stage and deployment, it could revolutionize the armoured battle of the future.

A related and similar weapon is the STAFF (Smart, target activated fire and forget) 155mm round. This employs an antenna in its nose which measures the radio waves reflected from the metal of the target. STAFF is fired in the general direction of the target and the projectile is then guided by an onboard computer. Like Copperhead, STAFF was designed specifically for the Central European battlefield in which NATO forces face large numbers of Warsaw Pact tank formations.

Some artillery is capable of firing nuclear as well as conventional rounds, a word concerning battlefield nuclear weapons, which are generally short-range missiles (or air delivered bombs). The missiles, such as the US Lance and Pershing 1 (now being replaced by the Pershing 2) have guidance systems, like the Soviet Scud B. Close support is provided for NATO forces by 155mm and 8in guns which can fire nuclear rounds – as can many of the Soviet weapons. Nuclear weapons are discussed in greater detail later in this chapter.

The fundamental mission of air defence is to protect installations and ground forces from direct air attack. To accomplish the mission, air defences are normally layered with radar-directed SAMs providing battlefield air defence against low-level attack from helicopters and attack aircraft. Tactical SAMs are employed against fighter-bombers and medium and high altitude offensive missions. An enemy air force must carry out defence suppression measures most effectively. The increase in capabilities of ground air defence systems has caused air forces to engage in more complicated and considerably more costly responses, and thus diverts them from their more primary offensive and ground support missions.

The best evidence of the effectiveness of modern air defences can be seen in the Arab-Israeli Wars. Prior to the 1973 War, the Israeli Air Force was conceded to have air supremacy in the Middle East. In the 1973 Yom Kippur War, the Egyptians manned a comprehensive Soviet designed air defence system and were able to degrade the IAF's effectiveness until the air defence cover over the Suez Canal and Sinai Desert were disintegrated in violent ground battles.

The Warsaw Pact has a dense air defence network made up of long-range search radar units, jamming units and both fixed-site and mobile defences. Mobile defences move with the first-echelon ground forces, the infantry and armour being protected from low-level attacks with the hand-carried SA-7 and the mobile ZSU-23-4 (23mm quad-barrel anti-aircraft guns). Behind these are the SA-6 and SA-8 low-altitude SAMs and the SA-4 mobile high-altitude SAMs. The Soviet air defence philosophy has been to compensate in technological lag with large numbers – Warsaw Pact air defences are effective because of their enormous volume.

However, more recently introduced Soviet systems are causing concern to NATO air staffs because of their technological sophistication. Even small WarPac units, for instance, can be armed with SA-9 Gaskins which have a 2-5 mile (3-8km) range and use an infra-red seeker, while the SA-10, introduced in 1978, is said to have characteristics similar to those of the latest US Patriot, with a range of 40 miles (64km), and an effective altitude of from about 1,000ft up to 100,000ft (300-30,000m). Such a system would be extremely effective against cruise missiles and low-flying strike aircraft such as the F-111 and Tornado.

Other new Soviet air defence weapons include the vehicle-mounted SA-11 and -13, the former using radar guidance, the latter infra-red homing and supposedly comparable to the American Chapparal. A new man-portable system is SA-14, which is reportedly designed to be comparable to the US Stinger and British Blowpipe.

NATO's air defence systems are thought to be more technologically advanced, but they are smaller in numbers. The NATO SAMs have better guidance systems and a higher kill ratio, and are also designed to be more resistant to ECM. The NATO hand-carried weapon, the US-designed Redeye,

**Above: US Patriot tactical air defence missile system makes extensive use of automation; during firing, only the Engagement Control Station is manned. A Patriot battalion comprises 765 men compared with 878 for Improved-Hawk.**

**Below: Soviet SA-10 air defence missile battery, comprising two launchers, two radar vehicles, a resupply vehicle and a command post. The Soviets have poured vast resources into air defence, and possess a great range of missile systems.**

**Oerlikon GDF-D03**
As the critical air threat has become fully apparent many air defence weapons have been developed. Once missiles seemed the full answer, but the tendency now is towards a gun/missile mix. Sophisticated gun systems are available, such as this GDF-D03. The drum magazines hold 430 rounds (including those in the feed system). Effective range is 4,374yds (4,000m). The 35mm guns have good anti-tank capability, with sub-calibre APDS-T rounds. The command/driving cab moves up and down.

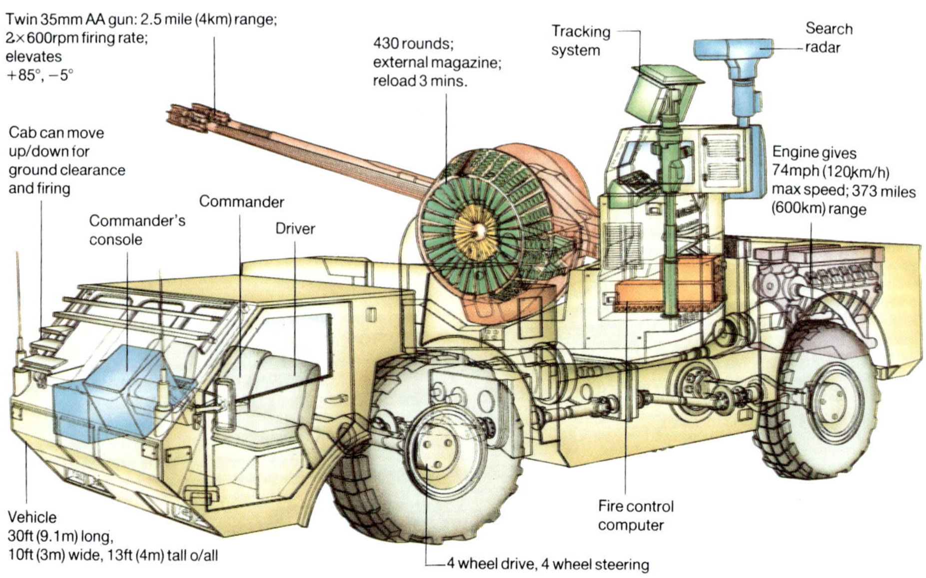

**The Patriot system**
Heart of the Patriot system fire unit is the AN/MPQ-53 radar. This combines in a single set the target search, detection, track and identification functions, as well as missile tracking and guidance, *and* ECCM! Autonomous in operation, it is remotely controlled by the Engagement Control Station computer. The Battery Engagement Control Station monitors readiness, allocates priorities to the radar, determines ECCM activity, computes guidance commands, and exchanges information automatically with battalion HQ. The Engagement Control Station provides the fire unit human interface, and is, in fact, the only manned element in the entire fire unit. The extensive automation of the system enables the human element in this battery command post to devote their attention to judgemental decision-making, which can include the final authorization for launch, where necessary. Also included in the fire unit is a twin 150kW power plant.

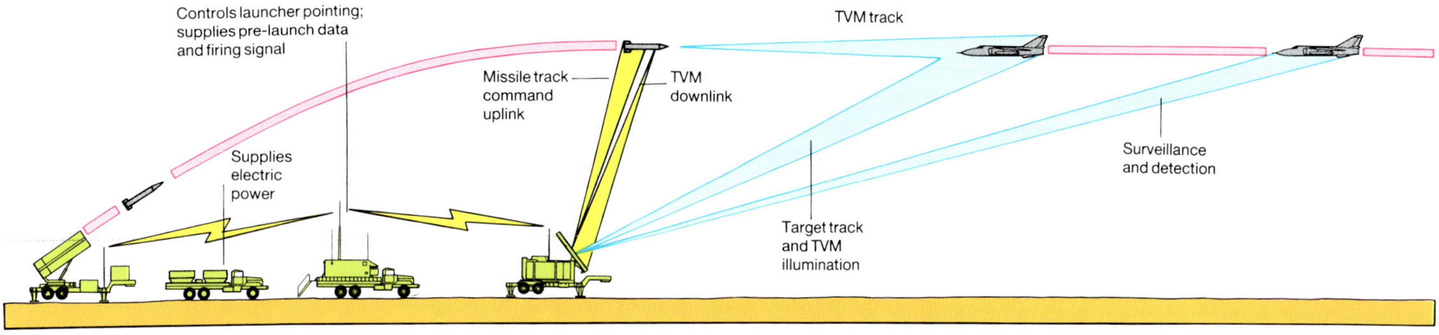

**Left: Following their success with the Seacat and Blowpipe missile systems, the British firm of Shorts have now produced Javelin, a very high velocity, shoulder-launched, air defence missile, using SACLOS guidance for head-on attack.**

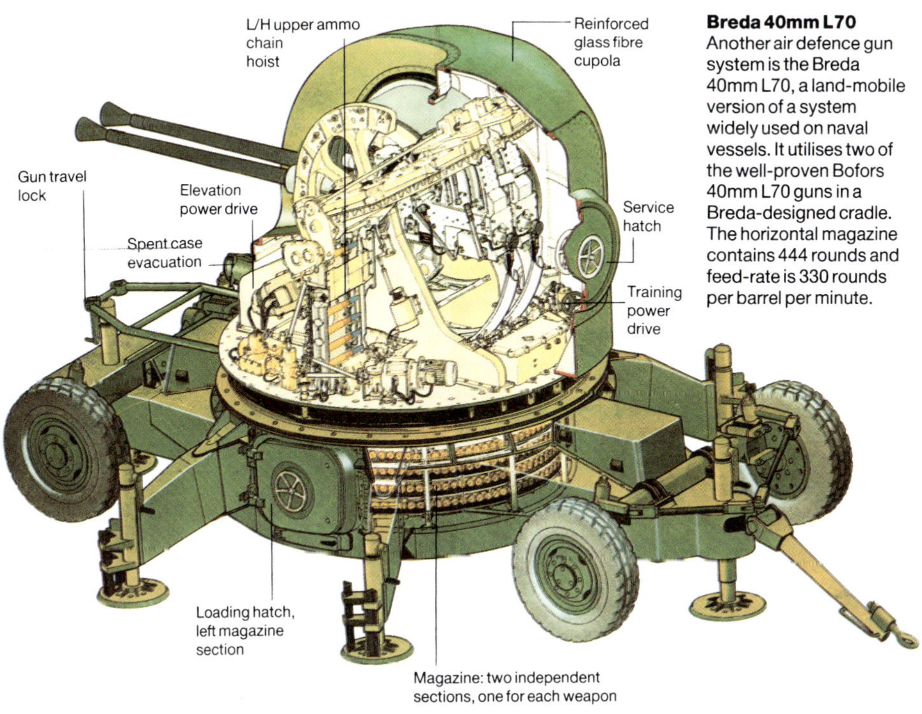


**Breda 40mm L70**
Another air defence gun system is the Breda 40mm L70, a land-mobile version of a system widely used on naval vessels. It utilises two of the well-proven Bofors 40mm L70 guns in a Breda-designed cradle. The horizontal magazine contains 444 rounds and feed-rate is 330 rounds per barrel per minute.

is now being replaced by the American Stinger and the very capable British Blowpipe, while the US Vulcan (a daylight, fair-weather system) is being replaced by the newer Division Air Defence System (DIVAD).

The Chaparral forward-area, low-altitude, mobile SAM has been operated in the field together with Vulcan. Originally, Chaparral was to be replaced by Roland 9, a low-level, all-weather SAM, but for the moment Chaparral is being improved and retained in service.

The French Crotale and British Rapier are other low-level mobile systems now in use. The new US-designed Patriot is scheduled to replace the low-to-medium altitude Hawk in the near future, it being designed to counter high-speed aircraft and missiles at all altitudes as well as to be resistant to jamming and other ECM.

When comparing the Soviet/Warsaw Pact air defences with those of the NATO forces, it is important to bear in mind that for a considerable period of time the NATO air forces were assumed capable of maintaining air superiority in Europe. However, while the NATO forces were improving their technological capabilities, the costs of those improvements meant there was less money to go toward the acquisition of aircraft, including the development of newer types – so the quantity of the forces did not increase incrementally. At the same time, the Soviets and their allies were greatly improving their counter-air capabilities (with, for example, ground-based air defences using radar-directed gunfire and guided missiles), while also improving their air forces from a fundamentally defensive role to one which could support conventional attack forces.

NATO's air forces will not therefore be able to carry out ground-attack missions without first defeating heavy ground-based air defences, while combating enemy air forces at the same time – a difficult task, as described in the chapter on Air Warfare.

With the advent of tactical nuclear weapons, the prospect of their use has permeated every aspect of military planning in all of the major nations. Military leaders and their staffs have envisaged that the employment of tactical nuclear weapons could accomplish the total destruction of large hostile ground forces elements, their command centres, logistics bases, supply depots and airfields as well. But these leaders have recognized that if these weapons should be used, their ability to accomplish rapidly what were, in earlier times, difficult and long-term missions, might also be accompanied by numerous problems – all of which would counterbalance possible military advantages.

The terrain and weather establish the basic physical setting for the land battle. Nuclear and chemical weapons, electronic warfare and obscurants such as smoke all compound the influence of terrain and weather.

The immediate effects of nuclear weapons are blast, thermal radiation, electromagnetic pulse (EMP) and initial nuclear radiation. These can all cause major personnel and material losses, and change the tempo and direction of the battle. Nuclear weapons have been used only twice in anger since their development – at Hiroshima and Nagasaki, Japan, in the closing days of World War II. While we have some knowledge of the effects of nuclear weapons on cities and their populations, these weapons have never been used against troops in battle. Tests and assessments give some indications of the effects on equipment, but no experience on the effects upon personnel.

**Above: Soviet Army SS-12 tactical nuclear missile is prepared for launching. Soviet Army formations possess many tactical nuclear weapons, but high-level political control is similar to that for NATO.**

**Below: The US Army's Pershing II tactical nuclear missile is exceptionally accurate and became controversial as it can reach targets in the western USSR. It is now being deployed in significant numbers.**

Should nuclear weapons ever be used on some future battlefield, while they offer the possibility of accomplishing quickly and completely major military tasks, their effects would be expected to produce many accompanying problems. Most of the problems would be encountered in the ancillary effects of the weapons since nuclear explosions can interfere with the operation of other (friendly) systems, such as communications and nearby aircraft, and can cause blindness or precipitate fallout on the users' own and allied forces, besides contaminating areas with radiation and debris. Military staffs envision that extensive use of nuclear weapons could cause enough problems as to counterbalance their possible military advantages. These questions have preoccupied army staffs more than anything else since 1945.

All the effects of nuclear weapons are dependent upon the height of the burst above the ground. The point on the ground under the burst is called *Ground Zero.* If the burst touches the ground it is a ground burst and if it does not then it is an air burst. (There is of course, an underground burst, but this does not figure significantly in tactical employment.)

About half the energy projected by a

nuclear explosion appears as a blast, in similar fashion to a conventional blast of dynamite or TNT. The difference is that of size: a one kiloton nuclear device, for example, is not large in nuclear terms, but it is the equivalent of a thousand tons of TNT. Had the Nagasaki atomic bomb (about 22 kilotons, compared with today's large strategic weapons which are calculated in megatonnage) been constructed using TNT, more than two thousand ten-ton vehicles would have been required to transport the necessary explosives! Nuclear blast effects are similar to those of conventional explosives, but in addition to the blast there is a shock wave transmitted through the ground like an earthquake tremor and which affects structures and roadways similar to the effects of an earthquake.

The thermal effects are caused by the intense brightness of the flash of the explosion, and its heat – over 1,000,000°C. The effects of flash depend upon whether the explosion occurs at night or during daylight hours, its duration (which, in turn, depends on its power), the weather conditions, a well as the general terrain. The effects of heat and light on men and equipment outside the immediate area of the explosion are difficult to predict.

The effects of radiation are both immediate and residual. It is calculated that with the smaller devices (up to about 20kT airbursts) only about 15 per cent of the overall casualties would be caused by radiation. Doses of radiation, however, are cumulative and since it is anticipated that doeses in excess of 200 rads would cause incapacity about a week after exposure and involve many fatalities, a soldier who was exposed to 200 rads in one day and 300 in successive days would probably have received a fatal dose. (The "rad" is the unit of energy absorption, which applies to all types of nuclear radiations, including alpha and beta particles and neutrons, as well as gamma rays.) Residual radiation results mainly from fallout, from the dust cloud spreading from the explosion. Militarily, residual radiation can deny to an enemy either transit through or use of a contaminated area.

The major electromagnetic effects resulting from a nuclear explosion are the electromagnetic pulse (EMP), which can damage unprotected electronic equipment, and the reaction on the atmosphere and ionosphere through which communications take place. Essentially, much of the equipment would be degraded, causing electronic command, control and communications to be most difficult, or impossible. The effects are greater on the high frequencies (3 to 30MHZ) than on the very high frequencies and ultra high frequencies (above 30MHZ) used for communications.

Obviously, then, nuclear weapons are tremendous force multipliers – that is, a small device possesses enormous lethal power – although they are relatively expensive to produce compared with the cost of conventional weapons. The attractiveness of nuclear weapons to military planners was based largely on their belief that the capability of the nuclear weapon and a diminished need for large numbers of troops could contribute to satisfactory military conclusion on the battlefield occurring more quickly and at less cost than with conventional weapons.

As has been the case with other developments, military planners have had to change traditional philosophies to take into account nuclear effects. To incapacitate or destroy an armoured vehicle a conventional one-ton bomb must detonate within about 30ft (9m) of the target. Since the accuracy of aerial bombing and conventional artillery (prior to the development of the modern PGMs) was considerably less than 100 per cent, several bomb loads or a lengthy artillery barrage would be required to blunt an enemy tank regiment's attack employing between 50 and 100 tanks and fighting vehicles (a condition NATO planners envisioned occurring in a Warsaw Pact attack). However, a single 100kT nuclear weapon could do enough damage to armoured vehicles within an area of about 2,200 yards (2,010m) in diameter to put a modern regiment in dispersed formation out of action. This would be the case even if the point of detonation were over a third of a

**Above: Pershing II warhead during the terminal phase of its journey. It is capable of penetrating to some depth before exploding, giving it the ability to attack underground command bunkers.**

**Below: Latest in the apparently unending series of new Soviet nuclear missiles is the SS-23, a truck-mounted and highly mobile battlefield system.**

mile (0.5km) in error. In addition to the armoured vehicles, the nuclear device would cause serious destruction to unprotected buildings and individuals within about 5 miles (8km) radius. Of course, fallout would cause additional damage for a longer period of time, and over a greater area.

While military planners were making their calculations about conditions on the nuclear battlefield they had in mind Central or Western Europe where an area of a 5-mile (8km) radius would probably include two or three villages, towns or urban areas with several thousand citizens in each one. When the planners realised that nuclear weapons would be used to defend against a massive (and presumably successful) conventional attack, they had to calculate that using only a few small nuclear weapons would not be likely to blunt or defeat the attack without the presence of considerable friendly conventional forces.

The effects of nuclear weapons has had two pervasive influences on modern military doctrine, tactics and plans. The first occurred in the early years following the development of tactical nuclear weapons when there were many theories about how the nuclear battlefield would work, but little knowledge and absolutely no experience. Planners designed forces and demanded equipment to fit their theories. Historically, one of the fundamental principles of warfare has been that of *concentration*. This means simply that in addition to the logistical and supply considerations, troops must be brought together in order to attack the enemy effectively or decisively. On the nuclear battlefield, however, it was theorised that concentrations of troops, weapons, supplies, etc., presented lucrative nuclear targets to an enemy. Thus, doctrine and tactics were developed which emphasised dispersion, Troops and their support were to be widely dispersed and hidden and brought together only when they could close quickly with the enemy in such a way that the enemy could use nuclear weapons only by risking his own forces. This tactic, called the "hugging" tactic, has been adopted by all major military forces.

Planners, at the same time, demanded the development of more precise, more accurate and smaller weapons so that today, it is possible, quite literally, to drop a relatively small nuclear weapon "down the hatch" of a tank. However, since a conventional weapon dropped down the hatch could accomplish the same results,

**Above: French nuclear weapons systems are becoming increasingly sophisticated. This is the planned Hades battlefield system with a 200 mile (320km) range. Comparison with the Soviet SS-23 is interesting.**

**The "neutron" bomb**
This diagram compares some of the effects of a conventional nuclear weapon (left half) with those of an enhanced radiation weapon ("neutron" bomb, right). The nuclear warhead has a yield of 10 kilotons and the neutron bomb one of 1 kiloton, and both are air bursts at a height of 330ft (100m), having been delivered by missile (blue trajectory). The area affected by the conventional nuclear weapon has been desolated by both blast and heat (red). But, inside the tank (foreground) the crew have survived and are capable of carrying on fighting. In contrast, the damaged area caused by the neutron bomb (red, on the right) is rather smaller. Electromagnetic Pulse (EMP) effects have destroyed the electronic components in the radar vehicle (foreground) and in the overflying aircraft. Most important of all, the radiation (yellow) has been greatly intensified on the right, killing the crews in the tanks and in the aircraft. It is a misconception to believe that neutron bombs "kill people and protect property", since there would be little point in saving property about to be occupied by the

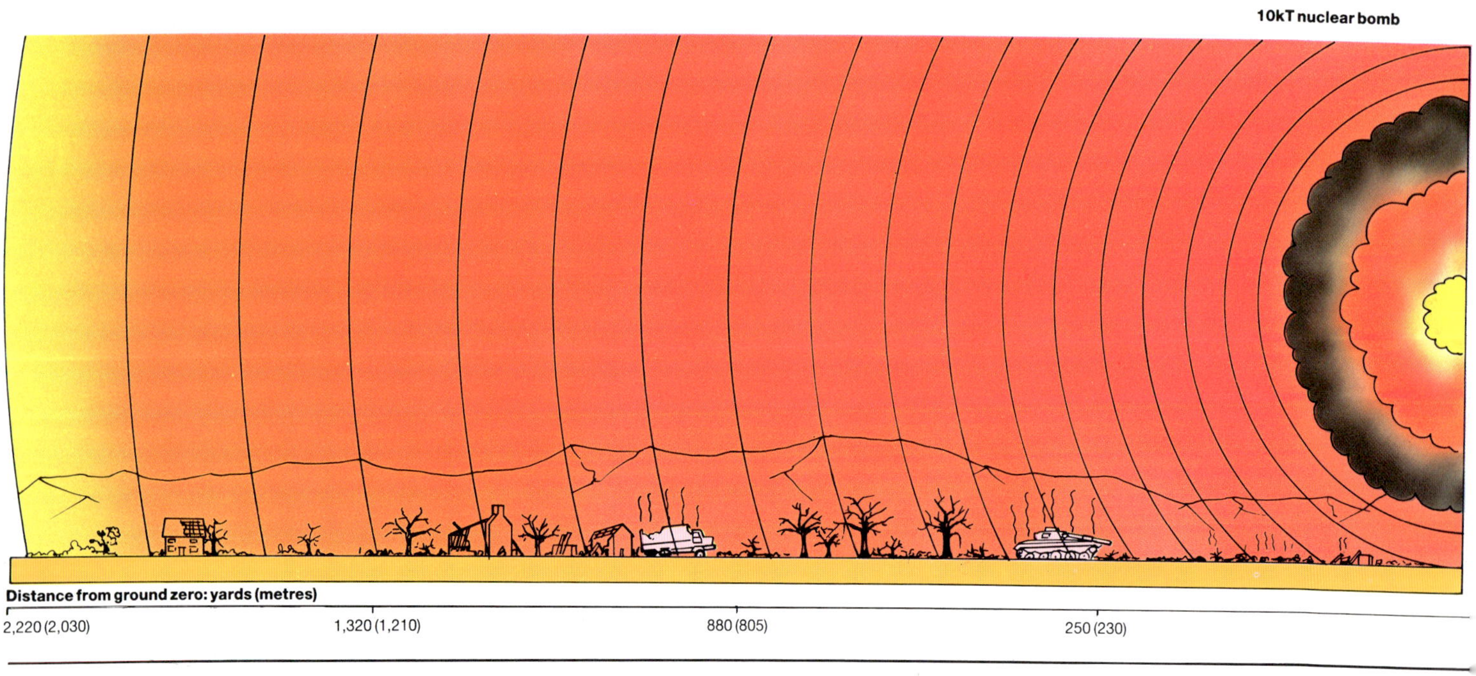

**Above: The US Ground-Launched Cruise Missile (GLCM) system is now in service in various NATO countries in Europe, despite intensive political campaigns in various countries to prevent deployment.**

military leaders do not see any advantage in using a nuclear device.

The problems of entering and moving on a nuclear battlefield combined together with all of the various theories of tactical nuclear weapons employment have led most contemporary military staff to conclude that any battlefield where nuclear weapons could be used at will militates against massive forces.

The idea of the nuclear battlefield may not be a practical one, but must be viewed only as a last, desperate measure taken in the face of impending defeat, but even then it has to be stated that Hitler, whose forces possessed some horrific nerve gases, did not use them even in the last desperate days leading to total defeat.

The current trend in battlefield nuclear weapons (since the late 1970s) has been in the development of smaller, cleaner weapons with more accurate and dual-capable delivery systems – systems capable of conventional as well as nuclear round delivery. Thus, it could be said that the most modern nuclear tactical weapons have come to resemble, in many ways, their conventional counterparts.

In the enhanced radiation weapon (the neutron bomb) some 80 per cent of the yield is released in the form of very high energy, deadly neutrons, which have a longer range than the neutrons released in standard fission weapons. It thus has a larger effective radius against personnel than normal fission weapons of the same yield, which kill primarily by blast and heat. In effect, it reduces physical damage and casualties resulting from blast but increases radiation damage. The weapon was developed by the United States, primarily as a response to the significant conventional threat posed by the large Soviet/Warsaw Pact armoured forces. A one kiloton neutron warhead would have the explosive equivalent of about one thousand tons of conventional explosives and would be expected to cause major destruction of the target area for about 1,640 yards (1,500m) in all directions from the point of explosion. The principal effect would be to release armour-penetrating radiation to kill or disable anyone within about 875 yards (800m). The major effect of the weapon is directed against the tank crew (and anyone else in the vicinity) rather than against the tank. Individuals within about 1,310 yards (1,200m) would receive radiation likely to be fatal within a few hours.

enemy! Rather, the purpose behind the neutron bomb is to kill immediately or disable the crews of tanks, aircraft and other manned weapons systems, which could not be destroyed by any other means except with individual counter-weapons. The NATO/WP numerical imbalance is such that NATO must find some new way of preventing defeat.

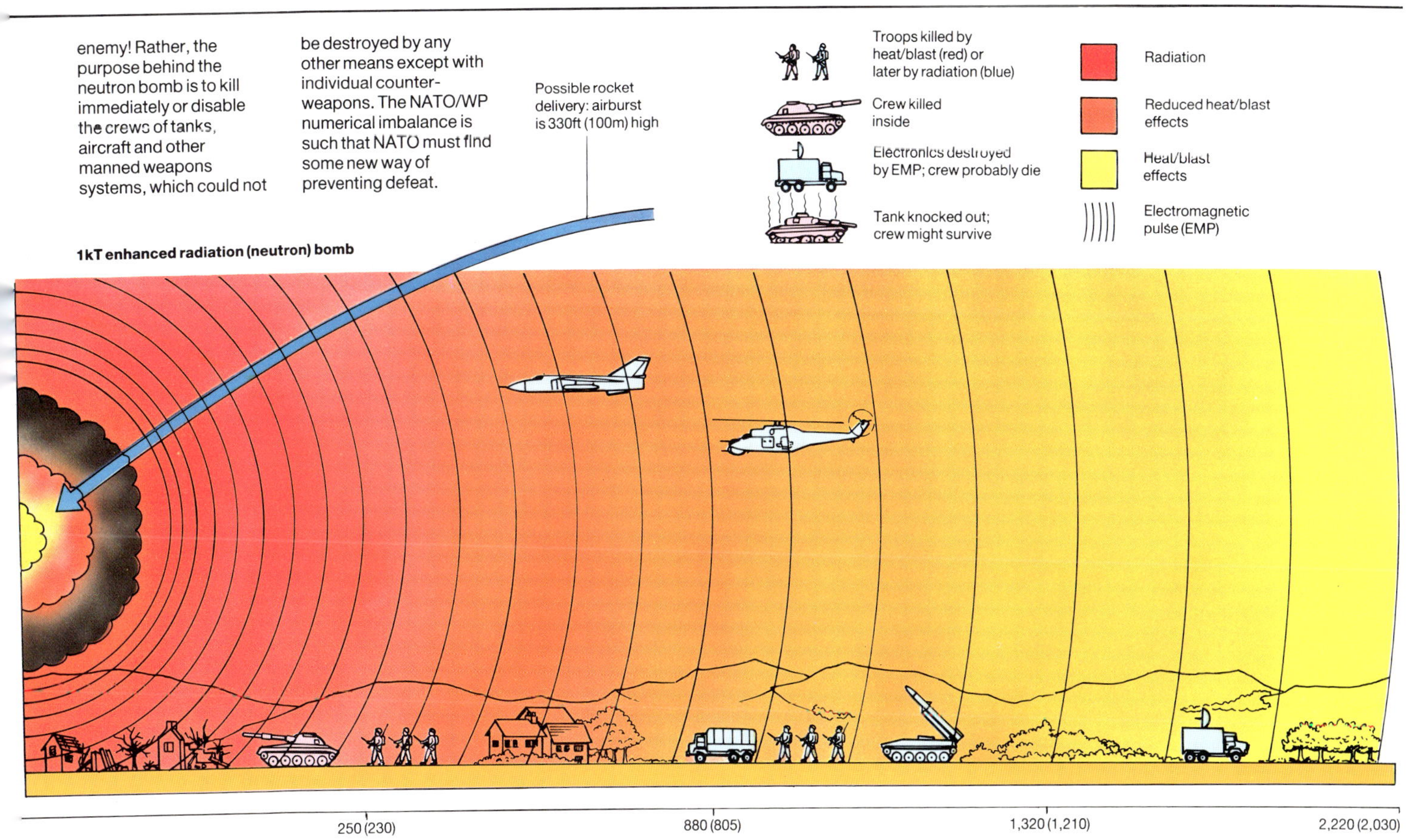

Chemical warfare (CW) is an emotive topic, but chemical weapons have been used on numerous occasions since the end of World War II, including the Indo-China war, the Egyptian campaign in the Yemen in the late 1950s and in the Soviet campaign in Afghanistan. In fact, in 1985 South Africans were reported to be using defoliants along their borders, while in 1983 and 1984 Soviet-supplied blister agents were used in the Iran/Iraq war. The terrible potential of CW has also been shown very clearly by the shocking consequences of the industrial accidents at Seveso in Italy and at the Union Carbide plant at Bhopal in India in 1984.

Modern chemical agents of military significance fall into five groups: nerve agents, incapacitating agents, toxins, psychological agents and defoliants. Older substances such as blister and choking agents are still found, although there is no known effort being devoted to further developments.

Contemporary nerve agents for use in general war are derivations of Tabun (GA), Sarin (GB) and Soman (GD), although these older agents may still be stocked in some countries. The Americans have a substance known as VX, but would like to manufacture a derivation of Sarin known as GB2. Earlier nerve agents were persistent, with consequential tactical problems, but the newer agents are non-persistent.

Incapacitating agents have a military use in anti-terrorist and anti-riot operations. These are normally disseminated in smoke form and have an immediate effect, but recovery is rapid. The main agents are CN (now little used by the military) and the widely used CS "gas".

Psychological agents have been tested by various armies. These derive their potency from psycho-chemicals such as lysergic acid (LSD). These hold out the possibility of so disorienting groups of soldiers that their positions could be taken over without a fight.

Toxins are chemical compounds derived from bacteriological organisms (eg, clostridium botulinum, or "botulin"). Some are fatal to humans, but do not induce epidemics, since they are chemical compounds rather than organic substances.

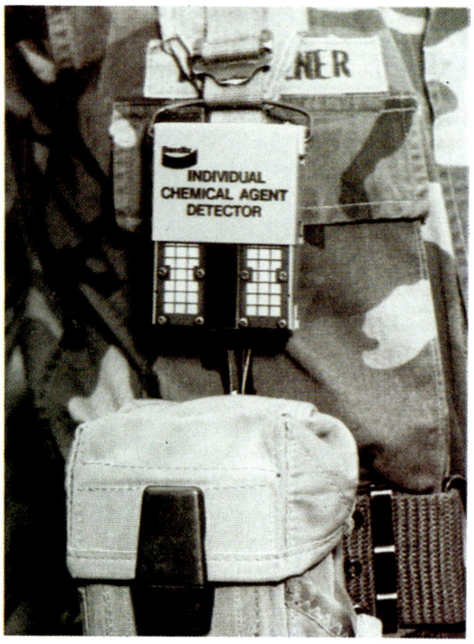


**Above: Bendix Individual Chemical Agent Detector employs electro-chemical phenomena which give alarm responses to G and V agents, mustard gas, lewisite, phosgene, hydrogen cyanide and cyanogen chloride.**

Other toxins have the same effects as psychological agents, being used, for example, to create acute terror symptoms.

Finally, there are the defoliants, which destroy plant life, either by causing unnaturally rapid growth or by killing leaves and shoots. Whilst having an application in some limited war situations, these are unlikely to have any major application in general war.

Chemical agents are stockpiled in liquid, gaseous or powder form and can be disseminated by aerial bombs, artillery shells, rockets, grenades, mines or by aerosol sprays from aircraft. Dissemination of these agents has never been an easy or precise art, and in the case of gaseous delivery is obviously very sensitive to wind conditions.

The USSR has long devoted considerable resources to CW and has a huge stockpile of assorted chemical weapons, the result of a continuous production. In the USA, however, political pressures have prevented production of any chemical munitions since 1969. Consequently, the stockpile of some 700,000 artillery shells, rockets and bombs has reached a stage where it has been described by General Bernard Rogers, SACEUR, as being on average 23 years old and substantially useless, with the exception of some 1,000 "Weteye" bombs. There are a number of 105mm chemical shells but, since few NATO armies retain 105mm artillery (the NATO standard is now 155mm), this is obviously of limited value. A further complication is that the current stocks are filled with the previous generation of chemical agents, which are unitary in construction and persistent in nature.

Meanwhile, the production plans and facilities for a new range of chemical munitions exist in the USA, but are held in abeyance pending Congressional authorization. The payload of these new-tech-

**Below: Modern armies regularly test chemical warfare detection aids and protective clothing, aware that chemical agents may cause large losses, shock and confusion among unprepared troops.**

Note: This section was prepared by Salamander editorial staff and does not purport to reflect the views of Colonel Friedman or any official organisation.

nology shells and bombs would be two independently non-lethal (albeit toxic) chemicals: difluoro and isopropyl alcohol amine. The production shell is filled only with difluoro, capped with a cardboard filler; the isopropyl alcohol amine is stored separately, being inserted into the shell or bomb only immediately prior to loading. Even then the two chemicals are in separate compartments and come into contact with each other only when the nose fuze sets off an expulsion charge which drives back a pusher plate, fracturing the membrane between the two components and allowing them to mix and form the deadly nerve agent GB-2. This is non-persistent and breaks down after a period of time, thus making decontamination by friendly troops passing through the area unnecessary.

After very considerable effort and financial expenditure virtually all NATO and Warsaw Pact military forces are now kitted out with protective clothing and respirators for use in a CW environment. Vehicles, buildings, ships and aircraft are proofed against chemical agents, usually by sealing the compartments and creating an overpressure. Much effort is also devoted to decontamination equipment for vehicles, aircraft and ships. However, it is generally agreed that protracted operations under such conditions are likely to be extremely difficult. Exercises have shown that wearing of CW clothing and respirators reduces the effectiveness of the wearer by some 40 to 50 per cent after a fairly short time, with a further and more marked deterioration after a few days. A consideration frequently overlooked is that no attempt has been made on either side of the Iron Curtain to supply NBC clothing to the civilian population at large which would be inescapably affected by the use of CW weapons in a conflict in Central Europe.

**Above: US Army sentry in full chemical warfare clothing during an exercise in Germany. Such clothing gives good protection but it is questionable as to how long wearers would remain efficient.**

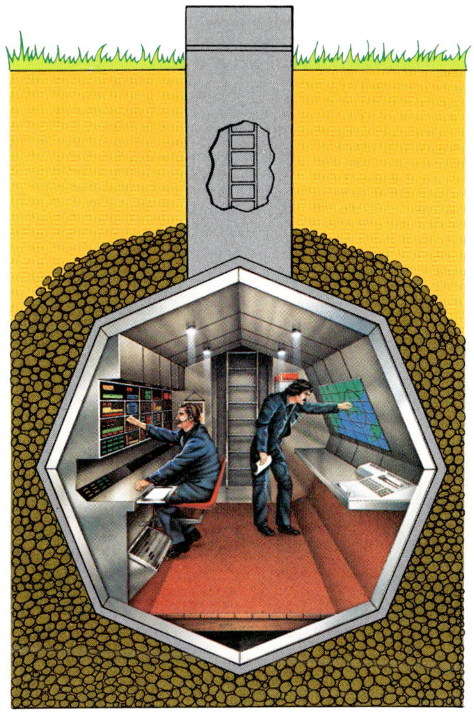

**Shelters**
The All-Purpose Defence (APD) shelter, produced by B. C. Barton, UK, is designed for occupation by personnel involved in military or other activities requiring protection under conventional war conditions, and is shown (left) as a command centre. It can obviously also be used as a store to protect sensitive equipment and supplies, and as a nuclear shelter. The shelter has an octagonal, pre-stressed outer cover, strengthened by a rigid framework. Sited below ground, it has an access-tower with an armour-plated entrance/exit hatch. Units, including service modules, can be interconnected as required to create an underground complex (below).

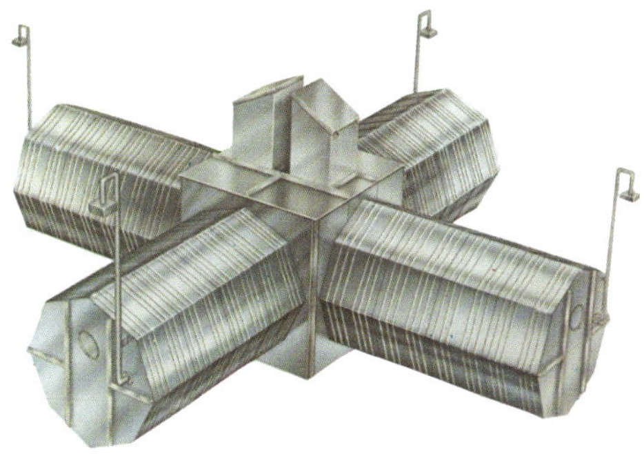

**Below: Soviet forces employ a variety of decontamination devices. This one is a surplus aircraft jet engine mounted on a trailer being used to decontaminate T-62 tanks on a Warsaw Pact exercise.**

**Below: US Army soldiers rush to don their chemical warfare clothing as the alarm is given during an exercise. Every sensible soldier dreads the thought that one day the alarm might be given for real.**

The imperative for any modern military commander is rapid reaction to an enemy concentration of mechanised forces. The ability of the commander to react in an effective and timely way depends upon his capability to discover the concentrations and dispositions of the opposing commander. What is required is easily remembered from the mnemonic aid known to every good non-commissioned officer: DRIL – *Detect, Recognise, Identify, Locate:*

Detect – Is there some activity there?
Recognise – What is it? A tank? An infantry unit?
Identify – It is an *enemy* tank; *friendly* troops.
Locate – The activity is at Map Reference xxxxx.

Most battlefield surveillance will be conducted by ground forces units. The principal area for the application of advanced technology, therefore, is to extend the ability, or improve it, of the ground observer to make his line-of-sight observations. The currently available means are: radar, thermal imaging, remotely piloted vehicles (RPVs) and unattended ground sensors (UGS).

Radar systems today can recognise some individual targets and are, overall, an excellent means for detecting targets. Many of the most advanced systems require the addition of a laser range finder to measure distances, however, and current production models do not measure speed. Radar systems also have a disadvantage in that

**Above: US Army Ground Surveillance Radar in Korea. Infantry units are now becoming rapidly more technically oriented as systems become more sophisticated.**

they are not passive systems (they emit an electronic signal) and are thus vulnerable to ECM.

Thermal imaging systems, instead of having a light-sensitive image projected, use a pattern of heat sensors which can "see" small differences in temperature. The operator is thus looking at a heat picture or graphic representation of heat reflections. These systems can be used as surveillance devices since most military targets emit heat. Thermal imaging can see through rain, smoke and fog, to some degree, and can produce an image adequate for identification. They have the added advantage that they can be used as a sight as well, and modern tanks, ATGMs and some other weapons systems are equipped with thermal imaging sights. However, these systems are also vulnerable to ECM and can be "spoofed" with decoys or blinded by false signals.

Image intensification systems used to take an image from a telescopic device and magnify not only the size of the image but also its intensity. Such systems can intensify a very dimly illuminated object by 40,000 times or more. An intensified image can reveal lights on the battlefield, particularly white lights at great distances, and can see objects illuminated only by starlight, and can detect infra-red devices from extreme distances, These systems which, like thermal imaging devices, are passive, are now being used to enhance vehicle drivers'/commanders' visibility, and also as sighting devices for a variety of weapons.

Remotely piloted vehicles are able to provide the ground commander with the ability to conduct surveillance above the line-of-sight. Since the Israeli successes with them in recent Middle-East conflicts, many nations are placing considerable emphasis on their development, production and deployment.

Most of the RPVs in production, or in train today, are small, propeller-driven pilotless aircraft (either fixed wing or rotary wing). Some are tethered (usually rotary wing). Some pilotless aircraft are not RPVs but are called drones, the basic difference being that the true RPV is

**Tank engagement**
A tank engagement using the Barr & Stroud DF3 fire control system. Against a stationary target the commander selects ammunition while the gunner acquires the target and flashes the laser (**1**). The micro-computer accepts the true range, calculates the ballistic aiming point and injects it into the field of view (**2**). The gunner then lays the gun and fires (**3**). Against a moving target a grid is inserted into the field of view (**4**), adjusts to match target speed and direction (**5**) and then follows the sequence (**1**) to (**3**).

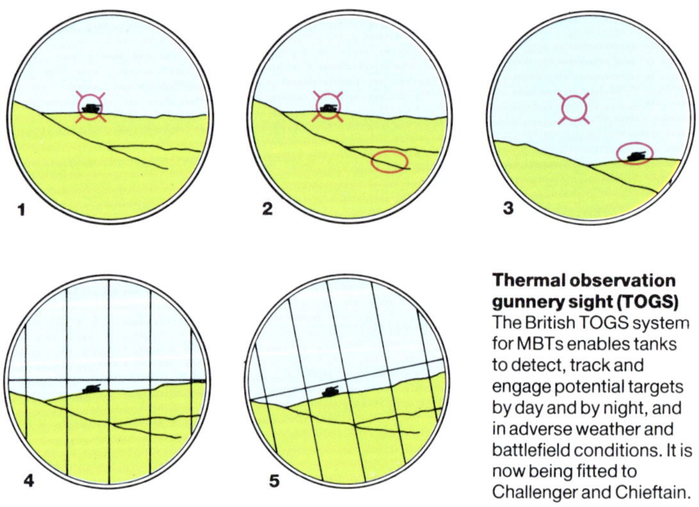

**Passive night vision**
The view through a Passive Night Vision Elbow Telescope, which uses a second generation, 25mm image intensifier. Being passive, it does not give away the tank's location, as did the earlier infra-red (IR) systems.

**Thermal observation gunnery sight (TOGS)**
The British TOGS system for MBTs enables tanks to detect, track and engage potential targets by day and by night, and in adverse weather and battlefield conditions. It is now being fitted to Challenger and Chieftain.

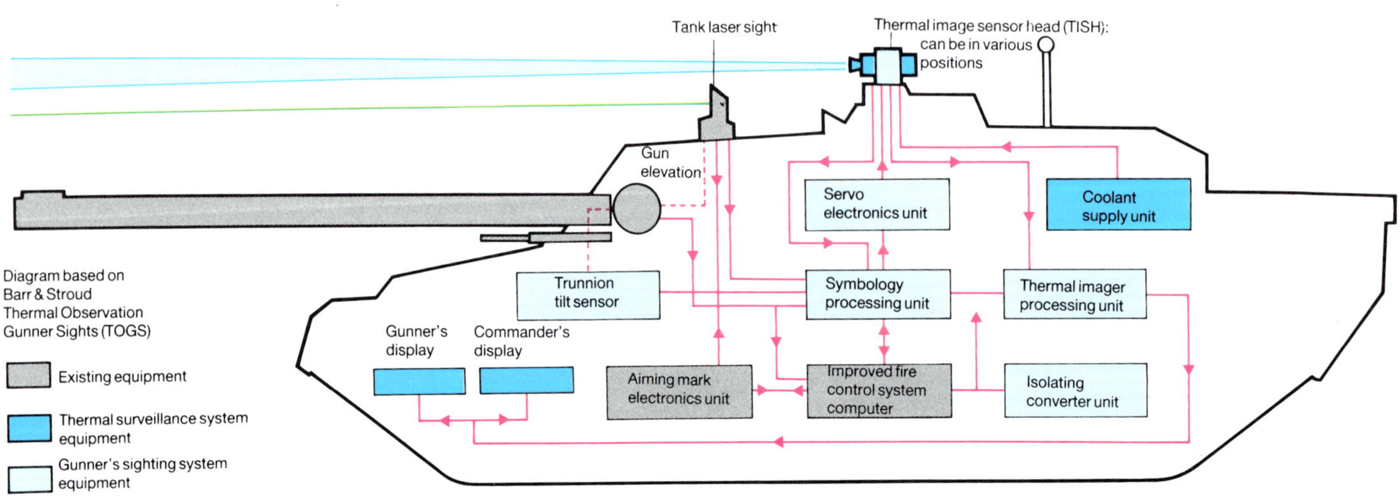

**Airborne surveillance drone system**
The low-level environment is becoming very hazardous for all aircraft, and increasing emphasis is being placed on drones for missions such as reconnaissance. One of the most effective of modern vehicles is the AN/USD-502. This is launched from a 4-tonne truck, by means of a strap-on rocket-booster, which is jettisoned after launch. Powered by its 240lb (108kg) turbojet the drone follows a programmed flight-path, with the sensors switched on at a pre-determined point. The drone carries a Zeiss camera and an infra-red line-scan sensor, with the latter capable of real-time data transmission back to the base site. This gives a substantial advantage over earlier systems in which the information was not available until the drone returned to base and the pictures were developed. Having completed its task the drone returns to base and is quickly readied for its next mission. Such unmanned systems are becoming more necessary on today's battlefield.

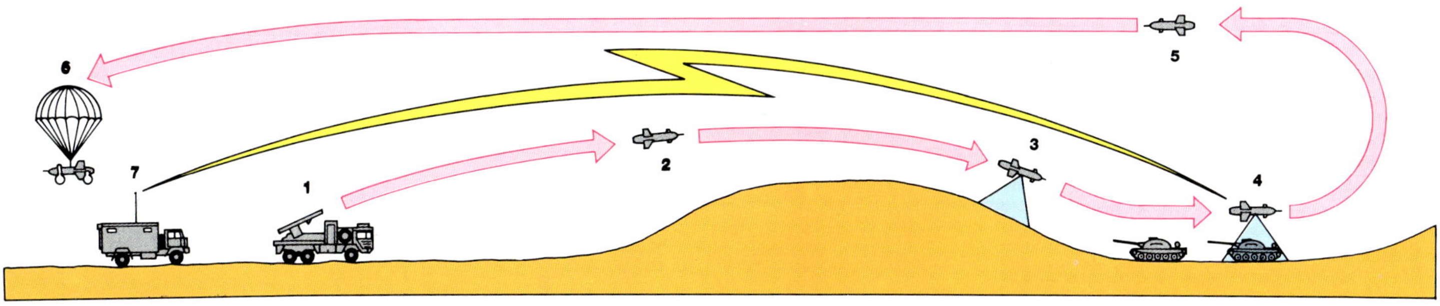


controlled from a distance by an electronic link while the drone is programmed to follow a pre-planned flight path.

The latest RPVs are designed to fly in hostile territory and locate targets, allow for the adjustment of artillery fire from remote distances and designate targets using lasers. RPVs can also perform reconnaissance, damage assessment, and other missions. Some models will be equipped with small television or forward looking infra-red night vision sensors.

The signals from the RPV are monitored by ground controllers. When a target is sighted, the controller centres a crosshair sighting device on the target image of the TV screen. At this point, the grid co-ordinates on the target can be determined, transmitted to an artillery unit, and the target can be engaged with artillery fire. The RPV can orbit the target area to adjust the artillery fire. If the controller desires to illiminate the target with a laser beam, he can do so by merely pressing a switch. Thereafter, laser-seeking weapons, such as the US Copperhead, or Hellfire or other laser-guided munitions fired into the general area of the target will home in on the laser spot. Once an engagement is completed the damage in the target area can be assessed from the picture on the TV tube. The RPV thus permits penetration of enemy zones protected by the most formidable defences or which are masked to friendly reconnaissance.

Most of the fixed-wing RPVs are launched by rail from a flatbed truck and can be recovered and reused after completion of a mission. The British currently are using a drone – MIDGE – which is launched from behind friendly lines and takes photographs which are available after return. The US is preparing to produce a fully advanced fixed-wing RPV for immediate deployment. Other nations are using the Israeli-developed Mastiff or Scout devices, and considerable interest has been shown in a Canadian tethered RPV which can provide data in real time from an elevated point and can be re-fueled in the air by means of the tether, thus providing extended operations. It is expected that the rapid advances of the newer micro-electronic technology will initiate a larger use of RPVs in the future.

Commanders can also obtain information from beyond their own line-of-sight capabilities (without risking their troops on reconnaissance missions) by using unattended ground sensors (UGS). These devices are seismic, thermal, magnetic or acoustic. At present, mostly thermal and seismic devices are in field use. UGS were widely used by the American forces in Vietnam where it was learned that they produced mixed results, including that some of the devices could be misled or (seismic in particular) reported *all* ground tremors. But it is possible now to produce devices which can discriminate between signals and report only those with military significance, or in accord with their programmed instructions. Such devices, planted at key points along possible enemy routes of advance or exposed flanks, could be used not only for target acquisition purposes but for intelligence-gathering as well. At the present time, the major expansion in UGS programmes is to develop additional means for their emplacement: today they are placed in position by aircraft, special artillery shells, or by hand.

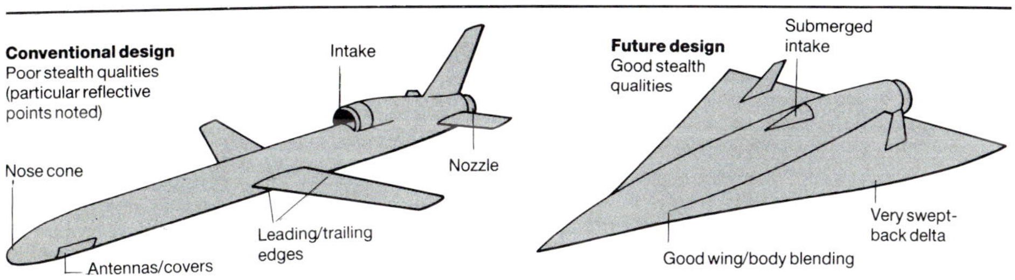


**Stealth for RPVs**
Even though RPVs and drones have so far been difficult to detect, it will soon be possible to detect them quickly and accurately, and thus destroy them. As with piloted aircraft, it will become necessary to employ stealth technology to extend their useful lives. Current drones and RPVs (left) give specular and cavity reflections and these will have to be overcome using modern design techniques and materials, as shown on the right.

**Left: The Lockheed Aquila is a small, propeller-driven RPV designed to locate targets, and to laser-designate targets for engagement by laser-seeking artillery shells or missiles, all at low cost.**

Perhaps in no other field of military application has advanced technology made its presence felt more than in command, control and communications ($C^3$).

It has always been necessary for combat commanders to assess the enemy situation in relation to its own, to develop a plan for battle and communicate the plan to his command. Once battle has commenced, the commander must thereafter be able to control the implementation of his plan and make rapid adjustments. The American Civil War (1860-65) is generally conceded to be the last major conflict in which the principal commanders could sit on a horse on some highpoint and personally control their forces in battle, and even then this was frequently not possible during that war.

There are two sides to this coin. While the commander might not be able directly to control the ebb and flow of the fighting, neither could his superiors supervise him from some distant capital. By World War II, national leaders could, and frequently did, conduct inter-allied business from great distances using modern and secure communications, sending detailed instructions to their military commanders in distant theatres of war. Just 20 years later, it was possible for an American President personally to converse with a platoon leader engaged in a firefight in Vietnam. This is not surprising, since he could also converse with astronauts in outer space.

Information flows from the forward point where one's own troops are in contact with those of the enemy – usually called the Forward Edge of the Battle Area (FEBA), or Forward Line of Own Troops (FLOT) – back to the various headquarters. Meanwhile command and control passes from the headquarters forward. During World War II, the armies in combat exercised their command through communications designed to follow the chain of command. Armies used land-line or (cable) communications, together with net radio and radio relay (microwave) systems.

Due to the complexity of modern military forces and because they must be widely dispersed, in recent years forces have tried to eliminate having a large number of vulnerable communications centres and have developed *area* systems. In such systems, communications centres and their associated equipment and troops are replaced by a matrix of trunk nodes which accomplish essentially the same functions as the communications centres but can, at the same time, do a lot more. These nodes can cover the operational area of the associated commands and are sited within "line-of-sight" of each other, a distance that varies according to terrain. The central component of the node is a vehicle (or shelter) containing a computer-controlled electronic switching device. This functions as an automatic exchange which can immediately connect one user (subscriber) to another.

The NATO nations have been developing a series of systems. The American system is called Tri-Tac, the British system, Ptarmigan, the French system, Rita and the German system, Autoko. All of these systems are less vulnerable than the systems they replaced; their secure speech facility and the ability to provide voice, teleprinter, facsimile and computer data links are all of great importance.

**Above: British Army communications experts testing Wavell command and control computer terminal, installed in a field car. Wavell will revolutionize command and control in 1 (British) Corps, Germany.**

Computer compatible communications are vital. Modern computers – particularly the newer micro-computers – are used to carry out such functions as routing military communications and assisting military staff officers as readily as they are to process airline ticketing and automobile sub-assembly manufacturing.

A number of major command and control computer systems are under development for the NATO armies, including Sigma (US), Heros (West Germany) and Sacra (France). The British Wavell 2 system is already in service with BAOR. All of these systems are being designed to be compatible with the Tactical Command Communications systems (US Tri-Trac, UK Ptarmigan, German Autoko and French Rita). The advanced systems make use of the latest micro-circuitry and digital communications. They are designed to provide rapid, secure and highly flexible control which will be able to incorporate improvement in later versions.

All function fundamentally in the same way. The basic concept is to store information and update it continuously, and display it whenever required. The newer devices can even project data on maps or overlays. A particular advantage is that the computer memories can up-date one another. A headquarters can pass the new data to every other unit through the area communications system.

The Soviets were quick to adopt the use of computers for their $C^3$ modernisation programme. Because of their philosophic preference for centralised planning and command, and their doctrinal approach to modern warfare, the Soviets reasoned that

**Tactical EW sensors**
An essential element of the modern land battle is the counter-$C^3$ battle, fought entirely in the electromagnetic spectrum. The hub of this battle is the EW operations centre, which has communications to the field HQ and to all the elements of the EW battle. Furthest forward are the Electronic Counter-Measures (ECM) vehicles, which provide high-powered jamming from just behind the Forward Edge of the Battle Area (FEBA). Also involved are Direction-Finders (DF) and radar ESM vehicles. Comint ESM vehicles intercept enemy communications and are located somewhat further from the FEBA. The whole system must be linked by its own secure communications.

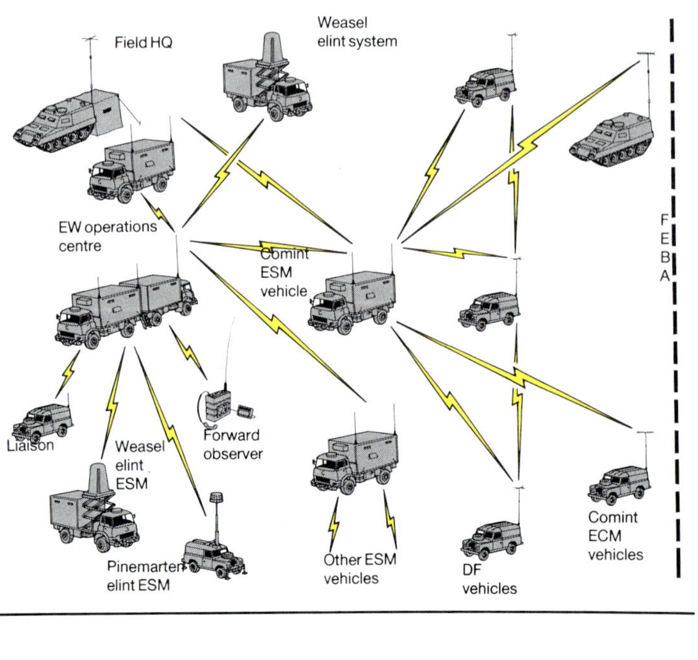

**Above: Israeli computerized command and control system. Most armies now find computerized systems essential to cope with the flood of information and the ever faster speed of movement on today's battlefield.**

computers could provide them with special benefits. They concluded that computers could be programmed with all the basic data concerning any area of conflict, then updated with recent particular data concerning weather, terrain and order of battle of an enemy, and that the computer could then optimise battle decisions. At present, published Soviet writings indicate that this system is still in the development stage, but it clearly indicates a possible direction for field use.

## Position location systems

After a major NATO exercise some years ago, one of the field commanders estimated that over two-thirds of the tactical communications activity during the exercise was for the purpose of reporting or finding the location of various units. A Position Location Reporting System (PLRS) is currently in production in the United States. This system can automatically provide combat commanders with accurate location data on deployed forces, and provide reliable navigation information together with a limited data communication capability to units.

PLRS consists of a master station at division artillery and up to 370 user units in manpack, vehicle and aircraft configurations distributed throughout a division or separate brigade manoeuvre and fire support elements. User units automatically transmit location data to the master station, where it is displayed instantly. This enables visual command and control of forces over a large area in a manner not possible since the time when commanders could see and hail all of their men from one place. Through the data communications capability of PLRS, users can send short preformatted or free-form messages between themselves and/or the master station to obtain automatic location and navigation information on themselves or other users. This feature significantly relieves the traffic on existing tactical communications previously commonly used for such purposes.

It is clear that with the employment of advanced concepts of command, control and communications, the widely dispersed, highly sophisticated armies of today and tomorrow have special tools which make their jobs simpler and more rapid in accomplishment. At the same time, the modern command, control and communications systems are more closely related to electronic warfare than ever before. While EW is the subject of a chapter of its own, it is necessary to point out that the commander who is not well defended against enemy EW by his own resources is in an extremely vulnerable condition; and the commander who cannot employ adequate EW resources against his enemy will find himself virtually deaf and blind. He needs both capabilities.

Victory in the land battles of the future will go to the commander who is able to avoid losing control of his forces before the time he is ready to engage the enemy, to concentrate his own forces at an appropriate point and advance, bringing to bear all of his arsenal of modern weapons against the most vulnerable points of his adversary. The commander who is unable to maintain control of his forces, to develop and execute a suitable plan vigorously and decisively will, as always, be doomed to defeat.

**Far left: British Weasel Electronic Intelligence (elint) system is an advanced automatic system for use in a dense radar environment. Left-hand screen shows graphical format of intercepted radars.**

**Left: The Hughes Position Location Reporting System (PLRS) gives details of the location of own or other units on request, thus answering one of a soldier's most frequent questions: "Where am I?"**

# NAVAL WARFARE

The 1980s and 1990s are likely to be among the most exciting years ever in naval technology, as a whole series of radical developments – and perhaps some major breakthroughs – look set to revolutionize many aspects of naval warfare. One event whose importance is difficult to overstress was the South Atlantic War of 1982 between the United Kingdom and Argentina, where a modern surface fleet deployed under real combat conditions for the first time since World War II. The anticipated major fleet action never took place, but the Royal Navy Task Force was heavily attacked and many important lessons were learned, which are being eagerly examined by all the world's navies.

The post-World War II period has seen the continued domination of the world's major fleets by the great aircraft carriers, and also the emergence of the missile as the pre-eminent ship-to-ship weapon. The past fifteen years, however, have also witnessed the emergence of the submarine as arguably the most important single type of naval warship, capable of world-wide deployments without replenishment or coming to the surface. During this period larger, more capable and quieter nuclear-powered missile-armed submarines have been introduced, culminating in the US Navy's Ohio class SSBNs of 18,000 tons displacement and 24 missiles, being completely outmatched by the Soviet Navy's Typhoon class with 20 missiles and, at 25,000 tons displacement, the largest submarine craft ever built by a very large margin.

Nuclear-powered attack submarines (SSNs) are faster, better armed and deeper diving then ever before, and both France and China are now building their own types. Even conventional submarines are improving steadily, although the great breakthrough of a non-nuclear replacement for the diesel engine/electric motor combination still seems a long way away. But the present position would alter dramatically if there was, at long last, to be a breakthrough in anti-submarine warfare (ASW) technology. SSBNs are currently the ultimate deterrent, hidden in the depths, able to strike at targets deep in the enemy's heartland from anywhere in the world's oceans. But can such an advantageous situation last more than another ten years? What then for the submarine?

On the surface the aircraft carrier reigns supreme. The US navy has its massive and powerful supercarriers (over 90,000 tons full load displacement and an air wing of about 90 combat aircraft), but these will be

**Above: USS *Georgia* (SSBN-729), one of the latest Ohio-class submarines to join the US fleet. With 24 Trident 1 SLBMs such submarines provide an almost invulnerable second-strike, counter-value capability – but for how long?**

**Below: Ships of the Standing Naval Force Atlantic (STANAVFORLANT) at sea symbolise the ability of NATO's navies to operate together, although true (and full) inter-operability of weapons and electronic systems still has some way to go.**

challenged in the early 1990s by the first of the Soviet supercarriers of some 75,000 tons displacement and with an air wing composed entirely of brand-new types of aircraft. Meanwhile, the French Navy intends to join the nuclear-powered carrier club by the end of the decade, and the British Royal Navy continues to develop the V/STOL carrier, a concept tested and proved in the South Atlantic War.

Other surface ships continue to develop at a rapid rate, but perhaps the biggest surprise has been the re-emergence of the battleship/battlecruiser. The US Navy has recommissioned its elderly Iowa class battleships in answer to the superb Soviet Kirov class battlecruisers, which are the most impressive capital ships to appear in the past 20 years.

Strategically the world's oceans are important as never before, carrying vast amounts of international commerce. Taking just oil as an example, 1,206 million tonnes were moved by sea in 1983, of which 899 million were destined for NATO countries, Japan and Australasia. Many other energy source materials are moved by sea (eg, bulk natural gas and coal), as are virtually all other raw materials. The global map, however, quickly shows that it is the West that depends for its survival upon the sea; for the Soviet Union maritime travel is simply a bonus. That is why the USSR did not really bother to build a major navy until the 1950s; even now its navy is mainly a means of posing an apparent threat to the West, and for im-

**Modern warships represent the most complex concentration of modern technology, possessing exceptional firepower, a multiplicity of sensors, and sophisticated command, control and communications systems. The 1982 South Atlantic War, however, taught many lessons to all navies.**

pressing the Third World. The oceans are critical to the West's survival; they are not to that of the USSR.

There have been many advances in naval tactics, weapons and systems, and in command and control of naval operations since the last major fleet actions in 1945. Naval forces have often been used since but normally only in limited affairs, until the deployment of the British Royal Navy Task Force to the South Atlantic in 1982. Although the expected major actions did not take place, the naval war was intense and crucial to the outcome of the campaign. Tactics and weapons were tried out which had previously only been tested in the slightly false atmosphere of exercises and simulators. But, in the disputed waters around the Falkland Islands, ships were sunk, aircraft shot down, many men killed and wounded, and hard lessons learned. The short campaign must therefore loom large in any assessment of the current and future state of naval technology.

However, while there have been remarkable developments in some areas, in others matters have changed but little. Submarine speeds underwater, for example, have trebled in the past thirty years, but surface ship speeds – at any rate with the traditional monohull – have increased only marginally: the British Task Force which went to the Falkland Islands in 1982 travelled no faster than did Admiral Sturdee when making the same journey, and for similar reasons, in 1914.

**Task Group protecting merchant convoy:** In any future war NATO's defence of Europe would depend upon the movement of men, stores and equipment across the Atlantic in vast quantities to reinforce those already "in-place". The Soviet naval and air threat is such that this great move could only take place using the convoy techniques of previous World Wars, but updated wherever possible to take account of the advances in modern technology. The schematic diagram below illustrating how such a task group might be deployed is not to scale, and it must be emphasised that this represents a very large area of sea. Further, although British symbols are used any NATO task group would use similar tactics. The use of many fixed- and rotary-wing aircraft is evident.

**TOTAL RESOURCES AVAILABLE**

| Warships | | Aircraft | | Weapons | |
|---|---|---|---|---|---|
| Helicopter Carrier | 1 | Sea Harrier | 5 | Exocet | 36 |
| ASW Frigate | 9 | Sea King | 10 | Sea Dart SAM | 7 |
| AD Frigate | 6 | Lynx | 24 | Sea Wolf SAM | 18 |
| Fleet Oiler | 1 | | | 4.5in Gun | 6 |
| SSN | 3 | | | Sub-Harpoon | Yes |

Fixed-wing, land-based, ASW aircraft; eg, P-3 Orion, Nimrod, Atlantique.

Long-range, heavy ASW helicopter, usually carrier-based; eg, Sea King.

Short-range, medium/light ASW helicopter, destroyer/frigate based; eg, Lynx.

Carrier-based, fighter aircraft on combat air patrol; eg, Sea Harrier.

Either SSNs or conventional submarines, eg, Valiant class.

High-speed container and passenger ships; convoy "commodore" commands.

Aircraft carrier, providing air cover and command-and-control facilities.

ASW destroyers/frigates for the anti-submarine screen, using own sensors/helicopters.

Inner screen of air defence destroyers/frigates for close air/missile defence.

Naval auxiliaries for Replenishment-at-Sea (RAS) of naval warships.

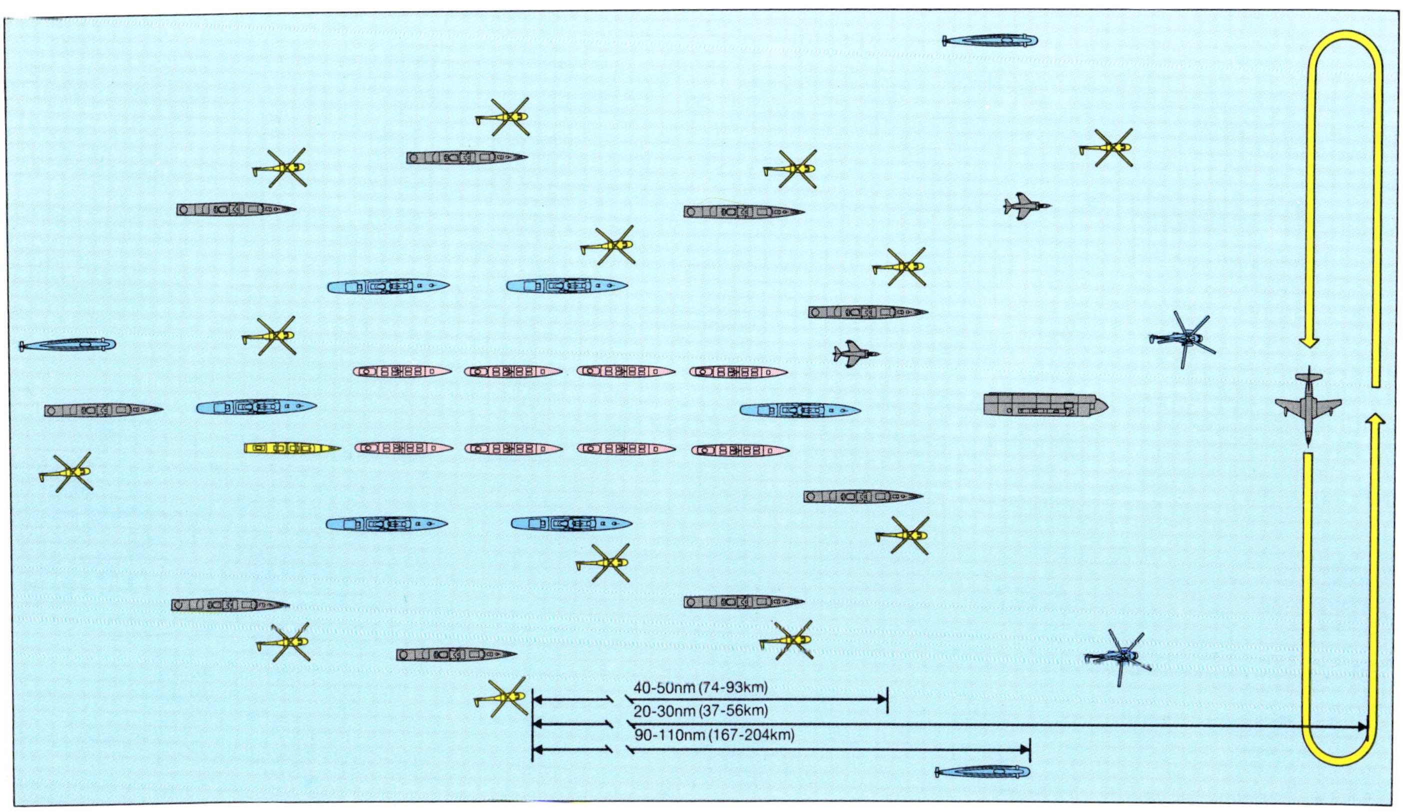

Surface warships have undergone unprecedented technological development since 1945, although the new ideas, concepts, weapons and systems were given little real combat testing until the South Atlantic War of 1982.

For many years the only really large surface combatants to be built were aircraft carriers. Virtually all battleships were scrapped in the 1950s and the largest ships to be built for many years thereafter were cruisers in the 8 to 10,000 tons class; instead, most navies concentrated on destroyers (4 to 6,000 tons), frigates (1 to 4,000 tons) and minor warships such as corvettes and fast patrol craft. But as happens so frequently with military equipment, size has gradually increased and the growing number of cruisers has been joined not only by the Soviet Kirov-class battlecruisers (25,000 tons), but also by refurbished World War II battleships of the Iowa-class (58,000 tons), which a prudent and far-sighted US Navy put into preservation in the 1950s rather than send them to the breakers yard.

Surface warships remain the high-visibility epitome of naval might and the world's surface fleets are gradually increasing in overall numbers, with individual units becoming ever more complex, sophisticated – and expensive. Surface warships are required to project sea power and have to be capable of anti-submarine (ASW), air defence (AD) and anti-surface roles, almost invariably in company with other combatants to form a task group. A warship combining all these capabilities will inevitably be large in order to carry all the sensors and weapons systems involved (and thus very expensive), but a ship which specialises in one particular role suffers tactical limitations, so that more vessels are required, leading to an overall greater fleet cost, even though unit cost may be less. This is the dilemma which all navies are facing, and to which they hope technology may somehow provide an answer.

It seems unlikely, however, that there will be a dramatic breakthrough in warship design or capability in the forseeable future, but rather a process of constant, if gradual, refinement. It seems improbable that surface warships will be able to remain undetected on the high seas, no matter what electronic countermeasures are taken. Ships' radar signatures may be reducible, for example, by use of radar-absorbent paint, not to avoid detection totally, but rather to enhance countermeasures against incoming missiles. Surveillance, both active and passive, will become more effective and longer ranged, and the integration of surveillance and target indication radars (with a 3-D facility) should lead not only to shorter weapon reaction times, but also to reduced top hamper. One device, widely used in other areas but seldom at sea is the Remotely Piloted Vehicle (RPV), which would seem to have considerable potential for long-range surveillance in support of maritime operations.

Great scope lies in the future development of command, control and communications systems, both within individual ships and between the ships of a task group. Fortunately the advent of digital

**Below: The Soviet Navy's Krivak class is an excellent example of contemporary warship design with a balanced armament and a well-integrated electronic outfit.**

**Surface warship weapons and sensors:** Surface warships represent a unique concentration of high technology packed into a relatively small and not invulnerable hull. In order to perform its assigned missions a warship must have a variety of weapons systems for both offensive and defensive tasks; for example, anti-submarine, anti-ship, air defence (against both aircraft and missiles). It must also have numerous sensors to detect and track air, surface and sub-surface targets, and control the weapons used to attack them. Finally, the ship needs a multiplicity of communications and ECM/ECCM/ESM systems. All this, plus the ship's own command-and-control and propulsion systems must be accommodated in a hull of finite size. There are many ways of achieving this; shown here is one Soviet answer: the battlecruiser *Frunze*.

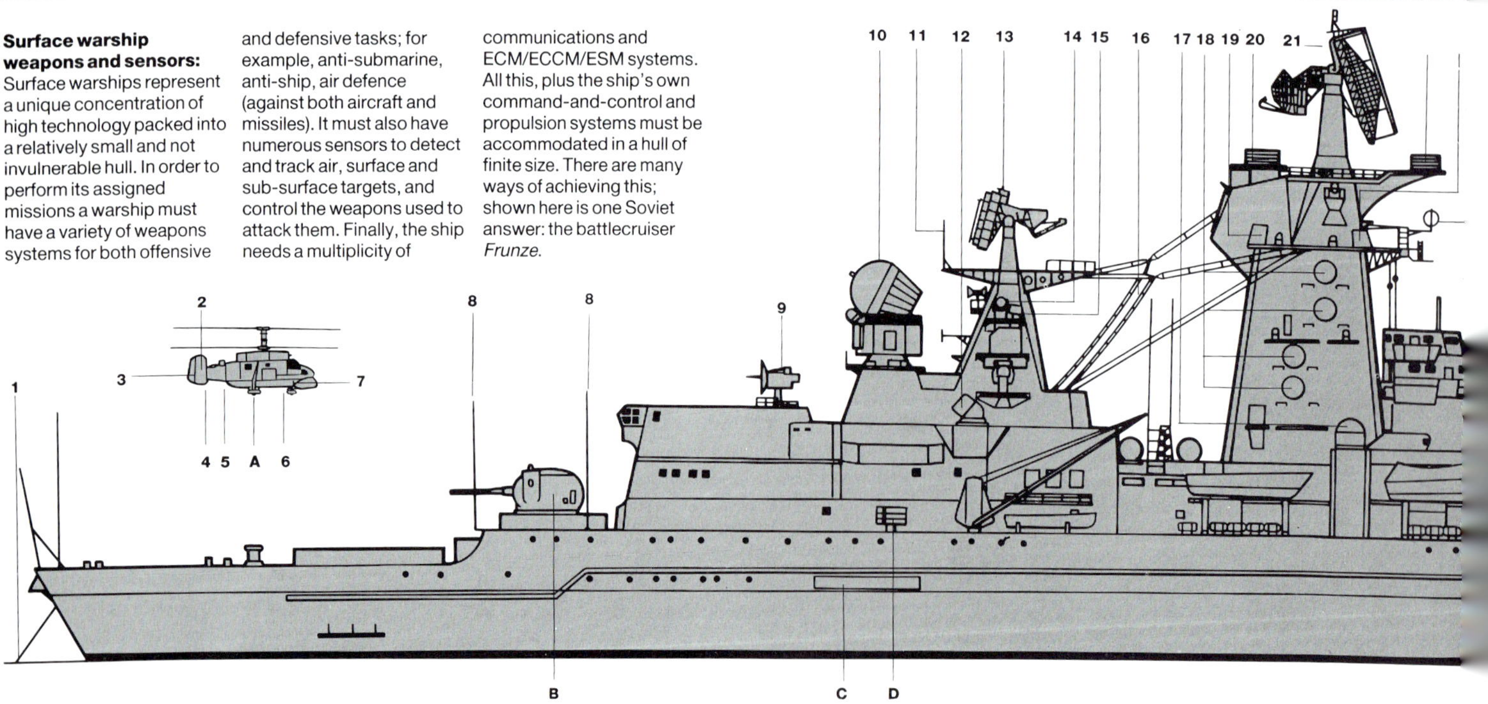


**Weapons and sensors of *Frunze***
The Soviet Kirov-class battlecruisers are the largest surface warships (apart from carriers) to have been built in recent years. This side view represents *Frunze*, second of the class. A major problem for naval designers is the integration of the numerous electronic and electrical systems on board a modern warship. The Soviets seem to be particularly good at this, *Frunze* showing a multiplicity of systems they fit into one hull. Weapon fits in Soviet warships tend to be complex. One field where they are setting the pace is in "hidden" launch systems, where missiles are contained in bins below deck and launched by removing the cover.

***Frunze*'s sensors:**
**1.** Stern-mounted, variable-depth, medium frequency sonar (VDS).
**2.** Electro-optical sensor.
**3.** Radio antenna.
**4.** MAD "bird".
**5.** Sonobuoys.
**6.** Search radar.
**7.** Homing antenna.
**8.** HF radio antennas.
**9.** Kite Screech fire control radar for 130mm twin guns.
**10.** Top Dome fire control radar for SA-N-6.
**11.** Cage Stalk broadband dipole radio antenna.
**12.** Bass Tilt for 30mm Gatling CIWS
**13.** Top Knot 3-D search radar.
**14.** "White light" searchlight.
**15.** Disc Cone.
**16.** Various folded-wire antennas.
**17.** Round House ECM antenna.
**18.** Side Globe ECM/ESM antennas.

systems enables a much greater degree of integration to take place, but this can be accompanied by data base distribution and duplication of facilities, both of which enhance survivability. One essential outcome should be a reduction in specialist and highly trained (and expensive) manpower, but a faster, more reliable and capable data-handling system will better enable ships to cope with the likely saturation attacks.

Weapons systems will probably continue to evolve gradually, with greater dependence upon missiles, although not to the total exclusion of guns (that lesson at least has been learnt), and it is inevitable that the West will follow the Soviet lead in using vertically launched missiles. Launching from hull-mounted silos gives increased protection to the missiles and magazines, and also reduces the complexity (and weight) of the installation.

Many of the advances in warship design will probably occur in apparently very mundane areas. Stabilisers, for example, have had a dramatic impact on crew effectiveness and weapon and sensor performance, while the simple expedient of covering in the bridge to protect the command staff from the elements led to a marked improvement in their stamina and effectiveness. Current attention is being paid to using multiplexing and data bus techniques to reduce the miles of cabling, thus reducing weight, complexity and maintenance requirements.

The South Atlantic War also showed that there is considerable scope for improvement in areas such as damage control and

**Above: US Navy ships (this is USS *John Young*) have fewer and less complex antennas than on Soviet ships, but this is because the equipment itself tends to be electronically more advanced.**

fire-fighting, which, in the halcyon (and cost-cutting) days of peace can all too easily be neglected. The aim of all navies will be to improve survivability, both by using less flammable materials and also by designing ships' interiors in a more sensible way.

The need to produce survivable sensors, control centres and weapons systems might even lead to a radical rethink of warship layout, with one set forward and the second aft in two separate, virtually self-efficient (but closely interlinked) operational areas. In such a scheme the helicopters would probably operate off the centre of the ship, which is in any case the most stable area.

One of the more worrying aspects of current designs for navies is the ever-escalating cost of surface warships. The three Ticonderoga class CGNs authorised in the Fiscal Year 1983 programme cost $1,600 million each, while the Arleigh Burke class is estimated to cost some $1.1 billion each (also in FY83 dollars); astronomic prices, but for very sophisticated ships in a wealthy nation. The British Type 42 destroyers currently cost some £85 million each (at 1980 prices) and even the "cheap" Type 23 is reported to be likely to cost some £75 million (although no firm figure has yet been published). No figures are available for Soviet ships, but the size and sophistication suggest that they will be proportionally as expensive to the Soviet economy as are their equivalents in the West. Unless, therefore, governments can provide more money for shipbuilding or technology can find ways of reducing

**System allocation:** This cut-away depicts the probable allocation of volume to the major systems on board *Frunze*. The unmarked areas are not empty, but are devoted to the many other requirements such as sailors' accommodation, ammunition, food, water and other consumables' storage.

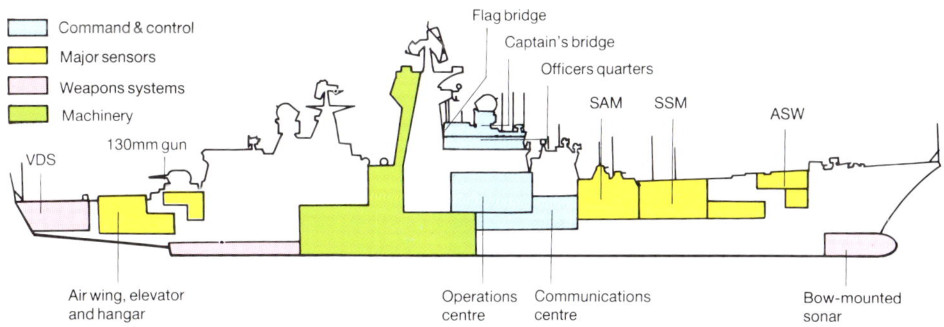


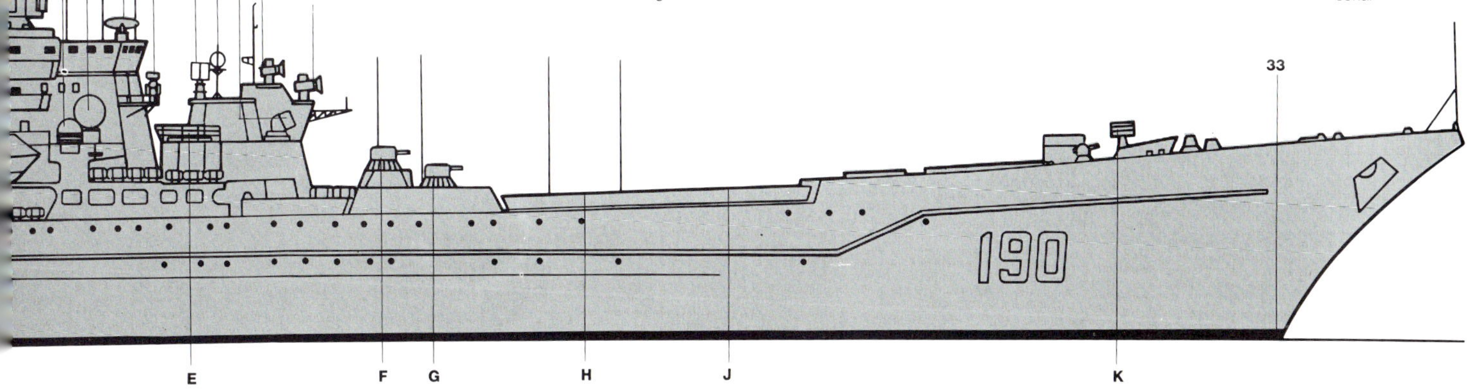


**10.** Rum Tub ESM antennas.
**20.** Round House ECM antenna.
**21.** Top Pair; combined Top Sail/Big Net radars for 3-D air surveillance.
**22.** DF Loop antenna.
**23.** Top Dome fire control radar for SA-N-6 missile.
**24.** Bob Tail electro-optical night/bad visibility surveillance system.
**25.** Large spherical radar, purpose not yet known.
**26.** Palm Frond navigation radar.
**27.** Electro-optical device.
**28.** DF Loop Type 4.
**29.** DF Loop Type 3.
**30.** Bass Tilt fire control radar for 30mm CIWS.
**31.** Eye Bowl fire control radar for SS-N-14 anti-ship missile system.
**32.** Second Eye Bowl antenna.
**33.** Low Frequency sonar, mounted in dome at foot of the stem.

***Frunze*'s weapons systems:**
**A.** Kamov Ka-25 Hormone or Ka-32 Helix ASW helicopter armed with homing torpedoes or depth bombs (HE/nuclear).
**B.** Twin 130mm gun turret (Kirov has two single 130mm turrets).
**C.** 21in (533mm) torpedo tubes; five per side behind doors.
**D.** RBU-1000 ASW rocket launcher; one per side.
**E.** SA-N-4 missile launcher; one per side. These launchers are normally retracted below deck level.
**F, G.** 30mm "Gatling" close-in weapons system.
**H.** 20 vertical launchers for the SS-N-19 anti-ship missile system.
**J.** 12 vertical launchers for SA-N-6.
**K.** RBU-6000 ASW rocket launcher.

(rather than increasing) costs there may be no alternative to a decrease in the number of ships that each navy can afford.

## Sensors

The modern warship is absolutely dependent upon its sensors, both active and passive. Radars are required to detect aircraft and surface targets. Air defence (AD) radars are invariably mounted as high as possible to give all-round coverage and to maximise their range. A modern 3-dimensional (3-D) AD radar antenna is large and heavy; the Soviet Top pair, for example, mounted atop the mast on the Kirov class battlecruiser, combines Top Sail and Big Net antennas, and is some 157ft (48m) above the water-line; the whole installation must weigh many tons.

Radar range depends upon power output and the height of the antenna and target, and may vary between 50 and 200 miles (80 and 322km). However, as was shown in the air-launched Exocet attacks upon HMS *Sheffield* in the South Atlantic War, a sea-skimming missile is very difficult to detect, even with sophisticated radar: the frontal area of an Exocet is 149sq in (962cm$^2$).

Some radars have to be supplemented by separate height-finding radars. Surface surveillance radars are also used, but are limited currently to the range of the horizon. Main improvements in this field are increased accuracy and definition, and detection of submarine periscopes is feasible with some sets.

At the forefront of air/surface radar technology is the US Navy's Aegis system, which is based on the AN/SPY-1A electronically-scanned fixed array radar, and which is now at sea on the Ticonderoga class cruisers. There are two pairs of fixed antennas on each ship (one pair forward, one pair aft), driven by an AN/UYK-1 digital computer to produce and steer multiple radar beams for target search, detection and tracking, thus obviating the need for a mechanically-rotating antenna. The system also provides target designation data for the missile illuminating radars, and tracks outgoing missiles. A lighter derivative, AN/SPY-1D, will be used on the Arleigh Burke class, but with only one transmitter and three illuminators, compared with two and four respectively on the Ticonderogas.

However efficient all these systems may be, they all suffer from one major shortcoming: as active transmitters they are easily detectable by hostile aircraft, ships and satellites. There are "quieter" electronic techniques (spread spectrum trans-

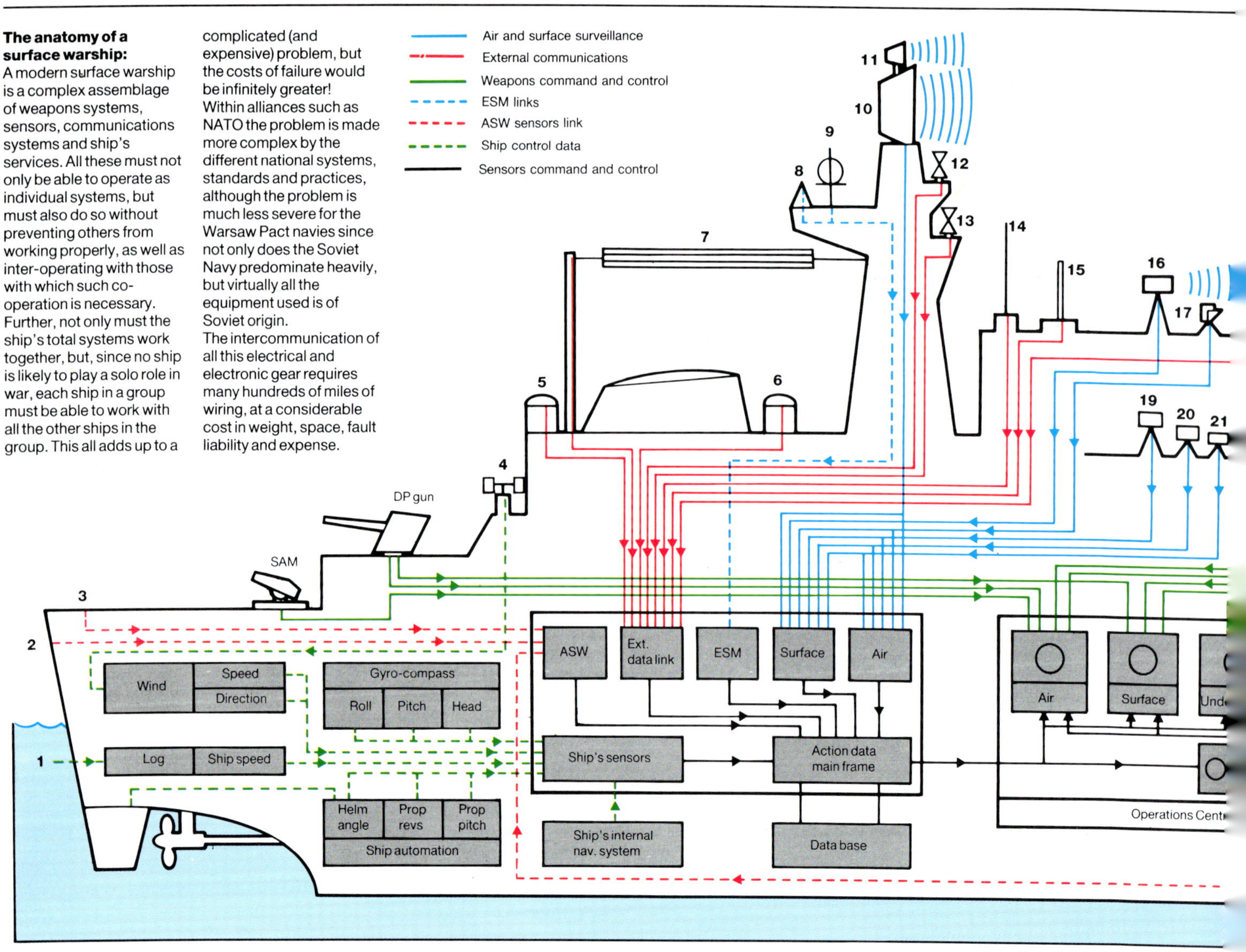


**The anatomy of a surface warship:** A modern surface warship is a complex assemblage of weapons systems, sensors, communications systems and ship's services. All these must not only be able to operate as individual systems, but must also do so without preventing others from working properly, as well as inter-operating with those with which such co-operation is necessary. Further, not only must the ship's total systems work together, but, since no ship is likely to play a solo role in war, each ship in a group must be able to work with all the other ships in the group. This all adds up to a complicated (and expensive) problem, but the costs of failure would be infinitely greater! Within alliances such as NATO the problem is made more complex by the different national systems, standards and practices, although the problem is much less severe for the Warsaw Pact navies since not only does the Soviet Navy predominate heavily, but virtually all the equipment used is of Soviet origin. The intercommunication of all this electrical and electronic gear requires many hundreds of miles of wiring, at a considerable cost in weight, space, fault liability and expense.

missions, for example) or procedural methods (such as switching radars on for only very brief periodic "looks") to make an enemy's task more difficult, but the only real solution lies in remote sensors or in passive systems. To this end electro-optical systems are being developed for some applications, but they are currently severely limited in range and by weather, and they are unlikely to provide adequate warning against, say, an incoming sea-skimmer travelling at Mach 1+.

Hull-mounted sonars are still the principal ASW sensors, and they can be used in either active or passive modes. These are supplemented in many ships by variable-depth sonars (VDS) which can be kept at a particular depth to detect submarines hiding below a thermal layer. Some surface warships are now also being fitted with towed arrays. One problem for surface ships is that hull-mounted sonars cannot be used at high speeds, while VDS and towed arrays limit speed and manoeuvrability.

This is exacerbated by the ever-increasing speed of some submarines: the Soviet Alfa class has demonstrated a speed of over 42 knots whilst shadowing NATO exercises, whereas most surface combatants have a maximum speed of some 32 knots in fair conditions, and considerably less in a seaway. This problem can be alleviated by a shipboard helicopter which can be used to detect and attack hostile submarines well out of range of the parent ship's weapons systems, and, in some cases, of its sensors as well. Virtually all destroyers and frigates now operate at least one helicopter for this reason, although the US Navy's Arleigh Burke class will have a helicopter landing deck but no hangar, its LAMPS III aircraft being required to use another ship's hangar facilities when necessary. This is an extraordinary reversal of an otherwise universal trend – even the latest built Soviet Krivak frigates have now overcome the one acknowledged shortcoming of this excellent design by having the full helicopter facilities which are missing on Krivak Is and IIs.

## Weapons fits

Weapons fits are the manifestation of national design philosophies, although appearances may sometimes be misleading. A study of contemporary surface warships suggests that Soviet naval staffs' top priority is the weapons systems, followed by propulsion, electronics, endurance and finally habitability, but that Western naval staffs give top priority to electronics and then to habitability, endurance, weapons and propulsion. However, Soviet naval weapons tend to be very obvious, as epitomised by the Slava class cruiser, which, quite literally, bristles with weaponry. The most striking feature is eight twin SS-N-12 SSM launchers lined up each side of the bridge, with a twin 130mm gun turret in the bows. AD armament comprises eight SA-N-6 and two SA-N-4 launchers, together with six 30mm gatlings, while ASM systems include two quintuple torpedo tubes, two RBU-6000 rocket launchers and a Hormone-B helicopter.

**Left: Air detection consoles in the spacious control centre of USS *Mount Whitney* (LCC-20). Effective integration of the electronic systems on modern warships poses a major design problem.**

**Above: The helicopter has become an invaluable extension of ships' sensor and weapons systems in the anti-submarine battle, where modern SSNs can travel faster than the hunting surface ships.**

**KEY**
**1.** Log measures distance.
**2.** Primary ASW inputs come from stern-mounted Variable Depth Sonar.
**3.** An essential input to the ASW equation is the ambient temperature gradient; this is supplied by periodic use of expendable bathythermo-graphs.
**4.** Anemometer gives wind-speed information.
**5.** Radio antenna for satellite communications, eg, to base and distant ships.
**6.** Satellite communications for ESM.
**7.** High-frequency (HF) links, eg, to base, and distant ships and aircraft.
**8, 9.** A whole range of devices conduct Electronic Support Measures (ESM), which involves monitoring the electromagnetic spectrum, classifying enemy transmissions and taking the appropriate countermeasures.
**10.** Airspace surveillance is conducted by a large radar, usually mounted at the highest point on the ship.
**11.** The airspace surveillance radar must include an Identification Friend-or-Foe (IFF) system.
**12.** Inputs from Airborne Early Warning (AEW) aircraft operating with the force.
**13.** Further inputs come in from ASW aircraft.
**14.** Input from force command ship, eg, aircraft carrier.
**15.** Submarines operating with a surface group can pass information by a variety of means; one is by use of a buoy, released at depth, and which transmits using a pre-recorded cassette when it reaches the surface.
**16.** Surface surveillance and navigation radar.
**17.** Visual information, eg, from the officer-of-the-watch.
**18.** ASW information from ship's helicopter.
**19, 20, 21.** Surface surveillance is making increasing use of modern techniques, eg, infrared, image intensification and thermal imaging.
**22.** Tucked away at the fore-foot is the major sonar.

For simplicity this diagram shows only one of most devices, and especially of the command-and-control systems within the hull. The South Atlantic War showed the shortcomings of such a concept, where a single operations room, which is the case in most current warships, could be lost to a single missile, thus leaving the ship with no stand-by. Much thought is now therefore being given to splitting the command-and-control functions and establishing duplicate data bases, so that the ship's nervous system cannot be totally disrupted by one enemy blow. It seems obvious, but took a war to prove it.

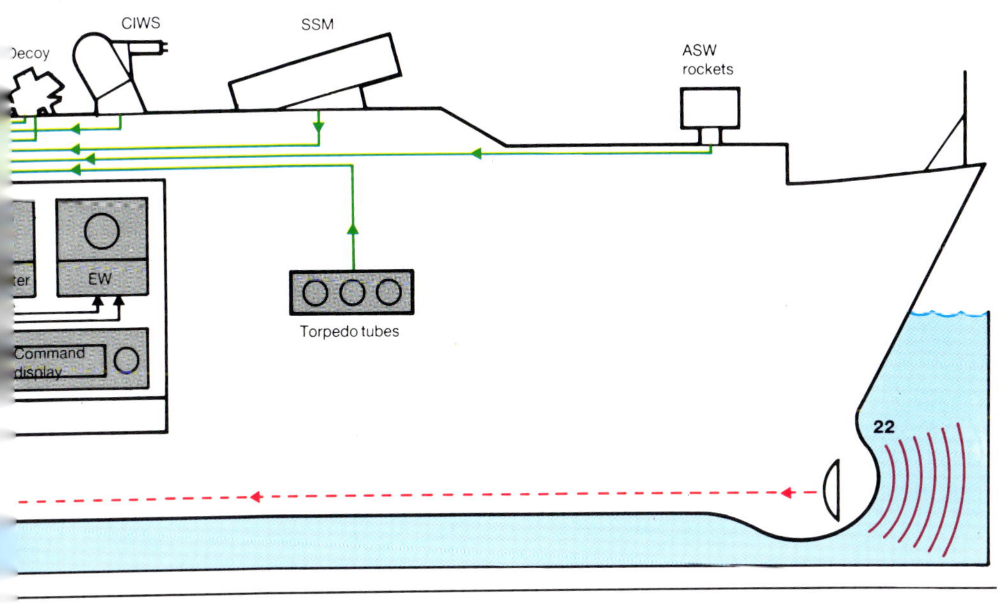

Every bit of deck-space on the Soviet ship seems to be taken up with something, while the equivalent US cruiser – USS *Virginia* – has large areas of totally free deck. *Virginia*'s surface weaponry includes two 5in (127mm) single gun turrets and two twin Harpoon launchers, with an AD armament of two twin mark 26 launchers (for Standard missiles) and two 20mm CIWS. ASW fit includes ASROC (fired from the same launchers as the Standard AAMs) and two triple torpedo tubes. No ASW helicopter facilities are installed on the first batch (CGN 38-41) but the "Modified Virginias" (CGN 42-45) will have hangarage for two LAMPS III.

Despite the apparent disparity in weaponry the Virginia class is as well armed as the Slava, and certainly has many more reloads, perhaps reflecting the different concepts of a future conflict: the Soviets aiming for an all-out, short-lived onslaught, while the US expects a more protracted campaign of resistance.

## Electronics

One of the most critical areas on a ship is that of electronics, which includes not just communications systems but also sensors (eg, radars) and other emitters; a close look at any modern warship reveals a plethora of antennas covering many wavebands, especially on Soviet ships. This obviously can lead to serious problems of mutual interference, to the point where some systems could become totally inoperable. The problems can be alleviated by designers, especially by careful siting. But this is not always sufficient and if mutual interference cannot be avoided it may be necessary to switch off one emitter while another is transmitting. This is inefficient and could lead to unfortunate consequences if, for example, a radar were to be switched off just as an incoming SSM comes into range. The integration of electronic systems is, however, proceeding apace and once again digital computers should provide the means for a solution to this problem.

## Propulsion

Propulsion is an area of considerable development. The oil-fired steam-turbine system is now found less frequently (only the Soviets use it in new construction, eg, Sovremmeny class cruisers) and modern developments centre upon nuclear, gas-turbine and diesel systems. Nuclear propulsion provides an answer to many problems, including that of fuel, and removes the need for space-consuming uptakes, downtakes and fuel tanks, as well as reducing the demands for replenishment by the fleet-train. The Nimitz class CVNs, for example, with two pressurized-water cooled reactors are estimated to be able to steam for some 800,000 to 1 million miles (1,287,000 to 1,600,000km) between core replacements. The smallest nuclear-powered surface warship currently in service is USS *Bainbridge* (8,592 tons), although there is no technical reason why smaller ships could not be built, and the way forward may be shown by the French Rubis class SSNs, which are powered by comparatively small reactors.

Major development is concentrating on the marine gas-turbine, and ships as large as the British Invincible class aircraft carriers (19,500 tons) are now powered exclusively by such means. Gas-turbines have a good power-to-weight ratio and are not large in themselves, although they tend to take up much valuable top-space with uptakes and downtakes to cope with their air demands and exhaust. Gas-turbines are expensive in first costs, but offer quick start possibilities, not available from other means of propulsion. Since they cannot reverse, current marine gas-turbines require either a reversing gearbox or a controllable propeller. The latter has a quicker response than a fixed-blade system, but is some 5 per cent less efficient when moving ahead.

Diesel engines are much used, especially in small combatants, but have heavy maintenance demands and can cause problems over removal routes for overhauls.

Various combinations of engines are used or under development, most of which are intended to reconcile the different propulsion requirements for high-speed and cruising. Combined Diesel or Gas-Turbine (COGOG) utilises diesels for cruising up to about 18 knots, giving good endurance, while the gas-turbine is used for quick starts and higher speeds. The German Navy uses this system in its Type 122 frigates in preference to the all-gas-turbine system used in the Dutch Kortenaar class, from which the Type 122 was developed. Such all gas-turbine systems (Combined Gas or Gas (COGOG)) use one type of turbine for cruise and another for high speed, each obviously optimised for its intended regime.

The Kortenaar's installation is typical, using two Rolls-Royce TM-3B Olympus (50,000shp) for high speed and two Rolls-Royce RM-1C Tynes (8,000shp) for cruising; the Kortenaar can accelerate from a standing start to 30 knots in under 2 minutes, a remarkable performance. A further combination is proposed for the Royal Navy's Type 23 frigate – Combined Diesel-Electric and Gas (CODLAG). In this design diesel generators will power elec-

**Above: The smooth lines and substantial armament of the French destroyer *Georges Leygues* (D-640) bespeak an efficient hull design, which is being utilised for both anti-submarine and anti-aircraft roles.**

**Below: USS *Ticonderoga* (CG-47) first of the cruisers taking the US Navy's AEGIS system to sea. The design of the broad-blade propellers shows the advances made in propulsion technology recently.**

**Above: The monohull has been supreme in ship design for centuries, but this is now being challenged by various new and revolutionary hull forms, including hovercraft, hydrofoils and catamarans.**

tric motors to give slow, very quiet operations for ASW, while the gas-turbines will be used for high-speed power.

## Hull design

The hull of a warship must be designed to accommodate the necessary sensors, weapons systems, propulsion machinery and crew; it must also be capable of allowing these to operate in all weather conditions and survive in a combat environment. Hull design is gradually evolving as new techniques and research are brought to bear.

Hydrofoil and hovercraft combatants are in production which are faster than monohulls and competitive with them in moderate sea-states, but with poor performance in very bad weather. Hovercraft have particular advantages in amphibious warfare and, because of their low underwater noise signature and fast speed, may have applications in coastal ASW and minehunting roles. Catamaran hulls are fast, quiet and as good in a seaway as a monohull; they also provide a large deck area and have a lower noise signature than a monohull. The Royal Australian Navy is building a 170-ton catamaran minehunter, while the US Navy has operated a research vessel (USS *Hayes*, 2,876 tons) and two Pigeon-class submarine rescue vessels for some years. These catamarans have been successful but the idea is not spreading fast; doubts remain over their suitability as major warships, especially in coping with asymmetric damage.

In general, therefore, the monohull remains supreme, largely because it is cheaper, well-understood and gives the best all-round performance. But there is still scope for refinement and there is much study of the advantages which might be offered by a shorter, beamier hull than is current practice. The "Osprey" design was originated in Britain, but the only craft to be built so far have come from Denmark. With a 505-ton displacement these ships have a length of 164ft (50m) and a beam of 34.5ft (10.5m): a beam-to-length ratio of 1:4.76, compared to the more usual 1:9 or higher. The US Navy appear to be following this trend with the new Arleigh Burke class (length 466ft/142m, beam 59ft/17.98m) compared with the current Spruance design (526.2ft/171.7m long, beam 55.1ft/16.8m), a beam-to-length ratio of 1:7.9 compared to 1:10.2. These shorter, beamier ships are claimed to confer better sea-keeping abilities as well as improved manoeuvrability, together with more usable space. The Osprey-class, for example, despite its low displacement, has a large helicopter deck.

Modern design and construction techniques have enabled shipbuilders to produce very efficient structures for surface warships. One major problem is the ever-increasing demand for equipment, especially sensors and antennas, to be as high as possible; coupled, in gas-turbine ships, with large downtakes and uptakes, this raises the metacentric height and reduces stability unless the superstructure can be lightened in compensation. One answer in the 1960s was to use aluminium in the upperworks, but the disastrous fire which occurred when the USS *Belknap* collided with the USS *John F Kennedy* in 1975 brought this solution into disrepute, and virtually every navy now uses steel again.

The effects of hits by missiles and bombs on Royal Navy ships in the South Atlantic War has led to a reconsideration of armour plating, at least for the more vulnerable areas, but at a weight of about 560lb (254kg) per square foot ($0.1m^2$) to defeat a 1,000lb (454kg) bomb, the effect on displacement is obvious. The US Navy's Arleigh Burke class is reported to have "unusual hull hardness, particularly around the ship's vital zones", which might well refer to an advanced form of armour possibly based on a ceramic construction, like the British Chobham tank armour.

A further form of passive protection is to disperse, and possibly duplicate, the ship's vital functions. The present practice of centralizing all weapon and sensor functions in one comprehensive operations room is not an ideal answer in combat, and a second, alternative "nerve centre" is required. Distributed data processing is also necessary, together with a replicated data base, to ensure that the ship's systems cannot be totally disabled by one hit on a solitary mainframe computer.

**Below: Artist's impression of the proposed Arleigh Burke class for the US Navy. The hull is notably shorter and beamier than its predecessors, leading to better sea-keeping and manoeuvrability.**

It is not surprising that there is more discussion about warships' armament than any other element in their system, since in all but a very few cases the weapons are the *raison d'être*. A surface warship needs to be able to attack and defeat other surface warships, submarine and aircraft, but to be able to mount all the sensors and weapons needed for all these roles would require a very large hull. Most ships of 10,000 tons or less are therefore optimised for one particular regime, with less than the optimum for others, relying on a task group to consist of a balanced mix of ships and give the necessary overall capability. Perhaps more than in any other environment, naval forces have concentrated on missile development for both offensive and defensive purposes and there can be little doubt that naval missiles have been developed to a higher point than in any other form of warfare.

## Anti-ship missiles (ASM)

ASMs first came to prominence when a Soviet-designed SS-N-2 missile from an Egyptian Navy patrol boat sank the Israeli destroyer *Eilat* in 1967. The impression then formed was reinforced in 1982 when Argentine Navy aircraft used the French Exocet missile in a series of attacks on British ships, leading to the sinking of HMS *Coventry* and the merchant ship *Atlantic Conveyor*. Although on both these occasions the missiles were actually air-launched, the Exocet is widely deployed in its shipborne form and in fact a County class destoyer was damaged by an Exocet launched from a naval mounting, albeit mounted on a wheeled trailer located near Port Stanley.

Despite the world-wide publicity generated by the loss of large ships to one missile each, a sober assessment shows that more Exocets were foiled than achieved hits, and that the hit on HMS *Glamorgan* did only limited damage. Further, the Exocets used by Iraq in the Gulf War against supertankers have rarely put the ships totally out of action.

The Exocet, like many other ASMs is subsonic and the new generation of fast-reaction SAMs and close-in weapons systems (CIWS) guns can almost certainly defeat such an attack under most circumstances. However, the latest ASMs, such as the Soviet AA-N-19, have speeds of up to Mach 2.5 and may well strain current defences beyond their limit.

An incoming Exocet flying at a height of 7 to 10ft (2 to 3m) and a speed of Mach 0.9+ could probably be detected by a frigate's surveillance radar at a range of some 10nm (18.5km), which is approximately 60 seconds before impact. Radar tracking would be initiated at a range of 3.4nm (6.3km) (ie, with 20 seconds to go), and a missile, such as Seawolf, launched when the incoming ASM was 2nm (3.74km) distant (ie, about 10-13 seconds before impact). This is dramatically short and requires computer control; indeed, several reports from the South Atlantic War stated that the first time ships' crews became aware of an imminent attack was when the defending Seawolf missiles were launched; without the computer, that would have been just too late!

However, if one of the newer missiles were travelling at Mach 2.5 and at the same sea-skimming height, the distances and timings would be very different; for example, first acquisition at 10nm (18.5km) would be only 19 seconds before impact.

As an offensive weapon the ASM forms an invaluable addition to a ship's armament, although no ship-on-ship engagement has yet taken place using only ASMs. The Soviet Kirov class battlecruiser mounts 20 S-N-19 ASMs in vertical launch tubes on the foredeck, although Soviet cruisers tend to mount their ASMs in deck-

**Below: The French Naval Crotale air-defence missile is a derivation of the successful land-based version. Like many naval surface-to-air missiles it has a secondary anti-ship capability.**

**Point defence:** A vital element in a modern warship's armoury is the air defence missile system – frequently, but not always, supplemented by a gun-based close-in weapons system (CIWS). Because of the inherent range limits on surveillance radars on ships at sea, air defence missiles must have rapid response times, high acceleration and good manoeuvrability. During the Falklands War of 1982 the first that ship crews knew of some incoming air attacks was when the Sea Wolfs launched under automatic control; human responses are too slow.

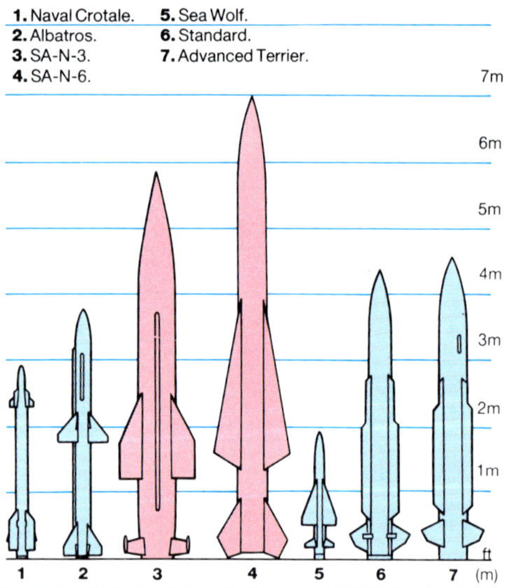

## SHIPBORNE SURFACE-TO-AIR MISSILES

| **Designation:** | **Naval Crotale** | **Albatros** | **SA-N-3 (Goblet)** | **SA-N-6** | **Sea Wolf** | **Standard (RIM-66A)** | **Advanced Terrier** |
|---|---|---|---|---|---|---|---|
| **Country:** | France | Italy | USSR | USSR | UK | USA | USA |
| **Type:** | Self-defence missile for ships against short-range airborne targets at medium and low altitude | Naval self-defence missile for use against medium and low altitude targets | Ship-borne air defence missile | Ship-borne air defence missile | Short range anti-aircraft and anti-missile system | Medium-range air defence missile system | Long-range air defence missile system |
| **Length:** ft(m) | 9.64 (2.94) | 12.1 (3.7) | 19.3 (5.9) | – | 6.2 (1.9) | 14.6 (4.47) | 14.7 (4.5) |
| **Diameter:** ft(m) | 6.2 (16) | 7.9 (20.3) | 15.7 (40) | – | 7.1 (18) | 13.5 (34.3) | 13.5 (1.17) |
| **Wingspan:** ft(m) | 21.2 (54) | 39.3 (100) | 0.47 (1.2) | – | 22 (55.9) | – | 42 (107) |
| **Range:** miles(km) | 11 (18) | – | 18.6 (30) | 37 (60) | 4 (6.4) | 11 (18) | 21.5 (35) |
| **Speed:** mph(km/h) | Mach 1.2 | Mach 2.5 | Mach 3 | Mach 6 | Mach 2 | Mach 2 | Mach 2.5 |
| **Guidance:** | Line-of-sight command and proportional navigation | Semi-active homing | Command guidance | Command guidance | Command line-and-sight | | Beam rider |
| **Warhead:** lb(kg) | 31 (14) HE with proximity fuse | | 88 (40) HE with proximity fuse | | 31 (14) HE with proximity fuse | | HE with proximity fuse or impact |
| **Remarks:** | Has secondary anti-ship capability | Designed to be integrated with gun batteries | | Few details yet available: fitted to *Kirov* | Now under development for vertical launch | Part of Aegis system | Constantly developed since entering service in 1956. |

mounted tubes set at about a 20° angle; in neither case does it appear that reloading at sea is possible. The SS-N-19 has an (estimated) range of some 310 miles (500km) and a speed in the region of Mach 2.5. External target detection and designation from aircraft is probably combined with on-board inertial guidance and homing systems.

The US Navy's principal ASM, the RGM-84A Harpoon utilizes very similar techniques, but has a turbojet cruise engine and its flight profile is much lower than that of the SS-N-19, but with a final manoeuvre phase to evade CIWS defences. No data inputs are required from the ship after launch; a computer guides the missile to the vicinity of the target where a frequency-agile homing radar takes over for the final phase.

The missile guidance system is programmed to aim for the centre of the hull where it will have most chance of damaging the target's vital services and, in particular, may hit the command centre. Some of the larger, anti-ship missiles, such as the Soviet SS-N-3 "Shaddock", have nuclear warheads. Most anti-ship missiles, however, use high-explosive warheads. Some use a hollow-charge warhead, which on impact against the ship's side is detonated and a jet of molten explosive projected into the interior of the ship where it causes secondary explosions and fires. Other missiles (for example, the US Navy's Harpoon) use a penetration blast warhead, with both proximity and impact-delay fuses.

Following the successful use of the Exocet missile in the South Atlantic War there have been a series of claims and counter-claims about the missile's performance. The Royal Navy has claimed that some of the warheads did not detonate, but that the missiles drove into the ships' interiors by the force of their impact, following which the unexpended fuel ignited, leading to fires in the ships. The manufacturers of the missile, not surprisingly, contest this.

The anti-ship missile is not, of course, infallible and various countermeasures have been devised. Chaff-launchers pro-

**Above: The British Sea Wolf SAM was proven in combat in the 1982 South Atlantic War; ships' crews often first knew of approaching targets when the Sea Wolf launched itself automatically.**

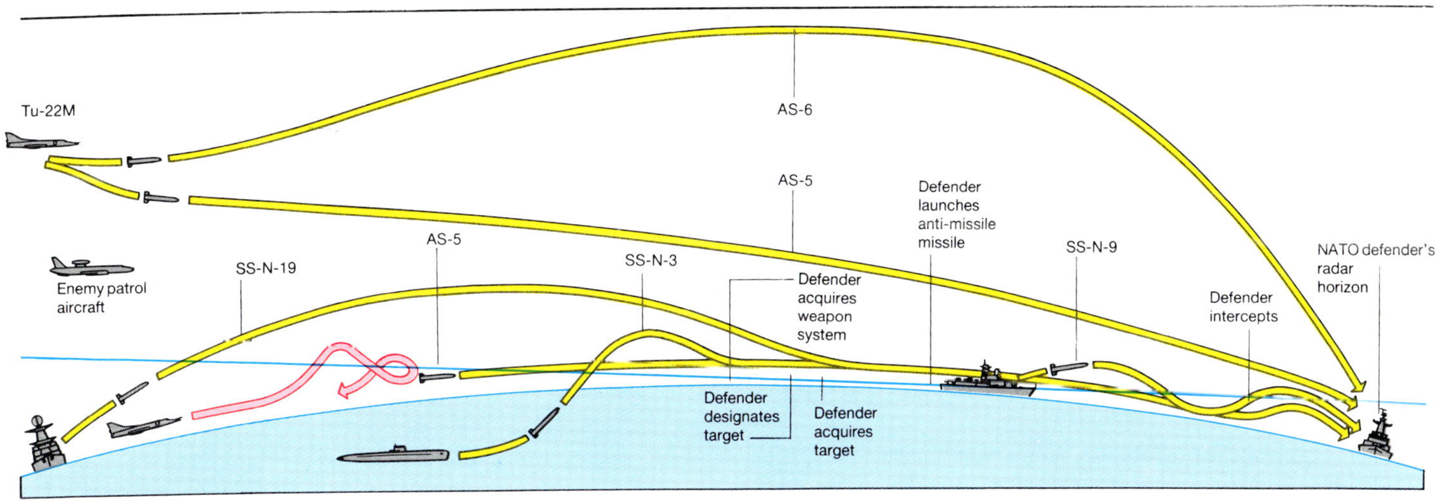


**Above: The missile threat**
The Soviet anti-ship missile threat to a NATO warship is serious. It ranges from aircraft-launched ASMs (eg, Tu-22M "Backfire" and AS-5 or AS-6), through ship-launched ASMs (eg, *Kirov* and SS-N-19) to submarine-launched missiles (eg, Charlie class and SS-N-9). A further hazard for the defence is the growing number of ASMs using the sea-skimming mode to minimise the radar warning time given to the target.

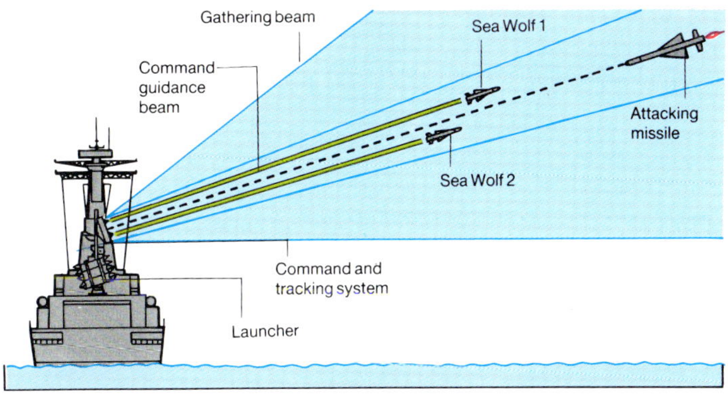


**Left: Sea Wolf defence**
The main elements of the Sea Wolf system, equally effective against missiles which dive on the ship or skim the wavetops. This system uses line-of-sight guidance with radar differential tracking or TV, both with radio command. Error signals, proportional to the missile deviation from the datum, are processed and coded correction signals sent to the missile to return it to the correct flight-path.

**Below: Decoys**
Another method of dealing with incoming ASMs is to "persuade" them to attack a decoy, which, to the missile detection system, appears to be a ship. One such system is the Plessey Shield which provides both "distraction" and "seduction" protection for naval and merchant ships of any size. In the distraction mode four separate chaff clouds are provided at ranges up to 1.25 miles (2km).

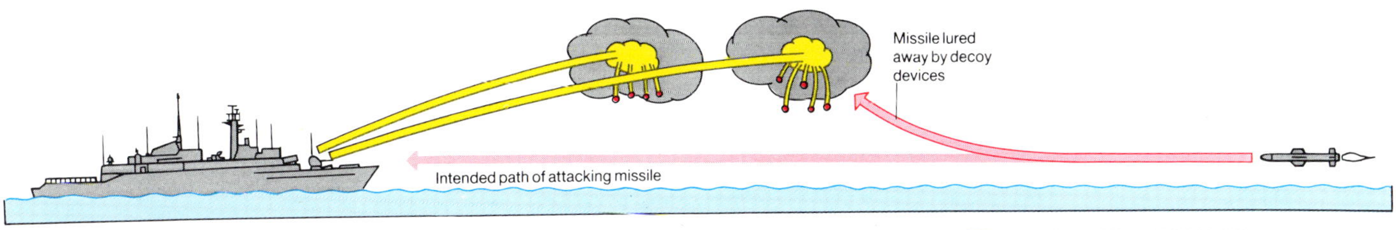

**Gun versus missile**
One method of dealing with incoming ASMs is by direct fire from high performance gun systems. However, an Exocet has a cross-sectional area of 149sq in (962sq cm) and to engage a target of this size, travelling at near super-sonic speed and just above wave level, and at sufficient distance from the ship, is a major undertaking. Many navies, however, have now produced CIWS to meet this requirement, such as the US Phalanx, the GEC/Hollandse Goalkeeper and Soviet 30mm "Gatling". The diagram below shows the triggering radius for Bofors PFHE Mk2 ammunition in the Breda Twin 40 L 70, which has to be minimised to adjust to sea clutter, a factor which is more critical the nearer the target ASM is to the wavetops. This demands exceptional accuracy from the proximity fuze fitted in the nose of each round, but this is now being achieved, leading to a return of interest in the gun.

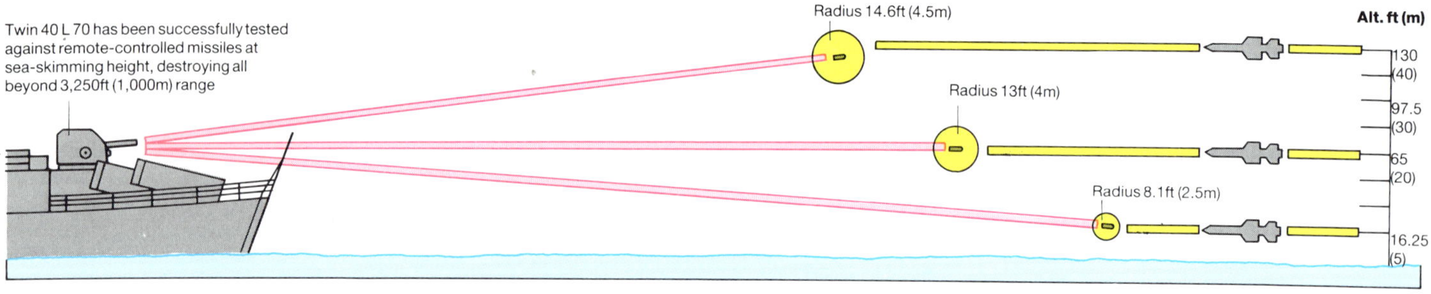


Twin 40 L 70 has been successfully tested against remote-controlled missiles at sea-skimming height, destroying all beyond 3,250ft (1,000m) range

vide a simple and cheap method of defence against incoming surface or air-launched ASMs, in which chaff-dispensing rockets are fired to form a radar decoy screen around the ship. Similarly, rockets are also used to launch infra-red decoys to confuse ASMs with that type of guidance system. These decoys usually take the form of infra-red flares equipped with both a parachute and a float, and which can simulate the radiant intensity of even a large ship. However, such methods are obviously not totally effective on all occasions, and must therefore be used in combination with quick-reaction guns and missiles, as described below.

## Guns

The naval missile school became so influential in some navies that powerful lobbies developed to do away with guns altogether and some warships in the 1970s appeared with only a token gun armament, eg, the British Type 22 with just two 40mm AA guns. The folly of such a proposition has repeatedly been shown and the need for at least one effective gun is now accepted. Fortunately modern naval guns are now extremely compact, and automated to the point where no turret crew is required. The US Navy's 5in/54 calibre Mark 45, for example, weighs 25 tonnes and fires 20 rounds per minute; total crew is six, none of whom are in the turret.

One of the most successful Western guns is the Oto Melara single 76mm/62 calibre gun, which has a firing rate variable between 10 and 85 rounds per minute. In service with 35 navies, the total installation weight is 7.5 tonnes. Apart from reducing weight and increasing automation, gun designers have managed to squeeze ever greater range out of the guns, and some guns can now fire rocket-assisted projectiles (RAP).

The threat from air attack and sea-skimming missiles has given rise to a requirement for close-in weapons systems (CIWS) for "last ditch" defence, in which extremely fast reaction time and a heavy volume of fire are of paramount importance. The US Navy has developed the Phalanx CIWS to meet this need. It marries the M61A1 Gatling gun to a very intelligent control system. The radar tracks the outgoing projectile and the incoming missile and seeks to eliminate the difference; the weapon has proved highly successful in service. The system is completely self-contained, unlike the 30mm CIWS on Soviet ships, which utilises the separately-mounted Bass Tilt radar. Nevertheless the Soviet CIWS, also a six-barrelled "Gatling" design, is an extremely neat installation and is fitted to numerous classes of Soviet ship. Another "stand-alone" CIWS is the seven-barel Goalkeeper, by Hollandse Signaal Apparat, which has a rate of fire of some 4,200 rounds per minute.

**Above: Italian Oto Melara 76mm/62 naval gun, here seen on a land-based test-rig, has been adopted by at least 35 navies; its firing rate is variable between 10 and 85 rounds per minute.**

**Above right: The South Atlantic War caused some major rethinking about ships' gun armaments, and there is a marked return to guns of various sizes especially in the 20 to 40mm calibre bracket.**

**Below: There is increasing emphasis in most navies on quick-reaction, rapid-firing Close-In Weapons Systems (CIWS), such as the US Phalanx, seen here, and a variety of excellent Soviet designs.**

**Oto 76/62 gun**
The Oto 76/62 is a good example of a modern, lightweight, dual-purpose naval gun system. The gun is remotely-controlled and can be laid rapidly on to fast-moving targets and then maintained in lay by a stabilization system. Feeding and loading systems are hydraulically operated through a self-contained power unit, and the firing rate is selectable from single rounds up to 120 rounds per minute. The revolving feed magazine contains 70 AA rounds (with another 10 in the feed system); surface ammo is manually loaded into the hoist. The gun is claimed to ensure a high kill probability against sea-skimming missiles and aircraft targets, including three simultaneous attacks by 3 targets in a 30° arc.

**Key:**
**1.** Rocking arms and loader drum; transfer rounds from the feeder hoist to the elevating mass.
**2.** Screw feeder hoist.
**3.** Revolving feed magazine, manually loaded and automatically feeding AA ammunition to the screw feeder hoist.
**4.** Presetting device to switch to surface ammunition loading mode.
**5.** Surface ammunition loading station.

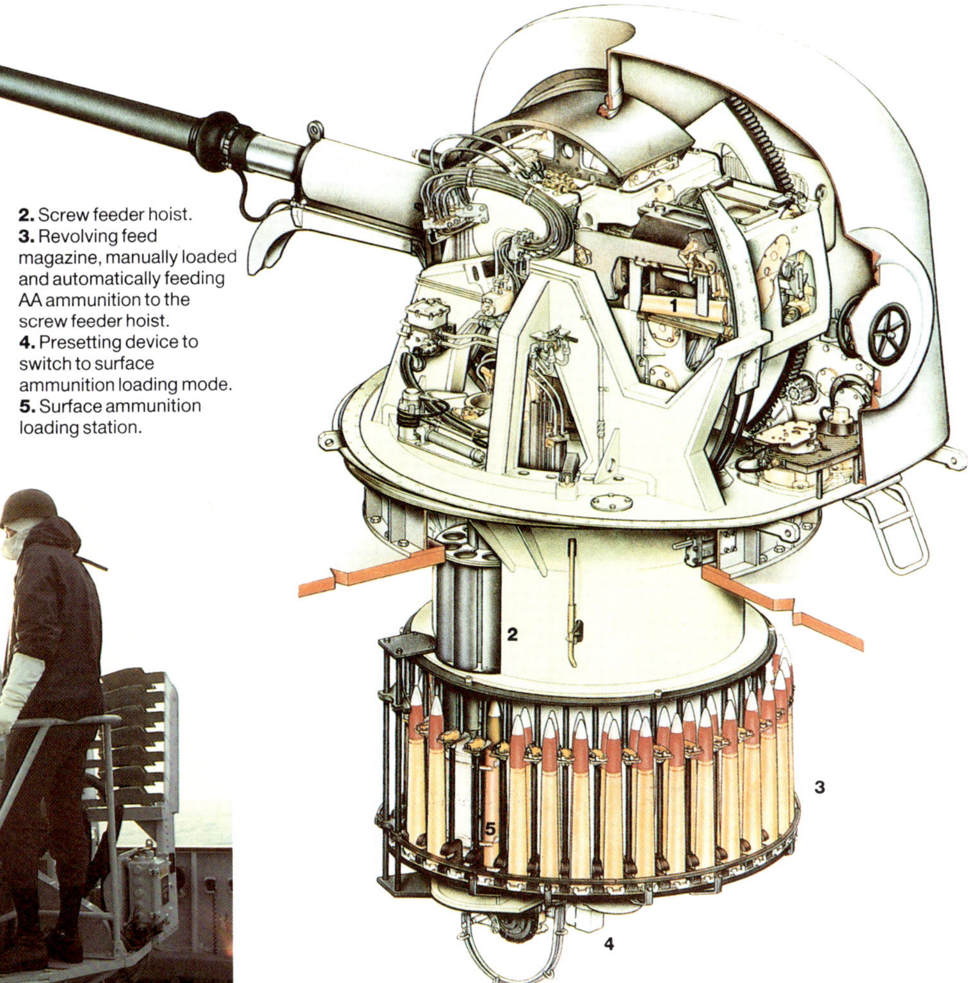

**Creusot-Loire 100mm Compact naval gun**
The revived interest in naval guns stems mainly from their high efficiency and the great degree of automation now achieved, which cuts down very considerably on the once-numerous gun crews. Turrets are rarely manned and even the feed system for ammunition now requires very few men to run. The Creusot-Loire Compact is a dual-purpose 100mm gun with a 90 rounds per minute rate of fire. The barrel has a double water-cooling system and has a life exceeding 3,000 rounds.

**Key:**
**1.** Mount power supply unit.
**2.** Compressed-air panel.
**3.** Elevation motor.
**4.** Train motor.
**5.** Mount electronic box.
**6.** Armature shunt box.
**7.** Compressed-air supply.
**8.** Water-cooling unit.
**9.** Hydraulic power unit.
**10.** Converter unit.
**11.** Magazine captain's box.
**12.** Magazine.
**13.** Hoist.
**14.** Remote control cabinet.
**15.** Processing cabinet.

Electric circuit
Water system
Hydraulic system

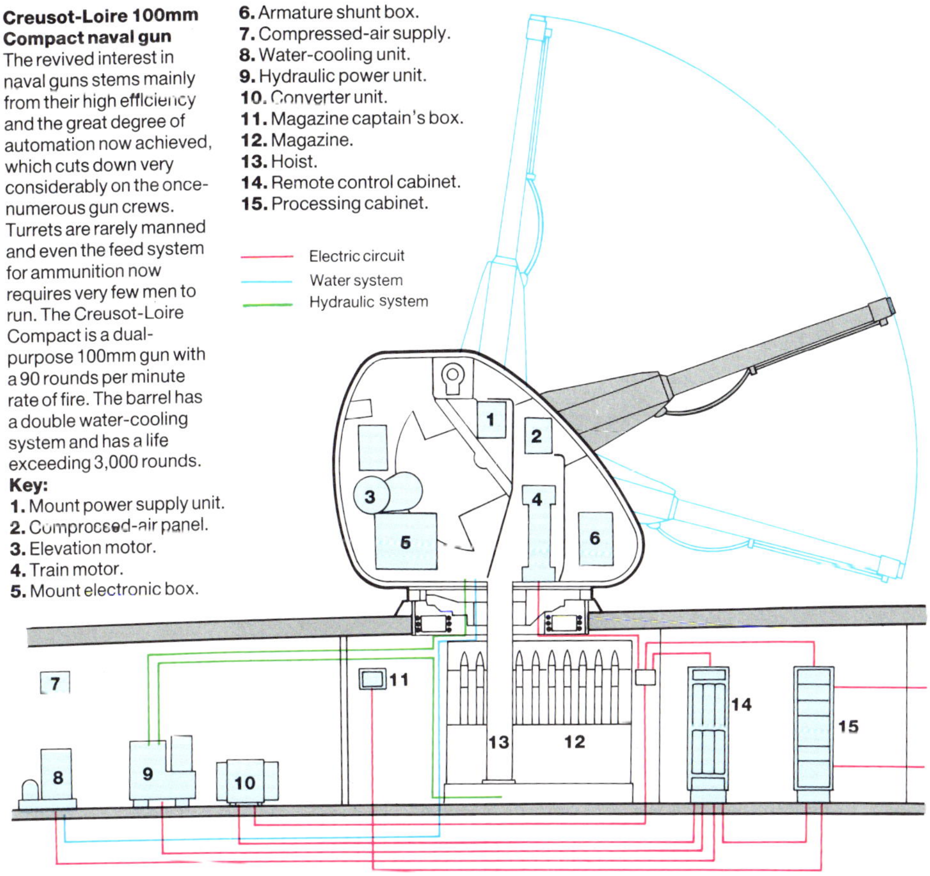

By far the largest gun currently in service is the US Navy's 16in (40.65cm), of which nine are mounted on each of the Iowa class battleships. These have a range of some 24.85 miles (40km). It is highly unlikely that guns of this calibre will ever be made again, and even the latest of the Soviet Kirov class battlecruisers mounts only one twin 5.1in (130mm) turret.

## Surface-to-air missiles

By far the most sophisticated air defence system currently at sea is the US Navy's Aegis, which is based on the AN/SPY-1A multi-function phased-array radar and uses the Standard SM-2 missile. This missile has a monopause receiver, an inertial reference unit for mid-course guidance, a two-way telemetry link for missile-position reporting, target position updating and guidance correction, and an on-board digital computer. The inertial reference unit enables the missile to navigate itself to the vicinity of the target where the semi-active radar-head takes over. The SM-2 is able to take full advantage of the Aegis' multiple target capability.

The British Seawolf is one of the most advanced anti-missile systems. The entire sequence is automatic and operations are timed in milliseconds. The Seawolf has destroyed an incoming 4.5in shell during tests; it did well under actual combat conditions in the South Atlantic War.

"The Fleet Carrier – the most impressive fighting machine the world has ever seen. This one ship can unleash a greater variety of lethal weapons with a greater destructive power, at longer range, than any man-of-war in history." These words by Admiral of the Fleet Lord Hill-Norton, former British Chief of the Defence Staff, accurately describe the current position of the aircraft carrier, the capital ship without parallel on today's oceans. The modern American carriers, with command centres rivalling those ashore, are capable of launching nuclear strikes and, with 90 of the very latest types of aircraft, have an air capability greater than that of all but the most sophisticated land-based air forces.

For many years the Western navies have dominated the aircraft carrier scene, with the US Navy maintaining a fleet of ever larger units, although the once significant British carrier force has dwindled away to almost nothing. The French maintain a two-carrier force but, these apart, there are a number of small (mostly ex-British) carriers scattered around the more sophisticated of the second-level navies.

Meanwhile, until recently, the Soviets, with no tradition of naval airpower, had nothing, despite the evidence from their writings that they considered the threat from nuclear-armed, carrier-borne aircraft to be of great significance to the Russian homeland, and numerous ship and submarine classes and weapons systems were developed to counter it. The Soviets themselves seem to have felt little need for carriers as their maritime ambitions apparently had not extended to blue-water power politics. But that position has dramatically changed and they are now posing an ever-growing threat to the West at sea. Admiral Gorshkov himself is on record as saying that: "The Soviet Navy will no longer be confined to its home waters, but will exploit the freedom of the seas and through its global presence in peacetime will spread Communist influence outside the borders of the USSR. Sea Power without Air Power is senseless."

## V/STOL less expensive

As in other naval spheres the cost of aircraft carriers is escalating rapidly; the next two US carriers of the Nimitz class will cost well over $2,000 million each for the ships alone, without the cost of the air wing. A number of navies have therefore been looking for a cheaper way to take airpower to sea, and in this the V/STOL (vertical/short takeoff and landing) aircraft seems to offer the only realistic answer.

There are currently four types of aircraft carrier. At the top end is the CTOL carrier (conventional take-off and landing), of which current examples range from the mighty American CVNs of 89,600 tons displacement down to the elderly carriers of about 20,000 tons in use in some navies. These ships operate air wings of a size proportional to the tonnage, and usually in a mix of strike, air defence, patrol and ASW types.

The giants of the current carriers are the US Navy's Nimitz class, each capable of carrying up to 85 aircraft. These will be countered by the new Soviet CVN of some 75,000 tons full load displacement, which is due for sea trials by the end of the decade. But the Soviet Navy is unlikely to be able to match the US Navy in numbers for perhaps 40 to 50 years. The French Navy is firmly committed to two new nuclear-powered carriers of some 50,000 tons (PAH-1, -2), but no other navies are known to have plans for CVNs at this time.

Second come the new medium-sized carriers with full-length flight-decks, but

Left: USS *Saratoga* (CV-60), like other fleet carriers, capable of unleashing a greater variety of lethal weapons with a greater destructive power, at longer range, than any man-of-war in history.

Below: The V/STOL Sea Harrier, seen here returning from a Combat Air Patrol (CAP) mission in the South Atlantic War, has given rise to a new breed of lighter, simpler and cheaper aircraft carriers.

Above: The first true aircraft carrier in the Soviet Navy, the four-ship Kiev-class is a precursor to the 80,000 ton fleet carriers (similar in concept to the US Navy's) now under construction.

which operate the V/STOL Sea Harrier or Yak-36 "Forger", together with large numbers of helicopters. These range in size from the Spanish *Principe de Asturias* (14,700 tons) to the British Invincible class (19,500 tons), with the Soviet Kiev class (42,000 tons) at the top end.

Similar in concept to these, but with a different primary purpose, are the third type, the amphibious assault ships, which have a large straight-through flight-deck for V/STOL aircraft and helicopters, but which also carry large marine forces (usually a battalion group). Epitome of this group is the US Navy's Tarawa class (39,300 tons), which can carry up to 19 of the huge CH-53 Sea Stallion helicopters, together with some 1,700 troops.

Fourth comes a small group of ships in the 8,000 to 20,000 tons category with half-length flight-decks aft for operating helicopters, or possibly AV-8A/C Harriers, operating in the VTOL mode. There are also the ever growing number of small warships with a stern flight-deck for one or two, usually ASW, helicopters.

The future of the large nuclear-powered supercarrier seems assured in the US and Soviet navies, with the Americans planning for a fleet of 15 and the USSR for at least two, and probably many more. These large carriers are certainly very impressive and carry significant air power; they are also extremely effective command, communications and control ($C^3$) centres. But they are also massively expensive and, despite their air wing, on-board armament and protective escorts, they are still vulnerable. They are certainly prime targets for submarine-, ship- and air-launched missile attacks, and could well also be targeted by land-based long-range missiles.

The British Royal Navy's experience with CVA-01 has shown that designs can become too expensive, and that the cost of large carriers, plus the concomitant cost of the specialised aircraft for the air wing, can exceed the budget. It would appear then that technology will be applied to producing much cheaper carriers, thus enabling more effective and more survivable carriers to join the smaller navies – and perhaps the larger navies, too.

For the air component of carrier fleets it seems there are two major areas for development. First is the yet more effective and capable V/STOL aircraft, and the second is the development of a viable fleet airborne AEW platform, which does not need a large CTOL aircraft carrier.

## Flight-deck design

The design of the flight-deck is obviously crucial to the efficient operation of an aircraft carrier. In this a number of innovations have played a vital role since the advent of the turbojet-powered aircraft, including the angled flight-deck, automated approach and landing aids, and the steam catapult. Further improvements, such as deck-edge lifts, represent sensible, evolutionary steps to increase usable deck areas. The abortive British CVA-01 design, although conceived in the 1960s, still probably represents the ultimate arrangement for economical and efficient use of deck-space.

The alternative approach is the straight-through flight-deck, but this is now only suitable for V/STOL aircraft. In this the major breakthrough has been the invention (by Commander Douglas Taylor, RN) of the "ski-jump", which like so many great inventions is cheap and truly simple. Not only is the jump very easy to construct, but no modifications whatsoever are required for the aircraft, while pilots find it safe and simple to use. Ski-jumps have already been fitted to the British carriers *Hermes* (12°), *Invincible* (7°), *Illustrious* (7°) and *Ark Royal* (12°), and to the Spanish *Principe de Asturias*. The US Marine Corps has bought three ramps for training at airfields but, so far as is known, no proposals have yet been made for a US Navy ship to be constructed with such a fitting.

The unique capabilities of the V/STOL Sea Harrier, combined with the ski-jump, have led to a wholly new concept in which a suitable merchant-ship could be converted into an aircraft carrier in a matter of days. The Shipborne Containerised Air--Defence System (SCADS) utilises some 230 standard International Standardisation Organisation (ISO) containers mounted on the deck of a 30,000-ton-plus container ship and topped by a specially prepared flight-deck. The containers can be fitted for roles such as missile launcher, fuel storage, aircraft maintenance, personnel accommodation, command post, and so on, to enable a thoroughly effective aircraft carrier to be produced. This is very similar to the US Navy's Arapaho system, which is currently being tested by the Royal Navy.

British Aerospace have even proposed doing away with a flight-deck altogether; their "Skyhook" design concept is a ship-mounted crane which would lift a Sea Harrier outboard and, when the pilot is ready, release it. Recovery requires the aircraft to hover in a notional 10ft (3m) cube area abreast of the crane position and moving forward at the same speed as the ship, whereupon the computer-controlled, space-stabilized crane locks on to the aircraft and lifts it on board. The system envisages all aircraft handling on the ship being done on mechanically operated trestles, which would not only lead to greater efficiency and more economic use of space, but could also much reduce the manpower requirements. Such a system could be mounted in a 5,000-ton hull, which could operate an air squadron of four Sea Harriers and two Sea Kings, thus giving the ship a capability far beyond that previously considered possible for such a small vessel.

An intermediate type flight-deck is used on board some cruiser-sized warships, mainly to enable a large number of helicopters to be operated. Such ships have a conventional superstructure and a large flight-deck aft, and include such types as the Soviet Moskva class (20,000 tons), the Italian *Vittorio Veneto* (8,850 tons) and the French *Jeanne d'Arc* (12,365 tons). Such designs enable a normal gun/missile armament to be mounted forward, thus considerably enhancing the ship's ability to look after itself, but the concept is confined to helicopters or VTOL fixed-wing aircraft, and would not be suitable for STOL operations.

## Aircraft on other ships

Finally, virtually every modern cruiser and destroyer, and most frigates now have flight decks and on-board facilities for one, if not two helicopters. This adds immeasurably to their ASW capabilities, since, as described in more detail elsewhere, the high speed of modern SSNs in comparison with ship and torpedo speeds means that the helicopter is essential to provide timely response at longer ranges. In addition, a number of modern anti-ship missile systems require an airborne relay/control station to give an Over-the-Horizon (OTH) capability. The smallest hull capable of accommodating a helicopter is the Osprey class of 505 tons, British-designed but built in Denmark, but these are fishery-protection vessels rather than warships, although, obviously, the Osprey design concept could be used for much larger warships. Surprisingly, the new US destroyer design – the Arleigh Burke – will have a flight-deck but no hangar, depending upon other ships in the task group for aircraft maintenance and repair facilities.

Most Western air-capable cruisers and destroyers have large and obvious hangars, whereas the Soviet Navy has developed a very neat (although possibly mechanically

**Length:** 591ft (180.2m).
**Beam:** 76.8ft (23.4m).
**Draught:** 22ft (6.7m).
**Flight deck:** 565×98ft (174×30.4m); no catapults; 6.5° ski-jump.
**Displacement:** 10,100 tons standard; 13,320 full load.
**Propulsion:** 4 2-shaft gas turbines, 80,000hp.
**Speed:** 30 knots (55.5km/h).
**Range:** 7,000nm (12,970km) at 20 knots (37km/h).
**Crew:** 550 plus 230 air group.
**Launched:** 1983.
**Commissioned:** 1985.

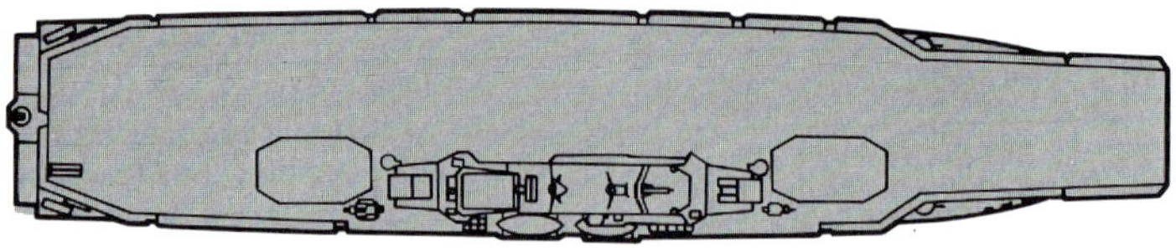

**Giuseppe Garibaldi** — **Italy**

**Air wing**
Accommodation for: 12 SH-3D helicopters in hangar, plus 4 on deck; or 10 Sea Harriers plus 1 SH-3D in hangar. (EH101 helicopters to replace SH-3Ds.)

**Armament**
3 sets of twin-mount 40/70 cannon plus Albatros launch rails and Aspide missiles for anti-aircraft and anti-missile; 4 double launch rails for Otomat Mk.2 anti-ship missiles; triple torpedo tubes.

**Length:** 643ft (196m).
**Beam:** 80ft (24.4m).
**Draught:** 30ft (9.1m).
**Flight deck:** 574×105ft (175×32m); 12° ski-jump; no catapults.
**Displacement:** 14,700 full load.
**Propulsion:** 2 gas turbines, 1 shaft; 46,000hp.
**Speed:** 26 knots (48km/h).
**Range:** 7,500nm (13,900km).
**Crew:** 774 ship, plus air group.
**Launched:** 1982.
**Commissioned:** Due 1986.

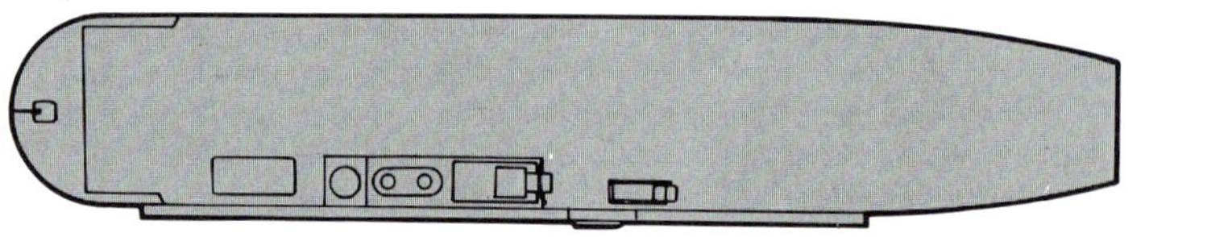

**Principe de Asturias** — **Spain**

**Air wing**
Accommodation for: 6 to 8 AV-8B (Harrier IIs) being built in US and UK; 6 to 8 Sea King helicopters; 8 AB 212 helicopters; maximum 20 aircraft.

**Armament**
Published so far: 4 12-barrel Meroka 20mm rapid-fire (CIWS) gun mounts; plus ECM chaff launchers. (Aircraft planned can of course carry wide range of air-to-air and ASW weapons.)

**Length:** 671ft (206.6m).
**Beam:** 90ft (27.5m).
**Draught:** 24ft (7.4m).
**Flight deck:** 545ft (167.8m) length; ski-jump varies from 7° to 15° with ship; no catapults.
**Displacement:** 16,000 tons standard; 19,500 full load.
**Propulsion:** 4 gas turbines, 2 shafts; 112,000hp.
**Speed:** 28 knots (52km/h).
**Range:** 5,000nm (9,265km) at 18 knots (33.3km/h).
**Crew:** 1,000 ship; plus air group.
**Launched:** 1977.
**Commissioned:** 1980.

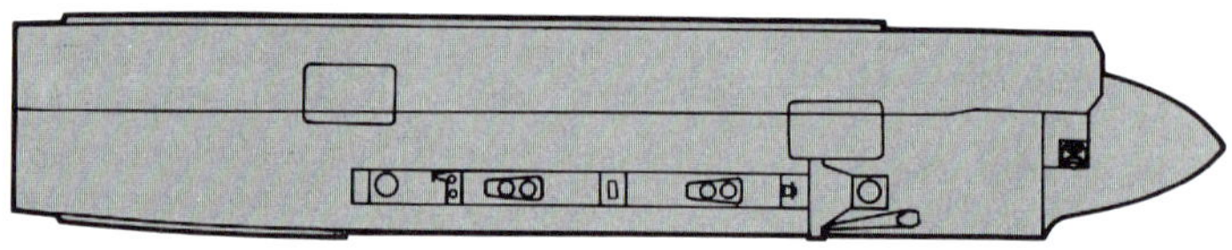

**Invincible** — **UK**

**Air wing**
*Invincible:* up to 10 Sea Harriers plus nine Sea King helicopters.
*Illustrious:* two extra Sea King AEW helicopters (to be 3 on all ships eventually).

**Armament:**
Two 20mm Phalanx CIWS; two 20mm cannon, plus Sea Dart missile installation, for anti-aircraft and anti-missile operation. (Sea Dart may also be used as an anti-ship weapon.)

Research project for use with BAe "SkyHook" system of launching Sea Harrier (and future V/STOL designs).
**Dimensions:** Small-to-medium ship with 250ft (77m) ski-jump flight deck plus drop-down hangar doors each side forming pads for additional launching of Sea Harriers direct from hangar deck level, via SkyHook "cranes".
**Displacement:** 5,800 tons.
**Propulsion:** Diesel.
**Speed:** 26 knots (48km/h).
**Range:** 4,500nm (8,340km) at 16 knots (30km/h).

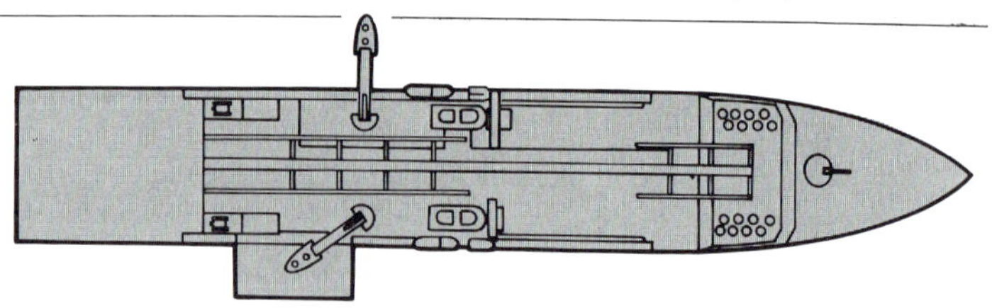

**"Vosper"** — **UK**

**Air wing**
At least five Sea Harriers (or future V/STOL fixed-wing aircraft), plus one helicopter for ASW or AEW, all in the hangar deck; possibly more on main deck.

**Armament**
This small escort carrier could be armed with a 100mm gun and two close-in weapon systems for anti-missile defence. It could also have extensive electronic warfare equipment and ASW sensors.

**Air power at sea**
Plan views are not drawn to scale, but an impression of scale is given by comparing all carriers to the Italian Garibaldi (top left). Getting air power to sea is critical to virtually all naval missions. Whereas 40 years ago the USA and the UK had the only effective carrier fleets, many more navies now have carriers, and the size and form of such ships are more varied than ever. The US Navy is aiming for 15 deployable carriers by 1990 with the majority being huge Nimitz class CVNs (top right). These have vast, clear, angled flightdecks, sidelifts and steam catapults. Their air wings are large and cover all necessary types. The US monopoly on super-carriers is being challenged by the USSR, who are building a 75,000 ton ship, believed to be called the *Kremlin*), for launching in 1987. The general layout, revealed by satellite pictures, is very similar to US practice. A total of 8 may be planned.

**Length:** 1,092ft (332.8m).
**Beam:** 251ft (76.4m).
**Draught:** 37ft (11.3m).
**Flight deck:** 1,092×252ft (332.8×76.8m), increasing by 5ft (1.5m) on 3rd ship, *Theodore Roosevelt*; 4 catapults.
**Displacement:** 81,600 tons standard; 91,400 full load.
**Propulsion:** 4-shaft nuclear, 2 reactors, 260,000hp.
**Speed:** 30 knots (55.5km/h).
**Crew:** 3,300 ship, plus 3,000 air wing.
**Launched:** 1972.
**Commissioned:** 1975.

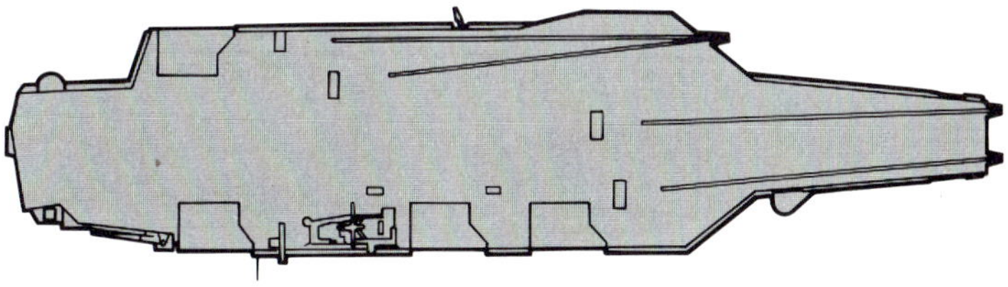

**Nimitz** **USA**

**Air wing**
24 F-14A Tomcat; 24 A-7E Corsair; 10 A-6E Intruder plus 4 KA-6D; 4 E-2C Hawkeye; 4 EA-6B Prowler; 10 S-3A Viking; plus 6 SH-3H Sea King helicopters.

**Armament**
3 Mk25 Basic Point Defence Missile System (BPDMS) launchers with Sea Sparrow missiles; being replaced by Mk29 launcher for NATO Sea Sparrow; plus 3 Phalanx CIWS guns.

Few estimates have been released of this aircraft carrier under construction in a Black Sea yard, except that it is assumed it will be similar to the large US Navy carriers, with a flight deck about 1,100ft (338m) long. Displacement is thought to be 75,000 tons. Sea trials could begin in 1988, and operational service within a few years from then.
*Note: Details based on Jane's Defence Weekly report by Robert Hutchinson and John W. R. Taylor.*

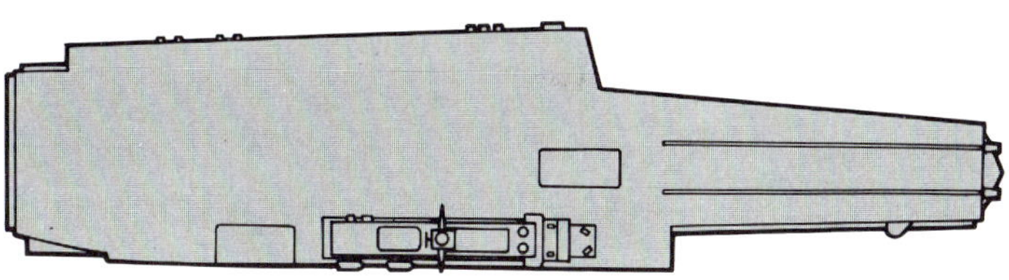

**"Kremlin"** **USSR**

**Air wing**
Possibly up to 70 tactical aircraft, including MiG-29 "Flanker" and "navalised" MiG-23 "Flogger" – but not Yak-36 "Forger" VTOL, apparently.

**Armament**
Vertically-launched missiles, possibly surface-to-air, in silos forward of island. No doubt the carrier (of which up to 8 may be planned?) will also have "CIWS" gun system for anti-missile defence.

**Length:** 899ft (274m).
**Beam:** 158ft (48m).
**Draught:** 33ft (10m).
**Flight deck:** 620×68ft (189×20.7m); 4.5° angle; no catapults.
**Displacement:** 33,000 tons standard; 38,000 full load.
**Propulsion:** 4-shaft geared turbines, 180,000hp.
**Speed:** 32 knots (59km/h).
**Range:** 13,500nm (25,000km) at 18 knots (33km/h).
**Crew:** 1,200 ship, plus air group.
**Launched:** 1972.
**Commissioned:** 1975.

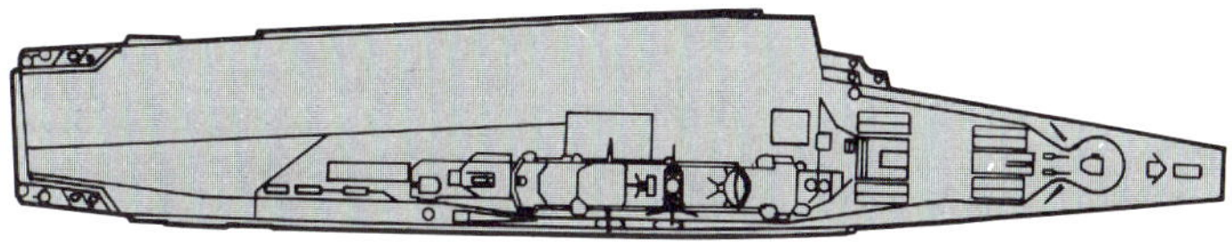

**Kiev** **USSR**

**Air wing**
12 Yak-36MP "Forger" V/STOL attack (2-seat trainers also carried); 14 Ka-25 "Hormone-A" helicopters for ASW; 6 Ka-25 "Hormone-B" for missile guidance.

**Armament:**
8 SS-N-12 SSM launchers; 2 twin SA-N-3, 2 twin SA-N-4 launchers; 2 twin 76.2mm DP guns; 8 30mm Gatling CIWS; 1 twin SUW-N-1 ASW launcher; 2 RBU-6000 launchers; 2×5 21in (533mm) torpedo tubes.

**Length:** 869.4ft (267m).
**Beam:** 104ft (32m).
**Draught:** 28ft (8.6m).
**Flight deck:** 543×97ft (167×30m); angled at 8°; two catapults.
**Displacement:** 27,307 tons standard; 32,780 full load.
**Propulsion:** 2 geared turbines, 2 shafts, 126,000hp.
**Speed:** 32 knots (59km/h).
**Range:** 7,500nm (13,900km) at 18 knots (33.3km/h).
**Crew:** 1,338.
**Launched:** 1957.
**Commissioned:** 1961.

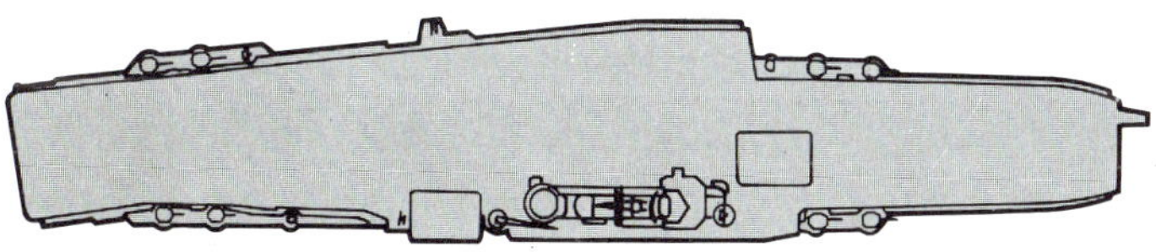

**Clemenceau** **France**

**Air wing**
40 attack aircraft (mainly Super Etendards); 2 Super Frelon and 2 Alouette III helicopters; Etendard IVP for reconnaissance; 1 Breguet Alizé ASW operations.

**Armament:**
Limited to 8 (100mm) automatic guns in single turrets, but of course Super Etendards are capable aircraft, and can carry very effective Exocet anti-ship missiles and other weapons. Air protection seems light.

The Soviet Kiev class comprises four units and is an innovative design which combines an angled flight-deck (for VTOL aircraft) with a heavy armament mounted on the forecastle. These ships are unlikely to be able to operate CTOL aircraft now being tested for the *Kremlin*.
The French Navy is building two CVNs, somewhat smaller than the US and Soviet super-carriers, but very capable nevertheless. In contrast there is a growing number of much smaller and cheaper carriers, designed, in the main, around the still unique capabilities of the Harrier/Sea Harrier. Similar in size and concept are the Italian *Garibaldi* and the Spanish *Principe de Asturias*, although the Italian Navy has not yet procured the Sea Harriers for which their ship is designed. First of the smaller V/STOL carriers was the RN's Invincible class, now fully proven in live combat. Still on the drawing board is BAe's "Harrier-carrier".

complicated) hangar, which opens up to admit the helicopter and then closes over it. Fitted on ships such as the Kara, Kresta and Udaloy classes, this hangar arrangement has many advantages, although working space for the aircraft engineers may well be very cramped.

Some navies have developed special devices to capture and control helicopters on such small flight decks. Most widely used of these is the Canadian "Bear-trap" device, which is mounted in the centre of the flight deck. Having engaged the trap while hovering some 10ft (3m) above the deck, the helicopter is then hauled down and held firmly on the deck. Other navies, although operating in the same waters as the Canadians, have not considered such a device necessary, proving that helicopters can land on even very small flight decks in almost any sea state.

## Roles for naval aircraft

Sea-based air power has five primary roles: anti-submarine warfare (ASW), strike/attack, air defence (AD), electronic warfare (EW) and airborne early warning (AEW). Airframes divide into three obvious categories: conventional take-off and landing (CTOL), fixed-wing vertical/short takeoff and landing (V/STOL) and helicopters.

ASW tasks are performed by CTO aircraft ranging from the four-place Lockheed S-3 Viking operating with the US Navy to the many types of helicopter such as the Soviet Kamov Ka-27 Helix and the US Sikorsky SH-60B Seahawk. The principal requirements of a CTOL ASW aircraft are to be able to transit economically and then spend as much time as possible on patrol, during which it must be able to detect, locate, identify and finally destroy submarine targets. This necessitates a variety of on-board sensors including radar (for detection and classification of surface targets), forward-looking infra-red (FLIR), magnetic anomaly detectors (MAD), sonobuoys and electronic countermeasures (ECM) equipment. All these sensors, plus inputs from the parent carrier, and from other ships and aircraft, produce so much information that considerable on-board processing power is necessary.

To attack its targets the aircraft also obviously needs a weapon load which ideally would include missiles for use against surface targets (Exocet, Harpoon, for example), as well as homing torpedoes, rockets and conventional/nuclear depth bombs. The only CTOL aircraft currently capable of performing the complete ASW task is the S-3 Viking. The Soviet Navy will probably produce a similar type of ASW aircraft when its new 75,000 ton carrier enters service.

An alternative to the CTOL ASW aircraft, albeit with less range/endurance, is the helicopter. This has been developed from a simple extension of the parent ship's sensors (with a real-time down-link) to a fully autonomous weapon system in its own right. Further, it has had the exceptional advantage of bringing an air capability to ships down to frigate size. The most advanced design in this category is the EHI EH-101, a British/Italian Sea King replacement. Not yet in service, this aircraft will be equipped with Ferranti Blue Kestrel search radar, Marconi sonar systems, Decca ESM (electronic support measures) and AN/AQS-81 MAD, all brought together by a Ferranti data-handling system.

Strike/attack roles are performed primarily by fixed-wing aircraft, using missiles, bombs or rockets. As so dramatically demonstrated in the South Atlantic War, a fast jet aircraft such as the French-built Super Etendard armed with Exocet sea-skimming missiles is a potentially lethal combination. Such an attack is feasible also by US aircraft armed with Harpoon missiles (Grumman A-6 Intruder, S-3 Viking), and the British Sea Harrier with Harpoon or Sea Eagle. The Soviet Navy currently has no capability in this area (from sea-based aircraft), but it must be presumed that they are developing a suitable aircraft/missile combination for service on the new carriers. Helicopters can also carry ASMs, and the French Super Frelon carrying AM.39 Exocets has been used by Iraq in the Gulf War with some success, while the British Lynx used Sea Skuas against (surfaced) submarines and surface ships in the South Atlantic War.

**Above: Grumman F-14A Tomcat of squadron VF-142 lands on board USS *Eisenhower* (CVN-69). This two-seat, multi-purpose aircraft is currently the most powerful naval fighter. This could change when the Soviet Navy's carriers go to sea.**

## Airborne air defence

Air defence by aircraft is a role confined to fixed-wing types, of which the technically most advanced is the Grumman F-14 Tomcat. Operating from the US Navy's larger CVNs, the F-14 also performs the reconnaissance role. The Sea Harrier demonstrated its air defence capabilities in the South Atlantic where it was responsible for shooting down in air-to-air combat eleven Mirages, eight A-4 Skyhawks, one Canberra, one Pucara and one C-130 Hercules. The only Soviet air defence aircraft is the Yak-36 "Forger", a fixed-wing VTOL machine which has limited capabilities.

In the area of specialised EW aircraft the only machine currently in service is the US Navy's Grumman EA-6 Prowler, a remarkable design with a comprehensive ECM/ESM capability. No other naval air arm has attempted to produce such a sophisticated (and extremely expensive) aircraft type.

In the South Atlantic War it was demonstrated that radar picket ships are an inadequate substitute for an airborne early-warning system. The only effective type currently in service is the Grumman E-2C Hawkeye, a twin-turboprop aircraft, which establishes a patrol some 200nm (370km) from its parent carrier, where its powerful APS-125 radar has considerable range. The British have produced an AEW version of the ubiquitous Westland Sea King helicopter, equipped with a Searchwater radar (as used on the Nimrod), mounted in an inflatable radome on the starboard side of the aircraft. Conceived in haste, this has proved an effective solution to the problem, although less economical than a fixed-wing CTOL aircraft.

The current Soviet practice of long-range land-based AEW for its naval task groups will doubtless prove vulnerable and inadequate in war and they will almost certainly have to develop a carrier-borne AEW aircraft to make their new task groups fully effective. The same also applies to the French Navy, whose new CVNs would inevitably be high-value targets in any naval conflict, and highly vulnerable without an AEW aircraft.

## The carrier in naval warfare

It is a dangerous over-simplification to regard the aircraft carrier simply as a mobile naval airfield, because it is very much more than that. Apart from the very small carriers, the large fleet carriers are

**Below: Grumman E-2C airborne early warning aircraft. Lack of AEW aircraft led to British losses off the Falklands in 1982, and the Soviet Navy also needs such aircraft to be fully effective.**

concentrations of air power, command and control facilities, and sensor technology without parallel. The carriers are invariably the largest units in their fleets, let alone in each individual task force, and are almost always the flagship. Their very size and the spaciousness of their accommodation enables them to provide the best facilities afloat for commanders and their staffs, with large arrays of communications and sensors. Further, they can carry large and very powerful computer complexes; for example, the American CVNs all house very powerful ASW computers, with real-time links not only to the accompanying ships in their task groups, but also, via satellite, back to US and NATO land-based facilities, such as the US Iliad 4.

It is quite clear that the Soviet Union has come to understand this, as is illustrated by the inexorable growth of their carrier capability. This almost certainly stems from the vulnerability they felt in the 1950s and 1960s to American nuclear strike aircraft on board the CVNs of the Forrestal and Kitty Hawk classes, and to which, at that time, they had no real answer. They have also seen how indispensable the carrier is to power projection, since air power is the vital ingredient in any threat to a distant power, or in an amphibious landing. However, even the vast experience and apparently bottomless purse of the Soviet Defence Ministry may find it difficult to produce simultaneously their first 75,000 ton fleet carrier (CVN) and their first carrier-borne fixed-wing attack, fighter, patrol and ASW aircraft.

Mighty as the carrier may appear, and despite all the protection provided by the variety of aircraft in its air wing and its accompanying ASW escorts (both surface and submarine), the aircraft carrier is nevertheless vulnerable. It is quite obviously a high-value target – in terms of naval warfare probably the highest value target of all – and any opponent will therefore inevitably seek to eliminate it at the earliest opportunity.

This was clearly illustrated in the South Atlantic War, where the Argentine carrier, *Vienticinco de Mayo*, was the Royal Navy's primary target, while the Argentines knew that if only they could damage or sink one, if not both, of the British carriers they would compel the Task Force to withdraw to Ascension Island. The British carriers had some very close shaves, and at one point it was only the diversion of an incoming Exocet on to the merchantman the *Atlantic Conveyor* (which it sank) that saved one of the carriers. The American carriers (and the new Soviet ones, too, when they appear) are enormous targets – the Nimitz class are 1,092ft (317m) long – and while their very size would enable them to absorb a lot of battle damage history suggests that no ship, however sophisticated, is totally invulnerable.

**Above: Sidewinder missile launched from a US Navy F/A-18 Hornet. This type is the current "state-of-the-art" combat fighter and Soviet designers have borrowed features for incorporation in the new "Fulcrum" and "Flanker" fighters.**

**Below: Exocet AM-39 anti-ship missile carried by a French Navy Super Etendard. This was the airframe/missile combination, used by the Argentine Navy, that led to British ships being lost off the Falklands in 1982.**

Faced with the high cost of these supercarriers, the concentration of valuable assets in one hull and their unavoidable vulnerability it is frequently suggested that the larger navies, and especially the US Navy, would be better advised to have a numerically greater fleet of much smaller carriers, say in the 30,000 to 40,000 ton range. However, there is no doubt that in the case of the supercarriers large size pays off. First of all, it enables each carrier to embark a suitably sized air wing with a mix of the various types of aircraft needed to attack the enemy and protect its own task group. (The dreadful problems which faced the British carriers in the South Atlantic War without an AEW aircraft underline this point only too clearly.) Secondly, with CVNs, the compact size of the power plant enables a substantial quantity of aircraft fuel and weaponry to be carried, while the absence of large down- and up-takes leads to a smaller superstructure, taking up less valuable space on the flight deck. It is for these reasons that the French Navy intends to replace its present CVAs with CVNs towards the end of this decade, with their size being as great as they can possibly afford.

A carrier fleet is thus an essential part of any navy with global pretensions. It will be very interesting to see how the East-West maritime strategic equation alters when the Soviet supercarriers join their fleet.

The submarine was for a long time a warship which fought primarily on the surface, retiring underwater to hide or when under actual attack. Sonar was an imprecise instrument and ASW weapons fairly ineffective. The considerable speed advantage held by surface vessels complicated the submariner's task, and the development in World War II of ship- and aircraft-mounted radar increasingly forced submarines beneath the surface, although the invention of the schnorkel enabled them to recharge their batteries at periscope depth. World War II also saw the appearance of the specialised ASW aircraft, which rapidly became one of the submarines' most effective foes.

Freedom from regular forays to the surface for life support and to recharge the batteries came with the nuclear-powered steam plant, conferring an endurance only effectively limited by physiological and psychological factors. Thus, the submarine became a true underwater weapons system, capable of world-wide deployment and able to operate at fleet speeds with surface task groups. Furthermore, submarines can now travel submerged at speeds equal to, and frequently exceeding, those of attacking surface vessels.

However, the conventionally-powered (diesel-electric) submarine continues to flourish; many hundreds operate with some 40 navies. Few navies can afford the increasingly expensive nuclear-powered submarines; only the US and French navies have announced their intention to go "all-nuclear" but the Soviet and British navies each have only one class of conventional boat in production.

Submerged submarines are difficult to detect – some more so than others – but there is no such thing as an "undetectable" submarine. All underwater craft have both acoustic and non-acoustic signatures, although many countries are conducting expensive research programmes into making detection of their submarines more difficult.

## The future

Perhaps the over-riding factor in future submarine design is that of cost, which is the end product of increasing growth in complexity, numbers of systems (both defensive and offensive) and size. Even with conventional diesel-electric submarines, apart from those designed for restricted waters such as the Baltic, size and cost are increasing. Many resolutions are made to produce smaller designs, but this somehow seldom seems to happen.

**Above: The French SSN, *Rubis* (S-601), is the first of a class of five. Its 2,670 tons submerged displacement is much less than any other SSN in commission and it clearly has a small nuclear reactor.**

**Below: The US Navy's Los Angeles class is armed with SUBROC, Tomahawk SLCM, Sub-Harpoon and torpedoes, making them the world's most heavily armed (and most expensive) hunter/killer submarines.**

The major research efforts are now going into making submarines yet quieter, faster and deeper diving. The emphasis on quietness does not necessarily make a submarine undetectable, but it does make the searcher work much harder probably to the extent of using active sonar and thus giving himself away. A real breakthrough would be to find a non-nuclear propulsion system which could free the submarine from the surface. It seems unlikely that there will be any major advances in battery design, although a process of continual refinement will obviously take place. Closed-cycle engines look a possibility, although no totally practical system has yet appeared, and one which produces a gaseous exhaust is in any case depth-limited since it could only be used as long as the gas pressure exceeded the water pressure.

Also under constant consideration, are fuel cells in which a fuel/oxidant reaction directly produces electricity; some (such as lithium/peroxide) even have a by-product of pure, potable water, with obvious benefits in a submarine. United Technologies in the USA are developing a fuel cell which it is hoped will produce (in a packaged system) outputs of up to 34,000hp (26mW). Although this is of a type unlikely to be suitable for submarine use, nevertheless there seems to be a reasonable prospect of viable fuel cells for submarines by the mid-1990s. Nothing is known of Soviet developments in this area, but in view of their huge investment in submarine technology there is no reason to think that they will lag behind the West.

## Nuclear-powered attack submarines

The first SSN was completed by the USA in 1954 after an unparalleled development effort. The Soviets have long since overtaken the US in sheer numbers, although

only now do Soviet SSNs possess the advanced features which have been common to all US boats since the early 1960s. The British, French and Chinese navies are the only others with SSNs.

The current type of US SSN, still under construction, is the Los Angeles class, which has a 50 per cent increase in displacement over earlier SSNs. In addition to the most advanced sensors and fire control equipment (now being retrofitted into the earlier Permit and Sturgeon boats) the Los Angeles class has regained the speed lost since the Skipjack class, through the installation of the more powerful S6G reactor.

Operating procedures in the Soviet Navy make it unlikely that there would be surface task groups against which American SSNs could operate, but the advent of the Kirov battlecruisers and of the fixed-wing attack carriers later in the decade may well change this. However, so far US SSNs have been designed to undertake three main roles: ASW hunter-killer, forward area attack and reconnaissance, and convoy/task force protection. In the Los Angeles class particular attention has been paid to quiet operation, the large hull making it easier to "cushion" the machinery. A major difference between the US and Soviet navies is that while the Soviets have gone for special-to-role submarines, such as guided-missile boats (eg, Charlie class SSGN) the Americans have tried to produce a single, multi-role vessel, such as the Los Angeles, and in this they have to a large degree been successful.

Even the considerable size of the Los Angeles is proving inadequate, however, and accommodation is so tight that apparently berths can be provided for only 95 of the nominal crew of 120 ratings. Much additional weaponry is taking up ever more internal space, despite the fact that the 12 Tomahawk missiles are now being put in external launch tubes rather than being stored in the torpedo room. The only reduction in space requirements is in automatic data processing (ADP) systems, due to technological advances leading to smaller mainframes.

The US Navy is therefore looking towards a new class of SSN, for initial production in 1989. One of the stated major design aims is to achieve even quieter operation, but Secretary of Defense Casper Weinberger has reported (FY85 Report to Congress) that the new submarine will carry more weapons, will mount more sensors and will be able to operate more effectively under ice. These comments, allied to the statement that "these improvements, (which) cannot be accommodated in the existing SSN-688 (Los Angeles) hull" imply that the new class will be larger, and most estimates lie in the 10,000 to 12,000 ton bracket.

**Above: The British Swiftsure class has demonstrated the value of pump-jets in which propellers are replaced by ducted, multi-blade rotors turning against stator vanes, to achieve much quieter running.**

**Below: Soviet Victor III SSN showing its towed array sonar pod atop the vertical rudder. Soviet SSNs are now much quieter and the latest "Uniform" class is causing NATO admirals considerable concern.**

## Soviet attack submarines

The current major class of SSN in Soviet service is the Victor class. The first boat appeared in 1968, and the class has continued in production, albeit with some modifications, well into the 1980s. The Alfa class, although produced in only very small numbers, was a revolutionary design constructed of titanium, with a liquid metal-cooled reactor and capable of more than 42 knots underwater. It is now being supplemented by the Mike class, also made of titanium. Soviet submarines, especially their SSNs, have always been noted for their noisy characteristics. However, a theory now being advanced in some quarters is that in more recent SSN classes this might be a deliberate peacetime "disinformation" ploy.

The first British SSN, HMS *Dreadnought*, had an American Skipjack propulsion unit, but subsequent classes have been all-British, although they have tended to follow the pattern of development of their US counterparts. Frequent deployments take place with surface task groups, and the designation Fleet Submarine makes it plain that the British see the role of their SSNs in the same light as the Americans. However, there are two major design differences: the British still keep the torpedo tubes well up in the bow while the Americans have theirs more or less amidships; and the British now use pump-jets as opposed to propellers.

Diving depth is obviously an important tactical consideration, although its major effect lies in giving a reasonable safety margin in high speed manoeuvres. Actual details are highly classified, but open source figures give a general indication for comparative purposes of normal operating depths: Los Angeles (SSN, USA) 1,475ft (450m); Victor III (SSN, USSR) 1,300ft (400m); Rubis (SSN, France) 980ft (300m). This depth performance is becoming more important as the ASW threat increases and is naturally leading to extensive R&D in materials.

The US navy has used a series of High-Yield (HY) steels for submarine construction; all classes since the *Thresher* have been built of HY80. (The number indicates the yield pressure of the steel; thus, HY80 can withstand a pressure of 80,000psi or 5,620kg/cm$^2$.) It was at one time planned to build the Los Angeles class of HY130, but it is now hoped to develop HY130 to an acceptable standard in time for use in the later boats of the next US Navy SSN class.

In other nations there is also much R&D into ever stronger materials. The Japanese use NS90 (NS = nickel steel) in their latest Yuushio class, while the French are using "Marel", a new high tension steel claimed to give a 50 per cent increase in diving depth. The USSR has continued to use steel for the majority of its submarines but, as referred to earlier, at least one class (Alfa) is known to be built of titanium. The Alfas are not only the world's fastest submarines (over 42 knots/74km/h), but also the deepest diving, with estimates varying between 2,000 and 3,000ft (607 and 914m).

Titanium is a very difficult material to weld, although the Soviets – long world leaders in metallurgy – seem to have solved this. Only six Alfas were built, but their successors would appear to be the newly revealed Mike class, also constructed of titanium. Titanium has an invaluable additional advantage in that it is non-magnetic and thus undetectable by devices such as airborne MAD and bottom-laid coils. Some years ago the West Germans attempted to achieve the same results by using non-magnetic steel for their Type 205, 206 and 207 submarines, which were intended primarily for service in the shallow waters of the Baltic. Early corrosion problems have long since been resolved but the idea has not been extended to larger designs.

## Manoeuvrability

Manoeuvrability is obviously of great importance to a submarine and again the USS Albacore has had a great effect upon all subsequent designs. The short hull improves manoeuvrability and during trials Albacore was able to turn at 3.2° per second, a figure probably well exceeded by the latest types.

The major control surfaces are the hydroplanes. Virtually all modern submarines have a cruciform tail empennage, with the US Los Angeles and Ohio classes being fitted with end plates to the horizontal hydroplanes, both to improve control and also as housings for hydrophones. The latest Swedish and Dutch classes, however, have an indexed X-shape; this configuration is less liable to physical damage (because it does not protrude beyond the outer edge of the hull) but requires computer-aided operation because of the complex relationship between steering and diving functions. Most nations use bow planes, placed as far forward as practicable, but the US Navy and some others, including Japan, use fin-mounted planes. While these may have advantages for manoeuvrability they need to be capable of rotation to the vertical for breaking through ice, which is not possible, for example, with the Los Angeles class.

## Propulsion

The greatest single advance in post-World War II submarine technology is without any doubt the advent of nuclear propulsion which has released a submarine from having to make regular forays to the surface to replenish its air supply and recharge its batteries. Nuclear propulsion confers virtually unlimited range: the USS *Nautilus*, for example, the first ever SSN, steamed some 150,000 miles (241,935km) on her third core, while HMS *Warspite* (a British SSN) recently completed a 16-week patrol in the South Atlantic, probably as long as any one crew can accomplish.

All British and US SSNs and SSBNs use pressurised-water reactors (PWR), with water acting as both coolant and moderator; this is a tried and tested technique which has proved very reliable. The way ahead for smaller and lighter reactors (or more power for the same size) lies in new coolants. Liquid sodium was tried in the

**Below: Despite the utilitarian and cluttered appearance inside this USN submarine the conditions for the crew are, in fact, greatly superior to those of even just twenty years ago.**

**Key:**
**1.** Sonar transducer array.
**2.** Transducer array access trunk and cable space.
**3.** Anchor windlass.
**4.** No. 1 main ballast tank, starboard.
**5.** No. 2 main ballast tank, starboard.
**6.** Forward hydroplanes operating gear.
**7.** Torpedo shutter.
**8.** Forward hydroplane, starboard.
**9.** Weapons embarking access covers.
**10.** Weapons embarking hatch.
**11.** Forward escape tower.
**12.** Indicator buoy cover.
**13.** Torpedo tube 21in (533mm).
**14.** Provision store.
**15.** Sonar transducer dome.
**16.** Crew accommodation.
**17.** Galley.
**18.** Torpedoes in stowage racks.
**19.** Navigation platform.
**20.** Attack periscope.
**21.** Surveillance periscope.
**22.** Radar mast and antenna.

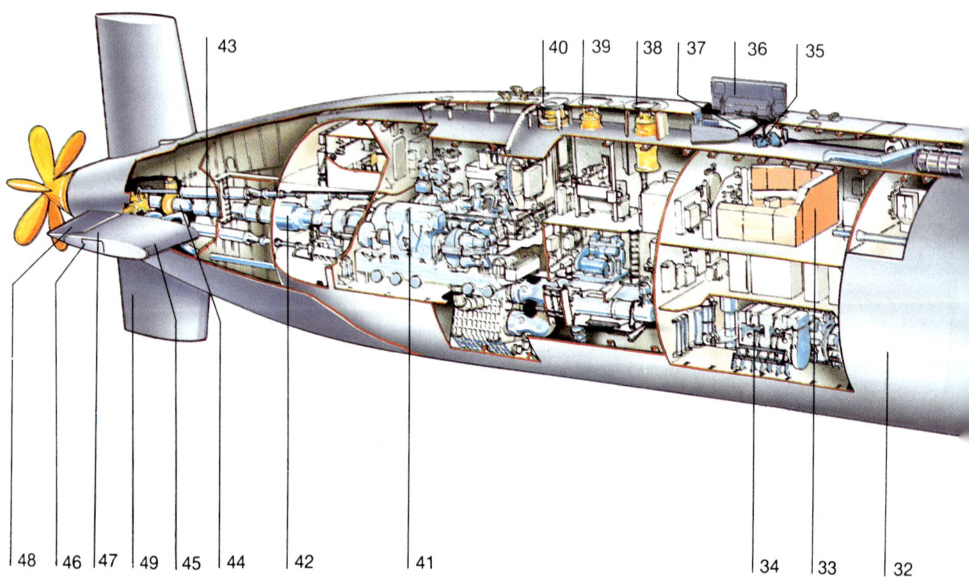

USS *Seawolf* in the late 1950s, but was eventually judged not reliable. Since then the only alternative appears to be some other form of liquid metal cooling, and it is mooted that the Soviet Alfa class uses such a system, possibly in a totally automated and unmanned engine room. It is also thought that such a system is used in the French Rubis class SSNs, which are by far the smallest SSNs in any navy.

A particular problem for nuclear-powered boats is that of machinery-generated noise, especially from gearing and rotating machinery, such as pumps, which must be kept running in all PWR systems. Machinery is mounted on rafts in most Western boats in an effort to isolate the vibrations from the hull, and turbo-electric drive has been tried (in USS *Lipscomb*) as has direct drive (USS *Jack*). Free-circulation was used in USS *Narwhal* and is believed to be used in the S8G reactor in the new US Ohio class SSBNs.

The final item in the drive-train is the propeller, which is one of the major causes of noise, and one of the most readily identifiable features of individual submarines. Modern submarine propellers have up to seven blades, usually of a scythe shape, and are designed to run at very low revolutions. But the latest British SSNs (Swiftsure and Trafalgar classes) are reported to use pump-jets, in which a ducted multi-bladed rotor turns against stator vanes, thus considerably complicating the ASW hunter's task.

**Below: Soviet Papa-class SSGN returning to port. The Soviet Navy has developed several specialised cruise missile classes specifically to deal with the threat posed by US carrier groups.**

## Submarine weapons

After World War II virtually every submarine abandoned its gun armament and for several decades the only weapon available to attack and patrol submarines was the torpedo. However, in numerous cases very sophisticated and advanced SSNs and SSBNs were at sea with torpedoes which had little better performance than those of World War II. The torpedo has had an extraordinary history in recent years with vast sums being expended by many nations, but with little overall advance in either performance or reliability, until recently. (This weapon is described in the section on ASW.)

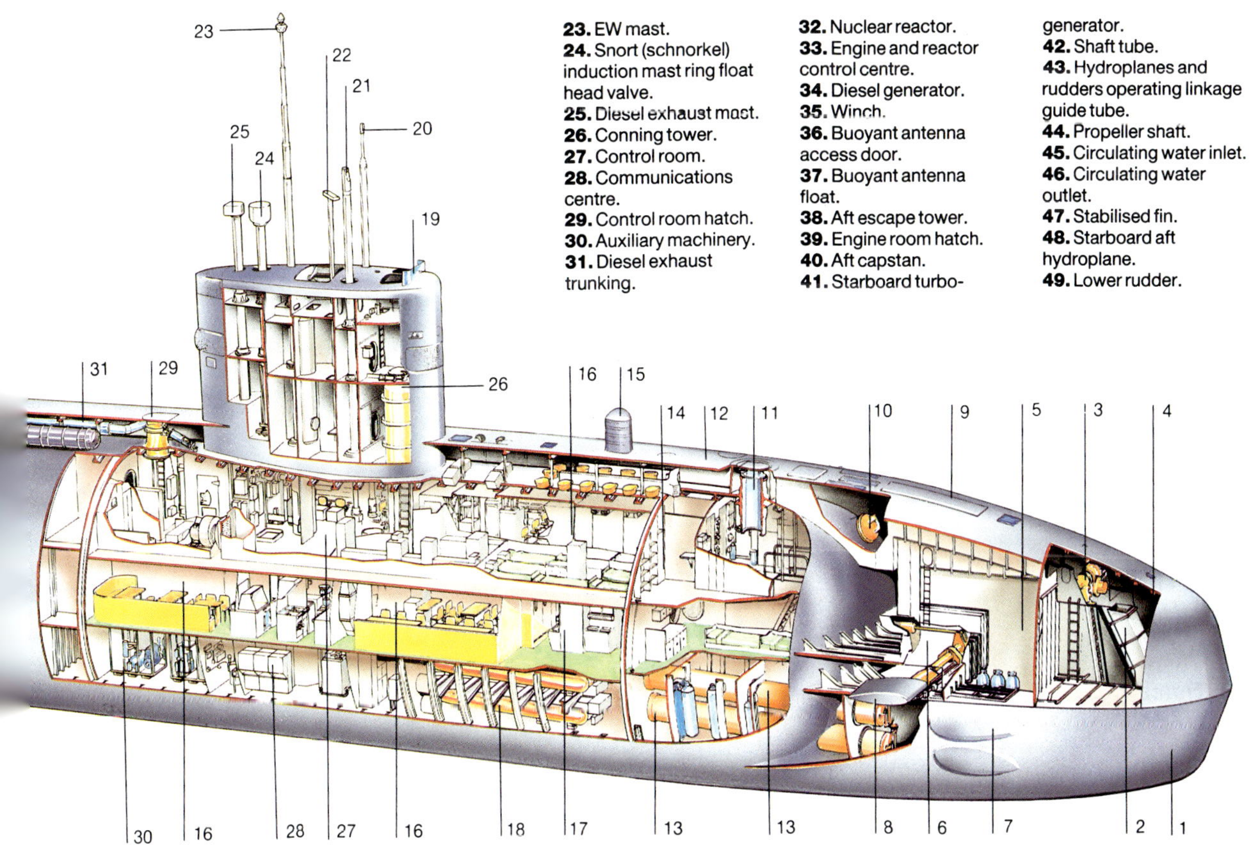

**23.** EW mast.
**24.** Snort (schnorkel) induction mast ring float head valve.
**25.** Diesel exhaust mast.
**26.** Conning tower.
**27.** Control room.
**28.** Communications centre.
**29.** Control room hatch.
**30.** Auxiliary machinery.
**31.** Diesel exhaust trunking.
**32.** Nuclear reactor.
**33.** Engine and reactor control centre.
**34.** Diesel generator.
**35.** Winch.
**36.** Buoyant antenna access door.
**37.** Buoyant antenna float.
**38.** Aft escape tower.
**39.** Engine room hatch.
**40.** Aft capstan.
**41.** Starboard turbo-generator.
**42.** Shaft tube.
**43.** Hydroplanes and rudders operating linkage guide tube.
**44.** Propeller shaft.
**45.** Circulating water inlet.
**46.** Circulating water outlet.
**47.** Stabilised fin.
**48.** Starboard aft hydroplane.
**49.** Lower rudder.

**Hunter-killer submarines**
Since their inception in the 1950s nuclear-powered hunter-killer submarines (SSN) have been intended mainly to seek and destroy other submarines, especially SSBNs. They have also been tasked with defending friendly SSBNs, and cooperating with surface task groups. One of the latest designs is the British Royal Navy's Trafalgar class, shown here. For security reasons the reactor compartment is not shown, and certain spaces within the submarine are not identified.
The relative spaciousness is noteworthy, with three decks running through the boat for most of its length. *Trafalgar*'s armament is a mix of 21 in (533mm) torpedoes and Sub-Harpoon, both of which are fired from the forward torpedo tubes. US Los Angeles SSNs also have vertically-launched Tomahawk, which are mounted in tubes in the space between the forward end of the pressure hull and the bow sonar.

Recently a variety of new weapons have become available to submarines, which can now be heavily armed and with much enhanced capability (for instance, nuclear or conventional strikes against targets far inland). The submarine-launched cruise missile (SLCM) has been around for some years and the Soviet Navy has long had specialised boats to carry them. The Juliett class diesel-electric cruise-missile submarines (SSG) and the nuclear-powered boats (SSGN – Echo, Charlie and Oscar classes) carry SLCMs mounted in bins, either amidships or in the bows. The missiles (SS-N-3, -7, -9 and -19) can be launched from underwater and have ranges up to 600 miles (966km); their most probable role would be to attack NATO task groups in the North Atlantic, especially those formed around attack aircraft carriers.

The US Navy took a different approach and instead of developing specialised submarines decided to produce weapons which could be launched from standard 21in (533mm) torpedo tubes. This was successfully applied to the Sub-Harpoon anti-ship missile and Tomahawk anti-ship (T-ASM) and land-attack (T-LAM) missile. But after vast expenditure on the Tomahawk programme it has been realised that these weapons will take up too much valuable space in the torpedo rooms and it has therefore been decided to fit them in vertically-mounted tubes inside the upper casing between the bow sonar and the forward-end of the pressure hull. This is a significant capability enhancement at little cost, but it is now irrelevant as to whether the missile can fit a torpedo tube or not, and thus much R&D money and effort would appear to have been wasted.

The Harpoon can attack ships out to 60 miles (97km) and the T-ASM to 250 miles (400km), while the T-LAM has a range of 1,367 miles (2,200km) and has both conventional and nuclear warheads. The main problem with T-LAM is that to exploit its capabilities fully it needs external target information, and a system known as Outlaw Shark is being deployed to achieve this. The deployment of Tomahawk in this role has increased the need for two-way communications with submarines, a notoriously difficult problem.

The only other submarine-launched anti-ship missile is the French SM39 Exocet, a new version of the very successful missile which achieved such a reputation in the South Atlantic War. Launched from a standard 21in (533mm) torpedo tube, the SM39 has a range of over 31 miles (50km).

An often forgotten capability of submarines is that of minelaying. Mine warfare is returning to importance and the submarine is in many ways the ideal minelayer, being able to lay both accurately and covertly, even though on the outward trip the mines take up valuable space on board.

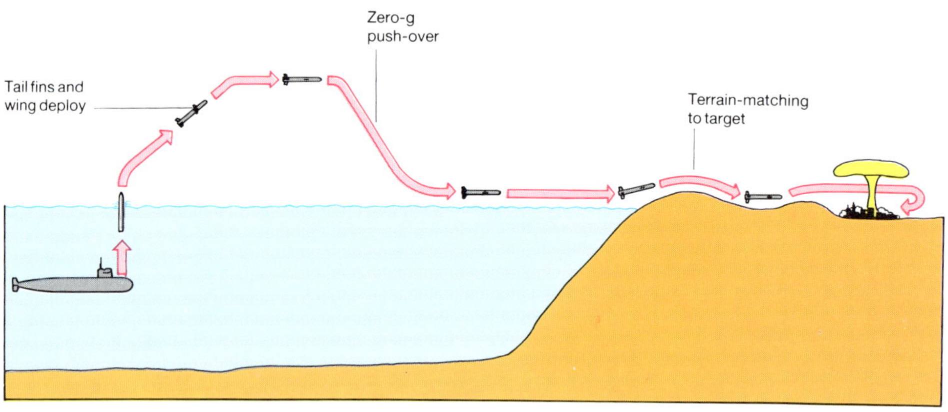


**SLCMs**
The SLCM is launched from a vertical tube (earlier development versions were launched from torpedo tubes). A lanyard attached to the submarine starts the boost motor and the missile then heads for and leaves the surface at an angle of 50°. Once in the air four tail fins spring open and the wings then deploy. The cruise engine starts up and the missile does a zero-g pushover to minimise the apogee and avoid detection by hostile radar. The missile then descends to very low altitude and heads landwards. The land-attack version's TERCOM (TERrain COMparison) matches the ground with data fed in prior to launch. This is especially important when crossing the coast to utilise the first position update, and is used thereafter when passing over selected small matrix areas of terrain to update the inertial system, which remains in command throughout.

**Below: A submarine-launched Harpoon anti-ship missile streaks away towards its target. This weapon is another significant addition to the submarine's increasing armoury, and it is being bought by a number of Western navies.**

## Sonars

Once submerged the submarine is totally dependent upon its sensors. The most important of these is sonar, which is described in the ASW section. Most submarines have a large sonar mounted in the bow, and many (for example, US SSNs and SSBNs) are also fitted with conformal arrays. Hydrophones (passive sonar) are also fitted, usually with arrays near the bow, amidships and near the stern, to give all-round coverage.

One important submarine sensor is that for listening to its own noises. Most navies use a surface ship to "inspect" their submarines as they leave port for noise emissions. But it is vital to continue such monitoring throughout the patrol to ensure that any new noise is detected and removed, otherwise it will act as a veritable beacon for ASW forces.

Just as important as the sensors themselves is the ability to assimilate and process their outputs, and in this advanced technology is making a major contribution. Not only are modern processors able to handle vast amounts of information, but they are also much smaller and easier to use. This problem of handling the ever-increasing amount of incoming information in such a timely manner that good use

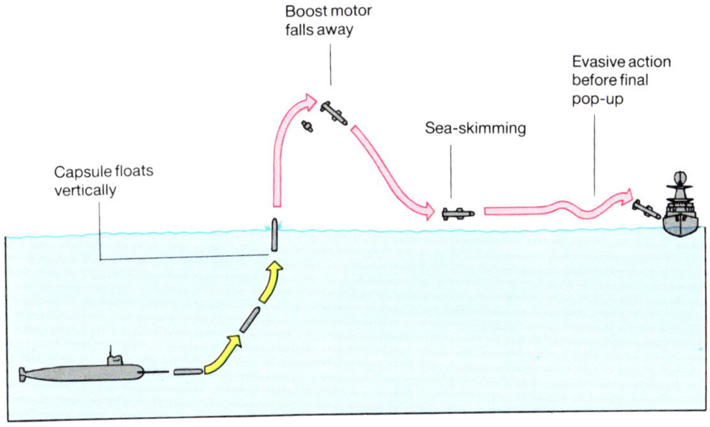


**Sub-Harpoon**
Sub-Harpoon gives suitably fitted SSNs and SSKs a highly effective anti-ship capability. The missile capsule fits a standard 21in (533mm) torpedo tube. On launch the capsule is ejected from the torpedo tube at some 50ft/sec (15.24m/sec) and is then steered to the surface by control fins. Nose and tail caps are then jettisoned, the boost rocket motor fires and the missile takes-off. Three seconds later the boost motor is jettisoned and the turbojet starts, whereupon the missile descends to its cruising altitude just above the waves. The guidance unit takes any necessary avoidance action close to the target before final impact.

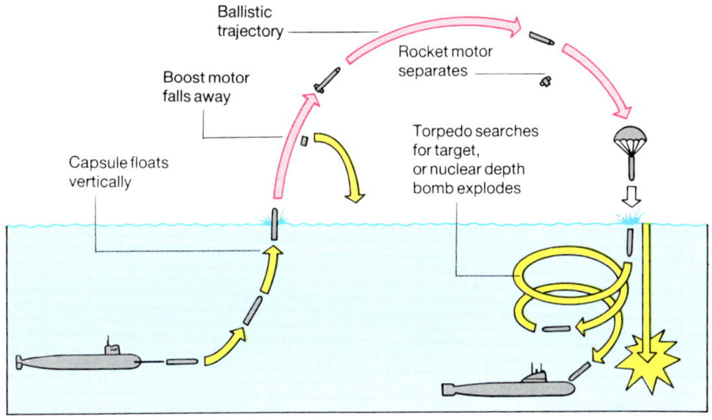


**ASW-SOW**
The ASW-Stand-Off Weapon gives submarines a stand-off capability against other submerged submarines. The missile is launched from the torpedo tube and floats to the surface, like Sub-Harpoon. Launch and lift-off are also similar. Once airborne, however, ASW-SOW follows a ballistic trajectory, dropping the rocket motor during the descent. The weapon completes its descent to the water by parachute. Two types of weapon may be carried: the Advanced Lightweight Torpedo (ALWT) or a nuclear depth bomb. The former carries out a normal search, before heading for the target; the latter simply descends to a pre-programmed depth.

**Above: French SM-39 submarine-launched anti-ship missile, showing the separation of the missile from the launching capsule in which it has been brought to the surface. This new version of the famous Exocet should repeat earlier successes.**

can be made of it is not confined to submarines.

Other navies have progressed from SSN to SSBN, but France, for political reasons, did it the other way round. The first French SSN was therefore not laid down until 1976, joining the fleet in 1982. This class, the Rubis or SNA, are the smallest SSNs to enter operational service. The Chinese Navy now has at least one SSN, although no pictures have yet been released to indicate its size or design.

## Conventional submarines

Despite earlier forecasts that the SSN would totally oust the conventional submarine, at least in the major navies, there are still a great number of the latter in service. Only the US Navy has attained a virtually all-nuclear status, although the French have announced their intention to follow the same path. The USSR has never announced its intentions in this area, but it still has many hundreds of conventional submarines in service and one new major class is in production. The British, who once wished to go all-nuclear, have 16 diesel-electric boats still in service, and recently announced plans to start construction of a new type (Vickers Type 2400 or Upholder class) at a rate of one a year.

The modern conventional submarine has certain advantages to offer, not least of which is economy. The latest Los Angeles SSNs in the FY84 programme will cost some $693 million each, while the British Trafalgar class cost £175 million at 1981 prices. These are deep-water boats but the costs are high and only a few navies can contemplate expenditure at that level. Furthermore, SSNs require large crews of highly trained professional sailors.

Against this, the conventional submarine is cheaper in first cost, more economical in operating costs, and requires a much smaller crew. Conventional boats are also very much quieter and much more suitable for patrol and reconnaissance operations, particularly clandestine missions. This latter is evidenced by the Soviet Whiskey class SS which grounded in November 1981 while surveying the Swedish Naval Base at Karlskrona, and by the missions undertaken by the British Oberon class submarine HMS *Onyx* in the South Atlantic War.

But conventional submarines are limited by battery life, and when surfacing or "snorting" are liable to detection. Nevertheless, there is still a place for them in every naval inventory except, apparently, for those of the USA and France.

## Submarine construction

The most significant development in submarine hull design was pioneered in the 1950s by the USS *Albacore* which, with a "teardrop" hull shape, cruciform tail empennage and single propeller, achieved an underwater speed of 33.5 knots and set the pattern for all subsequent hull designs. However, contemporary hull designs are less "fat" than the Albacore because a parallel mid-body is equally efficient if the forward and after ends are properly designed, and a tubular body is much easier (and cheaper) to construct.

Much research has been (and is being) devoted to the subject of hydrodynamic efficiency. Laminar flow has been tried, offering apparent drag reductions of over 50 per cent, but the theories were thwarted by the impurities of seawater. Streamlining has been the rule since the World War II German Type XXI boats, but great attention is now being paid to the actual nature of the surface. Special hull paints can reduce friction and thus either increase speed or reduce the amount of power needed for a given speed. Releasing polymers around the hull might also increase the speed for short periods. It has been reported that the exterior of Soviet submarines is coated in a compliant covering, derived from research into marine animals such as dolphins and killer whales, which combines suction and boundary-layer pressure equalisation. Further development, derived from the same research may lead to devices to modify shapes to match the boat's speed by, for example, progressively retracting the hydroplanes as speed increases. These researches may also account for the markedly different shape of the fins of the latest Soviet submarines which are not only smaller but also appear more akin to the fin of a marine animal than do those on Western submarines.

One of the most critical areas of advanced technology is in the murky world of anti-submarine warfare (ASW), where billions of dollars and roubles are being spent in a desperate and extremely expensive search to find an infallible method of detecting a submarine lurking in the depths of the ocean. Moving submarines have characteristics and they also cause effects, virtually all of which can be utilised to detect their presence. All major navies have large-scale research and development programmes to refine existing, and discover new, methods of detection, localisation and categorisation of submarines so that this can be done more accurately.

Submarine operational profiles can be divided into three main patterns, directly related to their type and role: nuclear-powered ballistic missile submarines (SSBN), nuclear-powered attack submarines (SSN), and conventionally-powered hunter-killer submarines (SSK). Submarines almost invariably start their patrols by leaving base on the surface, where they can obviously be detected by direct visual means, and by satellite or reconnaissance aircraft using photographic or electronic observation. (The only known exception to this is in Sweden where submarines are able to emerge directly from their pens cut into rock cliffs through a submerged tunnel into the sea, without allowing themselves to be seen on the surface.) All submarines must then pass over the shallows of the continental shelf, where they are relatively easy to detect, before reaching the open ocean. Some nations' submarines, particularly those of the USSR, must transit through "choke points" – passages restricted either in depth or breadth – where detection is particularly easy. Once in the deeper ocean, however, the operational profiles of the three main types differ.

SSBNs, once clear of the continental shelf, tend to move deep and fast to their patrol areas, taking precautions to avoid being trailed by SSNs. In patrol areas FBMSs cruise at about 3 knots (5.56km/h), varying their depth to match prevailing oceanic conditions and hide themselves from detection. The major vulnerability of FBMSs is their need to communicate with national command authorities and the requirement, at least with the present generation, to update periodically the Ships Inertial Navigation System (SINS). The primary means of communicating with a totally submerged submarine is by Very Low Frequency (3 to 30KHz) radio, but external antennas are essential for reception. Not surprisingly, major efforts are being made to develop new systems of communication and navigation which will overcome any inherently dangerous need to approach the surface.

The faster and more agile SSNs operate routinely at somewhat greater depths, but they, too, need to communicate with their base, or with the task groups with which they are serving. Radio transmissions by a submarine are immediately detectable by electronic surveillance, which will seek to analyse the content of the signal as well as to pinpoint the site of the transmitter. One way of overcoming this is to remote the transmitter, such as is done with the American AN/BRT-1, a buoy containing a cassette recorder and radio sender, which can be released by a submarine and which, when on the surface, can broadcast a message up to 4 minutes in duration to an aircraft or a ship. A preset time delay of 5 to 60 minutes enables the submarine to be some distance away before transmissions start; for example, at 20 knots a submerged SSN could be anywhere in an area of 1,664 square miles (4,310km$^2$) within one hour.

The greatest problem for SSKs is that they must routinely surface to obtain air for running the diesels which recharge their batteries, and also to expel the diesel exhaust. This can be achieved by exposing only the head of the schnorkel tube, but even this is a relatively easy target for modern radars and infra-red sensors. Further, the exhaust fumes from the schnorkel can be detected by "sniffers" mounted on most types of ASW aircraft. Thus, SSKs are faced with the paradox that, although submerged they are the quietest of all submarines and the most difficult to detect, they are inherently vulnerable by this inescapable requirement to approach the surface at regular intervals.

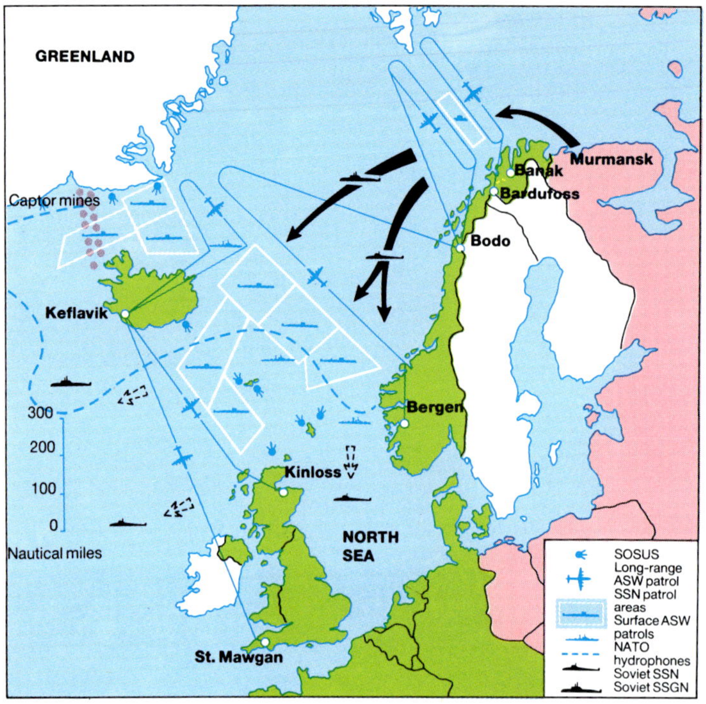


**The GIUK Gap**
One of the great technology battles is fought every day by NATO and Warsaw Pact forces in the Greenland-Iceland-UK (GIUK) gap. Any Soviet Navy ship or submarine wanting to leave northern waters must transit this gap. NATO wants to monitor this movement, not only for general intelligence purposes, but also to detect any "surge" in activity which might indicate warlike preparations.
The gap (a "choke point" in naval terms) is covered by a series of layers in which ASW aircraft, fixed monitoring devices (eg, SOSUS) and submarines (both SSN and SSK) carry out patrols. This map, taken from a Soviet source, shows how the Soviet Navy sees the problem and they are seeking methods of thwarting the NATO searchers, mainly through stealth technology.

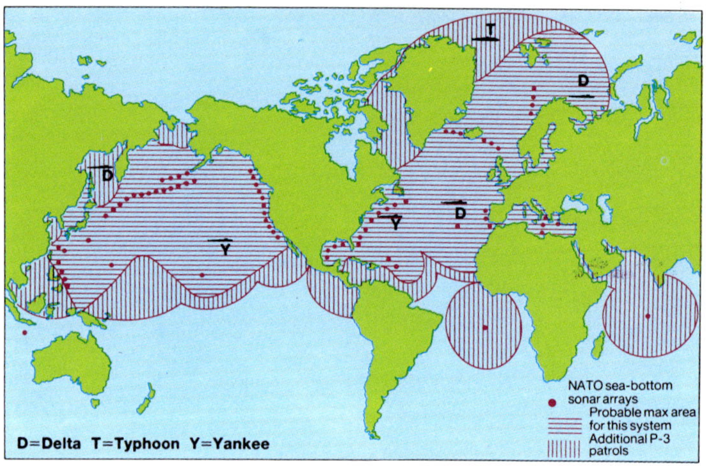


**US underwater detection capability**
The US and its allies have a wide coverage of the world's oceans to detect Soviet submarine activity. This map shows just the USN's own capability, using its static submerged systems of sonar arrays and its large fleet of P-3 Orion ASW patrol aircraft. Also shown are the known deployment areas for Soviet SSBNs.
The survivability of the static arrays in wartime is questionable as the Soviets know where they are in some detail. However, they are still valuable for peacetime use.

## The oceans

ASW can be divided into four capability areas according to the platform type and mission: submarine ASW, surface ASW, airborne ASW and surveillance ASW (ie, by radar and other sensors). All of these, however, must face up to a set of similar problems, the greatest of which is the nature of the ocean itself.

The ocean is virtually opaque to most forms of radiant energy, except for acoustic energy, which can travel great distances underwater. Sound transmissions such as those from a ship's sonar will be reflected by solid objects and the echoes detected by sensitive receivers, although such detection is but the start in a laborious process of identification and classification. The whole process is made more difficult by the complexity of the ocean environment, with its dynamic nature equivalent in many ways to that of "weather" in the atmosphere. This nature is difficult and cannot easily be predicted or characterised, a problem exacerbated by the fact that only a few of the oceans' phenomena, such as surface-wave activity, ice, tidal effects and local weather, can be observed by the human eye.

**Above: A SOSUS sensor, part of the US system, being towed into position prior to being lowered to the seabed. SOSUS is a good sensor system, but its wartime survivability is now open to some doubt.**

**Below: Soviet Navy "Hormone" ASW helicopter using a dunking sonar. Such helicopters are excellent ASW platforms and, using passive sonars, are very difficult for submarines to detect.**

The two best known of the oceans' characteristics are that they contain many dissolved chemicals, including, of course, the archetypical salt (sodium chloride), and that pressure increases with depth. But there are many further factors such as the variation in salinity (haloclines), variations in temperature (thermoclines), subsurface currents, counter-currents and waves, the effects of ocean-bed topography and the nature of the bottom, and the existence of macro- and micro-organisms. Finally, man-made noises affect matters; for example, other ships, oil drilling and harbour noises. All these have their effects upon the acoustical and optical properties of the ocean environment.

## ASW engagements

A typical ASW engagement will go through five stages: search, contact, approach, attack, close-combat and disengagement. In the search stage the hunting platform (airborne, surface warship, submarine) carries out a patrol, either in a general search pattern or, if a component of a task group, in a designated area. In the latter case both surface ships and submarines are faced with the problem that the speed of the group may well be in excess of the optimum ASW search speed. In this stage the hunter must try to avoid detection and will thus use passive sensors, coupled with inputs from other platforms (eg, satellites, SOSUS).

Contact comprises two elements: detection and classification. Detection is the indication that a possible target is actually in the search area, usually accompanied by a general bearing. Classification is a refining process, confirming that the contact really is a submarine (ie, not a merchant ship, a whale, or other harmless object) and then seeking actually to identify it. Examination of the contact's acoustic signature and comparison with the hunter's data bank can lead not only to a broad classification *(one of ours or one of theirs?)*, but also definition of type, class, and even, under certain circumstances, an individual submarine.

The approach stage begins with tracking, again normally accomplished by passive means to avoid alerting the target, followed by localization, when the hunter manoeuvres into an attack position. In the attack stage the hunter must be able to launch its weapons with a fair degree of confidence that the target will be hit; the primary reason for uncertainty for most navies at the moment lies in weapon (especially torpedo) performance, rather than sensors. Thus, the general submarine practice is to fire the torpedoes at about half their theoretical effective range (nominally some 10 to 12nm/11.5 to 13.8 miles/18.5 to 24km)), which is relatively close indeed. Even though a torpedo now carries its own acoustic sensors, and in many cases has a wire-guidance link back to its launcher, it has many limitations which affect the torpedo's chances of a successful engagement.

If the initial attack is a failure a close-combat phase will ensue in which the protagonists seek to eliminate one another. Such a submarine-versus-submarine engagement resembles two fighter aircraft wheeling and manoeuvering in a dogfight, identifying and evading hostile weapons while trying all the time to deliver the fatal blow themselves. This is the one phase of ASW in which the mechanical excellence of the submarine is a primary factor.

The final stage – disengagement – takes place in a return to the previous conditions of quiet, undetected operation.

## Active sonar

Active sonar devices transmit acoustic pulses in the audio frequency band (approximately 5 to 20kHz), the pulse rate being variable between about 12.5 to 700 milliseconds. Such variations in frequency and pulse rate are necessary to enable adjustments to be made to suit the prevailing oceanic conditions. Active sonars are used in submarines and surface vessels, and are also used in air-deployed sonobuoys. They are also used in torpedoes, using somewhat higher frequencies (typically, 20 to 35kHz) where the shorter range is offset by that greater spatial resolution.

One characteristic of all underwater-mounted transducers is self-noise, which is generated by the relative movement between the acoustic transmitter-receiver and the water surrounding it. A further complication is that transmission power is limited by the cavitation effect, in which gaseous bubbles appear on the emitting surface. But the greatest problem for sonar is the complex variation in the prevailing conditions of the ocean (as described above), and to maximise a platform's sonar capabilities it is frequently necessary to carry various different sonar sets, optimised for different regimes.

By its very nature, an active sonar system reveals its presence, thus enabling the target to detect it and take evasive action. As regards airborne sonars, it is very difficult for a submerged submarine to detect the aircraft's presence until the moment when an active sonar in a sonobuoy starts to transmit.

Active sonar systems comprise a large number of transducers mounted in an array. Surface ships use two-dimensional arrays (flat circles or rectangles); a typical modern system, the US Navy's AN/SQS-26 has 576 transducer elements in a cylindrical array, housed in a large bulbous dome at the foot of the stem. Such bow-mounted sonars are usually indicated by a sharply overhanging bow, as on most US Navy

ASW ships, whereas a less acute bow angle (eg, on British warships) normally indicates a sonar dome mounted in the keel. Cylindrical arrays are usually used in the bows of submarines. All these arrays are fixed and beams are formed electronically to give directional resolution, while Doppler shift in the return signal gives moving target indication (MTI). But a very slow moving target (such as an SSBN on patrol) with a low Doppler in a noisy or high reverberation environment, is notoriously difficult to detect.

The most effective current submarine active sonar systems are the AN/BQQ-5 installed in the US Navy Los Angeles class SSNs and the British Type 2020 now being fitted to the Trafalgar class. The AN/BQQ-5 has a bow-mounted spherical array which, using the Submarine Active Detection System (SADS) upgrade, is integrated with other onboard systems, such as the mine detection and avoidance sonar (MIDAS), under-ice systems and the forward-mounted conformal array.

Further developments in active sonar centre upon maximising its excellent detection capabilities while trying to reduce its characteristic and revealing signature. One method being tried in the United States is spread-spectrum transmission, in which the signal energy is spread over a wide range of frequencies in a pattern known only to the transmitter and receiver, thus, hopefully, losing the signal in the general oceanic "noise." This system has a further benefit in that it enhances the probability of overcoming the fluctuations in the acoustic path inherent in the oceanic environment.

**Above: An engineering model of the US Anti-Submarine Warfare – Stand-Off Weapon (ASW-SOW) is tested to ensure that it reaches the surface in the correct attitude when launched from a submarine.**

Parametric sonar has been the subject of much research. This system depends upon the mixing of two high frequencies to produce a difference frequency (higher minus lower), which is then selected for transmission. By careful selection of the two original frequencies a suitable sonar signal (100 to 1000Hz) can be produced, but this has a narrow band-width similar to that of the original higher frequencies, thus giving much better spatial resolution than can be obtained with a normal low frequency beam. Although used in minehunting and navigation, the use of parametric sonar has been restricted by problems over electronic scanning and in reducing the transducers to a reasonable size.

## Passive sonar

Rapidly growing in importance is passive sonar which, because it makes no transmissions, ensures that it does not prejudice the hunter's position. The detector is the hydrophone, a very sensitive listening device optimised for submarine noises, and which can be assembled in a variety of arrays according to the particular task, although the only substantial differences lie in the signal processing techniques. Narrowband processing is an extremely sophisticated technique which requires spectrum analysers and great computer power to produce the information, although the current revolution in microprocessors is easing this problem. The USA is clearly ahead of the world, and especially of the USSR, in this area, which is of crucial importance in detecting and analysing slow-moving targets.

Broad-band processing looks at the full spectrum of incoming signals and separates constant noises (eg, submarine flow-noise) from random oceanic noises. It thus tends to be used for the initial detection of a submarine target and for analysing the movement of the target relative to the searcher. Its main value lies where the noises emitted by the target exceed the ambient noise level and it enables rapid target detection to be made as well as relatively accurate target bearings. New techniques such as transient acoustic processing (detection and analysis of sudden, brief noises such as weapon launch or random machine noise) are adding to the effectiveness of broad-band processing.

**ASW missiles**
One of the critical ASW problems is dealing with targets detected at some distance; a torpedo will take too long to reach the target and will be detected en route. One answer is to use an airborne carrier to transport the torpedo to the vicinity of the target. Such carriers can either be rockets or winged, turbojet-powered devices, but all end by depositing the torpedo, usually by parachute, into the sea. Also of advantage is that the target has no knowledge of the imminence of attack until the torpedo actually enters the water.

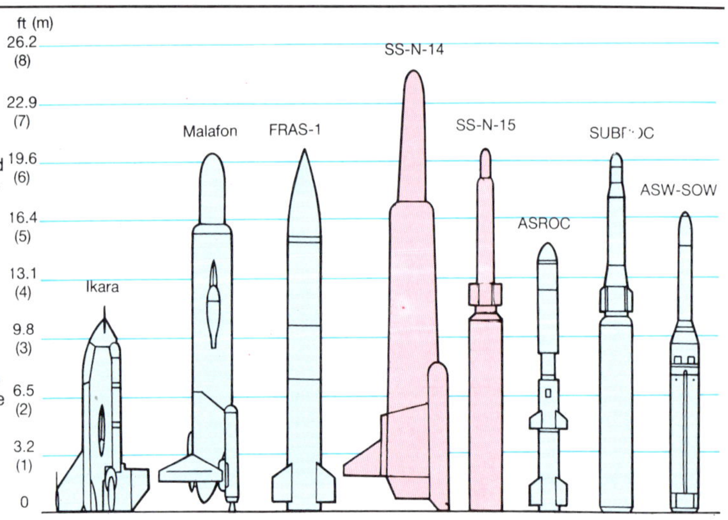


## ANTI-SUBMARINE WEAPONS

| **Designation:** | **Ikara** | **Malafon** | **FRAS-1** | **SS-N-14** | **SS-N-15/16** | **ASROC** | **SUBROC** | **ASW-SOW** |
|---|---|---|---|---|---|---|---|---|
| **Country:** | Australia | France | USSR | USSR | USSR | USA | USA | USA |
| **Type:** | Rocket-propelled, winged torpedo carrier | Rocket-boosted, winged torpedo carrier | Rocket-propelled, nuclear depth-bomb carrier | Rocket-propelled, winged torpedo/ depth bomb carrier | Submarine-launched, rocket-propelled missile | Ship-launched, rocket-propelled missile | Submarine-launched, rocket-propelled missile | Ship or submarine-launched, rocket-propelled missile |
| **Length:** ft(m) | 11.2 (3.42) | 20.1 (6.15) | 20.3 (6.2) | 24.9 (7.6) | – | 15.0 (4.6) | 20.5 (6.25) | 17.0 (5.2) |
| **Diameter:** ft(m) | N/A | 2.13 (0.65) | 1.80 (0.55) | – | – | 1.04 (0.32) | 1.73 (0.53) | 1.57 (0.48) |
| **Wingspan:** ft(m) | 4.98 (1.52) | 10.8 (3.3) | N/A | – | – | ??? (0.84) | N/A | 2.75 (0.84) |
| **Range:** miles(km) | 12.4 (20) | 8.0 (13) | 18.6 (30) | 34 (55) | – | 1.2-6.2 (2-10) | 35 (56) | 31 (50) |
| **Speed:** mph(km/h) | Mach 0.9 | 516 (830) | High subsonic | High subsonic | – | High subsonic | Mach 1+ | Mach 1+ |
| **Guidance:** | Radio command | Radio command | None | Radio command | None | None | Inertial | Inertial |
| **Payload:** | Lightweight torpedo (eg Mk44) HE | Lightweight acoustic torpedo HE | Depth-bomb Nuclear | Homing torpedo or nuclear depth-bomb HE or nuclear | Homing torpedo or nuclear depth-bomb HE or nuclear | Mk46 acoustic torpedo or nuclear depth-bomb HE or nuclear | Nuclear depth-bomb Nuclear (1kT) | Homing torpedo or nuclear depth-bomb HE or nuclear |
| **Remarks:** | | Has secondary anti-ship role | FRAS = Free-Rocket Anti-Submarine | May have anti-ship role | Similar to ASROC | | Launched from standard torpedo tube | |

Passive hull-borne sonars normally use the same arrays as the active systems, and with a spherical submarine array using digital steering a coverage of some 270° horizontally and 50° vertically can be obtained. Towed arrays are now being used by both submarines and surface ships and comprise a large number of hydrophones (up to several hundred in some cases). A typical towed array is part of the AN/BQQ-5 system used aboard the US Navy's Los Angeles class SSNs. The cable is 2,624ft (800m) long and 0.37in (9.5mm) in diameter, and is tapered at both ends to reduce drag. The array contains the hydrophones and electronics, which includes a multiplexer to reduce the wiring running through the towing cable. One advantage of towed arrays for submarines is that they provide a rearward-looking capability not possible with a bow-mounted sonar.

**Above: Underwater sensor technology is improving rapidly as shown by this Klein 100kHz Side Scan taken off the Canadian coast. The ship profile can be clearly recognised as that of an elderly Liberty ship.**

Not surprisingly, towed arrays limit the manoeuvrability of the towing ship as the actual array needs to be straight to obtain a coherent signal. Furthermore, speed has to be limited since the array is liable to oscillate above a certain speed, and this can give rise to false readings. For the towing craft (whether surface ship or submarine) following a relatively straight course at a constant slow speed makes the vessel itself very vulnerable. The problem is heightened for a surface ship by the need for separation from the main body of a task group to maximise sonar performance, although as shown with British radar picket ships in the South Atlantic in 1982, such isolated single ships are highly vulnerable to air attack. Finally, towed arrays themselves may also give rise to detectable noise, either from water-flow over the array, or by the towing cable resonating at its "natural" frequency.

For surface ships, stowage of towed arrays is mainly a matter of meeting the space requirements for the equipment on the quarterdeck; but for submarines the problem is more acute and the Los Angeles SSNs stow the array in a tube between the pressure hull and the outer casing, with the winch in the forward ballast tank. Earlier classes have to have the cable clipped on as they leave harbour on patrol (although this has the benefit of reducing the number of arrays needed).

## Non-acoustic sensors

A submerged submarine moving through the water leaves a wake, which is detectable by active sonar. This turbulence, conical in shape, eventually reaches the surface well astern of the boat where it causes minute variations in the wave pattern. Incredibly, it seems possible to detect this phenomenon at considerable ranges, and both the USA and USSR are experimenting with Over-the-Horizon Backscatter (OTH-B) radar for this purpose. Depending upon the depth of the submarine and the prevailing oceanic conditions it may also be possible that the wake turbulence will force colder water to rise and mix with warmer water at the surface; this causes a temperature differential detectable by satellite or aircraft-borne sensors.

When a submarine is travelling near the surface there is a tiny but perceptible rise in the surface-level of the water above the hull; this rise is potentially detectable by satellites. The USSR is known to be interested in this technique and the USA has at least one satellite (SEASAT) which has a radio altimeter with a vertical resolution of 3.9in (10cm) that could be used in this role.

Submarines are also detectable by the electrical and magnetic fields they create. The Soviet writer V. Nikyaylin said in 1962 that there are electro-chemical processes on submarine hulls which generate varying electrical potentials with currents flowing between them using seawater as the conducting medium. The rate of change of the consequential electrical and electromagnetic fields is detectable by sensitive devices, one of which could be made up of very large seabed electric coils.

Most submarine hulls are mainly ferrous bodies which move through the lines of force of the Earth's natural magnetic field, creating a magnetic "anomaly" which is detectable, especially by an airborne detector (Magnetic Anomaly Detector – MAD). Such detector units are mounted in extensions behind the tail in fixed-wing aircraft or in an aerodynamic body ("bird") towed behind a helicopter on a cable. All advanced ASW aircraft are fitted with such devices, including the US P-3 Orion and S-3 Viking, British Nimrod, French Atlantique and Soviet Il-38 ("May").

MAD techniques are not suitable for area searches, having a maximum detection

**Searching for an SSBN**
As seawater is a complex medium it is first necessary to establish ambient criteria. Satellites monitor weather (**1**), seastate(**2**), oceanographic data (**3**) and thermal variations (**4**). Seabed surveillance systems (**5**) identify all unwanted noise sources, eg, fish. Another satellite (**6**) relates solar activity to natural variations in the Earth's magnetic field. Merchant vessels (**7**) are eliminated by voluntary reporting (**8**) or by satellite tracking (**9**). The first step in tracking SSBNs is by satellite photography (**10**) as they leave port and by electronic monitoring by satellite (**11**) and land-based stations (**12**). ASW aircraft (**13**) use a combination of detectors including MAD, sonobuoys (**14**), thermal measurements (**15**) and forward-looking infra-red (FLIR) (**16**). Rapidly Deployed Surveillance System (**17**) is air delivered. Surface ships depend mainly on sonar (**18**) and towed arrays (**19**). Attack submarines depend upon sonar, both hull-mounted and towed (**20**). Vital passive devices are Sound Surveillance Systems (SOSUS) (**5**) and large coils have also been laid on the seabed to monitor electrical field variations (**21**). Finally, traces of a submarine's passage can be detected by Over-the-Horizon (Backscatter) radar (**22**). All this information is fed back to a huge computer (**23**).

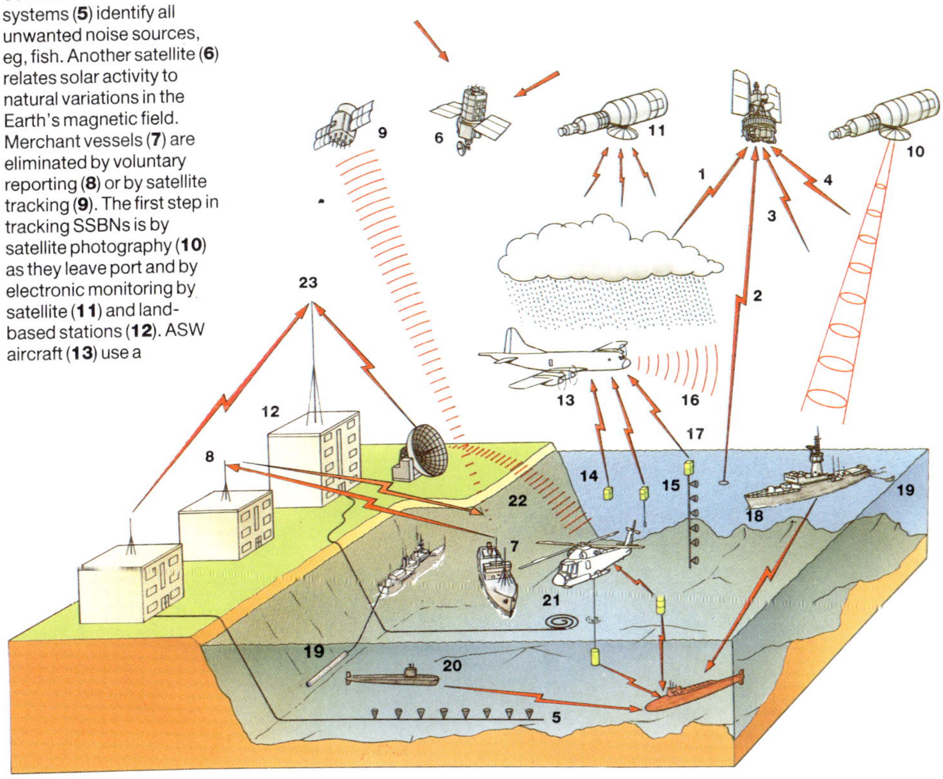

range of only some 1,000ft (305m), but they are invaluable for the precise locating of underwater targets detected by other means, such as SOSUS. However, the use of titanium hulls (as in the Soviet Alfa class) introduces a new factor since this is a non-magnetic material and cannot be detected by MAD sensors. The days of this widely used technique may be limited.

## Underwater optics

Sea water is an efficient light absorber, but there is a "window" in the blue-green portion of the visible spectrum. Efforts are being made to exploit this for communications purposes with blue-green lasers, utilising satellites to project the beam on to the "target" area. The laser could be mounted on the satellite, but the power, reliability and life requirements may preclude this for the foreseeable future. However, another possibility lies in a ground-based laser transmitting a beam which is bounced off an orbiting reflector on to the desired oceanic area. All the major relevant areas could be covered by three symetrically placed geosynchronous, equatorially orbiting reflectors. This requires great precision in laser target acquisition, pointing and tracking, but the technology for this already exists with the NASA space telescope which can be pointed with an accuracy of 25 nanorads (that is, with an error of not more than 1ft (0.305m) in 7,500 miles (12,069km), an accuracy of $2.5\times10^{-6}$).

A similar concept could be used to detect submarines, either by detecting minute return reflections from a satellite-borne sensor, or by mounting a blue-green laser and suitable detectors in an ASW aircraft. In the latter role this could well be a replacement for MAD.

## ASW weapons

The archetypical World War II ASW weapon, the depth charge, is now little used by surface ships because to allow a modern submarine within launcher range – about 2 miles (3.2km) – would be extremely hazardous. However, depth charges are used by ASW helicopters; a good example is the British Mark 11 which contains 180lb (81.6kg) of torpex. Far more effective, but with major political implications (and therefore tactical limitations) is the nuclear depth bomb, exemplified by the US B57, which can be deployed on S-3 Viking, P-3 Orion and SH-3 Sea King, and which has a reported yield in the 5 to 20kT range.

Mines have a major role in ASW, although the days of the moored or bottom-sitting mine are probably numbered. Far more effective is the US Navy CAPTOR

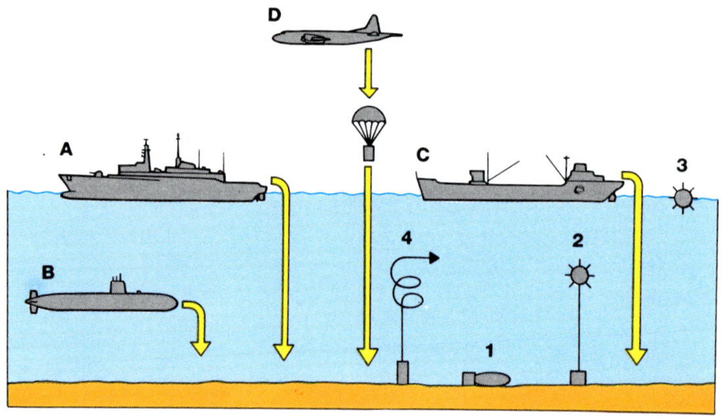


**Minehunting**
The scene depicts a minehunting operation by an Eridan "Tripartite" mine hunter. The system consists of a number of elements, the first of which is a hull-mounted sonar system for the detection and classification of mines. The information from the radar is processed and displayed to the commander. Location and navigation is achieved by means of radar, radio navigation, doppler log and autopilot. Finally, mine identification and neutralization is made by remotely-controlled underwater vehicles, divers, or by mechanical sweeping gear. Of particular value on the Eridan class is the PAP 104 system, consisting of two wire-guided, remotely controlled vehicles which have an on-board TV camera and 27 explosive charges. This French system can operate at up to 5,470 yards (5,000m) from the mother ship and at a depth of 395ft (120m). To reduce acoustical detection and to improve the very precise handling required during minehunting operations the Eridan has two separate and independent propulsion systems. The first is a 1,370kW diesel. The second is an "auxiliary" system based on an Astazou IVB gas-turbine alternator to drive the "active rudders" and the associated bow thrusters; a second gas-turbine powers the ship's services; the third is on stand-by. Complement is 49 with full accommodation and facilities on board.

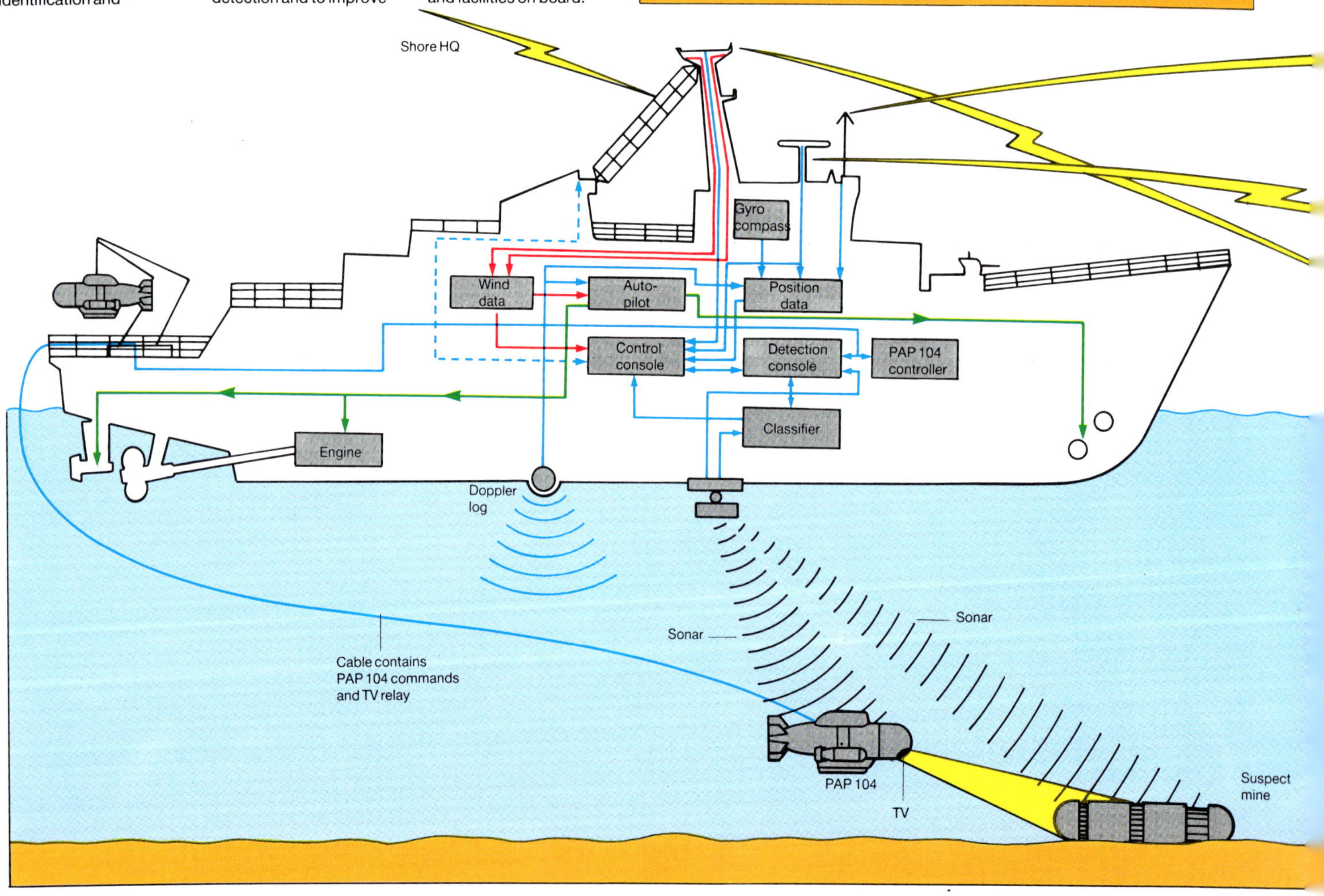

**Above: The remotely controlled vehicle is an essential element of the PAP 104 minehunting system, part of NATO's riposte to the growing Soviet mine threat.**

**Minehunting systems**
Mine warfare was a neglected art for several decades, but is now returning into vogue. There are probably more minesweepers/hunters being built now than for many years.
Mines can be delivered by warship (**A**) and submarine (**B**). Although there have been specialised minelayers in the past, both surface and submarine, today this activity is invariably undertaken as a secondary role. Mines can also be laid by merchant ships (**C**) as was done by the Libyans in the Red Sea in 1983. Finally, aircraft can be used (**D**); a rapid and effective method in times of tension.
Mines can either lie on the seabed (**1**), be moored to float at a specified depth (**2**), float on the surface (**3**), or be launched from a seabed capsule (**4**). Activation can be by means of proximity, impact, acoustic or magnetic fuzes.

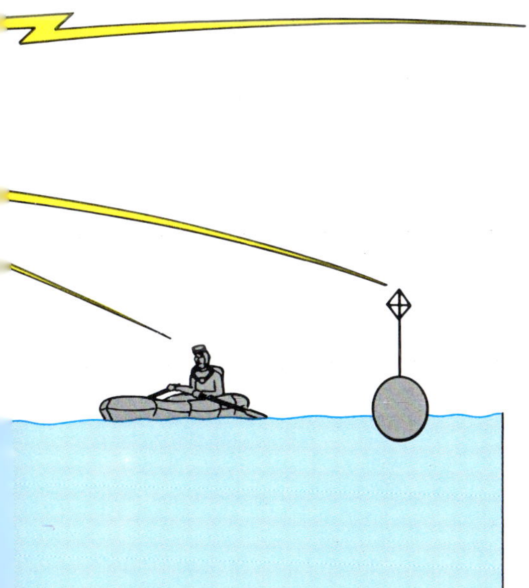

(EnCAPsulated TORpedo) which is designed exclusively for attacking submarines. CAPTOR consists of a Mark 46 torpedo housed in a tube, and can be laid by submarines, aircraft and surface ships. It sits on the ocean floor and monitors all passing maritime traffic using passive sonar; this has a range of some 3,000ft (1,000m) and is gated to exclude surface traffic. On identification of a submarine target the active sonar system is switched on, optimum launch time computed and the torpedo launched. No Identification Friend or Foe (IFF) system is fitted, and thus friendly submarines must be kept clear of any CAPTOR minefields.

## Stand-off weapons

To overcome the problem of reaching the target submarine before it can itself attack its hunter a number of stand-off weapons have been developed which deliver a depth-charge or torpedo by a carrier missile or rocket. The US ASROC (Anti-Submarine Rocket) consists of a Mark 46 acoustic homing torpedo with a strap-on rocket motor. Somewhat similar weapons are made in France (Malafon), Australia (Ikara) and the USSR (FRAS-1, SS-N-14). Ranges are not great; for example, that of the ASROC is estimated to be 1.25 to 6.2 miles (2 to 10km). The US SUBROC (Submarine Rocket) and Soviet SS-N-15/16 provide submarines with a similar capability, but fired underwater from standard torpedo tubes, before driving to the surface and then up into the air.

The US Navy intends to produce a common successor to SUBROC and ASROC, to be known as the Anti-Submarine Warfare Stand-Off Weapon (ASW-SOW). This is designed to combat the threat posed by such submarines as the Soviet Alfa and Sierra classes, and estimates of its range vary from 35 to 100 miles (56 to 160km). Vertically launched from surface ships, and expelled from standard torpedo tubes in submarines, the system comprises a common missile carrying either the Advanced Lightweight Torpedo (ALWT) or a nuclear depth charge. The latter would almost certainly be a new common device to replace the current W55, B57 and W44 nuclear warheads all currently in service with the US Navy.

A vexed question for most navies is the torpedo, which has been the subject of some extremely expensive (and often abortive) development programmes. The current generation have a top speed which is now being equalled (even surpassed) by that of their quarry; the US Mark 46, for example, has a speed of 40 knots (46mph, 74km/h) which is exactly that of the Soviet Alfa SSN; a stern chase is thus completely out of the question.

The problem of lack of speed and range for torpedoes has been partially answered for surface warships by the use of helicopters and stand-off delivery systems (eg, ASROC, Ikara) and, in the case of submarines, by SUBROC. There is still a need, however, for a much faster torpedo. In the US this has led to the Advanced Capability (ADCAP) programme for the Mark 48 torpedo which will raise its speed to 55 knots (63mph, 102km/h), and the British Tigerfish is reported to be capable of about the same speed. The next generation British torpedo, Spearfish, however, uses a gas turbine and a pump-jet to attain even higher speeds

The Stingray guidance system is complex. On launch an active/passive sonar transducer works in the passive mode, any change to active being automatic once the target indicates that it has become aware of the attack. Guidance wires link the torpedo to the launching submarine, but not only are commands fed to the torpedo but sonar information is also passed back to the submarine's system.

Warhead design is also becoming important as submarine hulls become ever stronger, the most difficult target being the enormous, double-hulled Soviet Typhoon class SSBN. Some form of directed-energy (hollow-charge) warhead could be necessary to penetrate such hulls.

However, in general terms the capabilities of current torpedoes lag behind those of the sonars that support them: they are slow, lacking in range and very noisy. Reports from the South Atlantic War of 1982, for example, abound with stories of both the Royal Navy and the Argentine Navy suffering from malfunctioning torpedoes. The USSR and some other nations are experimenting with electric motors, which at the moment are quiet, but slow and lacking in range; increases in either can only be obtained by more battery power, which means bigger torpedoes or a smaller warhead.

Torpedo sonar, too, needs to be improved and the US Defense Advanced Projects Agency (DARPA) is working on this, possibly heading towards a torpedo with on-board signal processing, coupled with an interactive fibre-optic link with the parent submarine. A further project is to adopt such a system for use with ASW aircraft, utilising a fibre-optic link from the torpedo to a buoy which has a radio uplink to the aircraft.

**Above: US Navy design concept for a ship-to-shore landing vessel. Advanced technology has not had as great an impact on amphibious warfare techniques as it has had upon other types of warfare.**

If so-called "blue-water" fleets are intended to establish dominance over the world's oceans and carrier-borne air forces to maintain air superiority over the fleets, then the final element for a major navy is the ability to land a balanced amphibious force and establish it securely ashore. Despite its importance there has been surprisingly little technical advance in amphibious warfare and, for example, the British soldiers and marines who landed in San Carlos Bay in the South Atlantic War of 1982 used ships and landing craft that were little different from those their fathers used in the Normandy landings in 1944. The only transport equipment present not available in 1944 were the helicopters.

The landing ships and craft of today are straightforward developments of those used in World War II; bigger perhaps, and slightly more sophisticated in some ways, but no different in principle. Even the US Tarawa class and the Soviet Ivan Rogovs are very direct descendants of World War II concepts, and amphibious vehicles like USMC LVTPs and Soviet PT-76s are but developments of the Duplex-Drive Sherman and the DUKW.

**Above: One of two air cushion test vehicles whose best attributes were incorporated in the US Navy's Landing-Craft Air-Cushion Vehicle (LCAC). The ACV has taken a long time to become militarily viable.**

## Air cushion vehicles

One significant development in amphibious warfare is the air cushion vehicle (ACV). From the tiny, inefficient prototype of 25 years ago the ACV is now a sophisticated craft, of which the Soviet Aist class (12 in service) is probably the current world leader. With an operating weight of some 220 tons (233.5 tonnes) and powered by two gas-turbines, the Aist can carry up to 150 troops and four light tanks (eg, PT-76) and is in full service with the Soviet Marines. The only Western ACV of note in service is the British missile-armed BH-7, with an operating weight of 50 tons (53.07 tonnes), of which six are in service in Iran and one in the UK. The ACV is one of those ideas which seems always to be on the edge of a breakthrough, but never actually makes it. The Soviet Navy has, however, shown the practicability of the idea, although even their craft are confined to a coastal role, and open-ocean ACVs still seem to be a long-way off.

In the USA trials over many years (including the use of ACVs in the Vietnam War) have led to the Landing-Craft Air Cushion Vehicle (LCAC), a 150 ton (159.2 tonne) craft capable of over 40 knots. Unlike the Soviet ACVs, the LCAC is a "surface effect ship" (SES) with two side structures stretching the length of the hull and an open well down the centre for the payload. No weapons are mounted since the LCAC is simply a sophisticated lifting-platform for carrying men, vehicles and weapons ashore; total procurement is planned to be 107. The large, civilian-operated ACVs used on the busy Anglo-French cross-Channel links have demonstrated the viability and reliability of the concept, but it would seem that navies, by and large, remain unconvinced.

Another idea which is a long time in reaching fruition is the wing-in-ground-effect (WIG) machine, which physically resembles an aircraft, but which is designed to remain within the ground effect regime. Persistent reports appear in Western journals concerning Soviet developed machines, accompanied by artist's drawings, but not one photograph (even of the grainy, silhouette-only) type has appeared. (But it was also many years after the Soviet Typhoon class SSBN entered service before a reasonable picture was

**Above: The USSR is always ready to adopt and adapt Western technology. There are many ACVs in service, such as the Soviet Navy Gus-class craft; far more than in the West where the concept was born.**

**Above: The first production LCAC was delivered to the US Navy in December 1984; a total of 107 have been ordered. The LCAC is in effect a powered platform, for amphibious over-the-beach delivery.**

**Below: The Sea Harrier has provided a major breakthrough in amphibious warfare technology, making immediate air support available at much less complexity and cost than with other fixed-wing types.**

published in the West, and that proved that many of the artist's impressions had been very close to the truth.) An amphibious *ekranoplan* is reported to be under development which is powered by a large, tail-mounted marine turboprop with contra-rotating blades and two bow-mounted turbojets with swivelling nozzles, similar to, but rather larger than, those fitted to the Yak-36 "Fencer". The craft is reportedly armed with two SS-N-22, the latest type of sea-skimming anti-ship missile. The craft is fully amphibious and clearly has great potential in landing operations by the Soviet Marines.

## Command, Control and Communications

Study of any amphibious landing operation reveals that one of the major problems is always the so-called C³. This particularly applies in the assault phase when the naval, land and air forces are all in the same fairly small geographical area, and all needing to communicate both within their own environment and to each of the other two elements. There is also a major requirement for staffs to be co-located to ensure effective co-operation. There have long been specialised headquarters ships to meet this requirement, this approach being epitomised by the USS *Blue Ridge*, a highly sophisticated centre for C³.

It can be considered wasteful to devote one hull entirely to such a role and most other navies allocate the space for staffs and communications in other ships, such as large assault ships (Landing Platform Docks/LPDs) like the British Fearless class LPDs. In this, advanced technology is playing a major role in that automatic data processing and communications equipment is becoming more capable, more efficient, and – most importantly – smaller, cheaper and more reliable. USS *Blue Ridge*, for example, has three computer systems, one each to support naval tactical data, naval intelligence and amphibious control, while the larger and newer Tarawa-class LHAs have one Integrated Tactical Amphibious Warfare Data System.

## Air support of landings

A critical element in an amphibious landing is air support. Initially, of course, this can be provided from aircraft carriers, but there is always a strong desire, especially from the land commanders, to get the aircraft ashore and on to a land base. With CTOL fixed-wing aircraft this has always been a difficulty, making it necessary to capture an existing airfield in the early stages. The V/STOL aircraft such as the Harrier has altered this, and the most enthusiastic proponent of the Harrier has been the US Marine Corps who, from the very start of the programme, realised the potential of this unique machine.

Although the Yak-36 "Forger" would obviously be used in a Soviet amphibious landing, there has been no evidence so far of the aircraft being operated away from the Kiev class carriers on a temporary landing site as is common practice for the Harrier. Indeed, its ability to operate from such hastily prepared landing sites ashore must be somewhat limited due to its lack of a rolling takeoff capability, which will doubtless be rectified in its successor.

# UNCONVENTIONAL WARFARE

The preceding chapters have described the effects of advanced technology upon the military forces intended primarily to fight in a general war on a global scale. However, such a war has not occurred since 1945, although there have been moments when it appeared to be a possibility. Nevertheless, the world, despite a theoretical state of "peace", has been riven by an apparently endless series of wars, some major, many minor.

The most common form of conflict has been revolutionary warfare (also known as insurgency, or guerrilla war), with campaigns having been fought in Malaya, Indochina (first against the French and then against the USA), Kenya, Cyprus, Aden, Algeria, Nicaragua, Bolivia and Cuba, to name but a few. Associated with this is terrorism, which can be either an adjunct of a widespread revolutionary war or, as seems to have become more prevalent in the past 10 to 15 years, a conflict-form of its own, with no apparent intention of expanding to the true "revolutionary" stage.

There have, of course, been many "conventional" or "low intensity" international conflicts, ranging from the Korean War of the early 1950s, through the various Arab–Israeli wars to the Iran–Iraq war (still continuing in 1985 as the longest war between two nations this century) and the Anglo-Argentine conflict of 1982. These conventional wars have tended to be fought within well-defined, if implicitly agreed, limits;

## SPECIAL FORCES

The past two decades have seen an almost unprecedented rise in the number of special forces, raised mainly to conduct peacetime operations which are, usually for political reasons, beyond the means of the police forces or the conventional military. Virtually all of these special forces would have a genuine role in general war, but there can be no doubt that it is their peacetime role which the governments concerned find of greatest value. These forces exist both as autonomous groups in their own right, such as the American "Delta", and as elite sub-groups of a larger body, such as the British Special Boat Squadron (SBS) which is a part of the Royal Marines.

The two largest of these groups are the United States Forces, now well over thirty battalions strong, and the Soviet Spetsnaz, which are believed to number about 30,000 in peacetime. The Americans also have three Ranger battalions and the group known as "Delta". The British, in addition to the SBS already mentioned, have the famous Special Air Service (SAS), who conducted the rescue operation which finished the Iranian Embassy siege in London in 1980. The West German special force is Grenzschutzgruppe 9 (GSG9), which is unusual in that it is part of the Border Police rather than a part of the army, as is the case in virtually every other nation.

These special forces tend to have highly specialised equipment requirements, and use advanced technology in a very specialised way. Their requirements are to be able to observe, usually at very close ranges, without themselves being seen, to communicate reliably and using the shortest possible time on the air, and to fight close-quarter battles in confined circumstances. To a large extent, however, and for very understandable reasons, their equipment is usually very highly classified, and like

**While major international wars remain the greatest threat to global stability, indeed to global survival, minor wars and terrorist "incidents" continue unabated, with an ever-present danger that they might escalate into a superpower confrontation. Here, too, technology is playing a vital role.**

they have, however, almost invariably been conducted by both sides using the weapons, organisations, tactics and technology intended for use in general war. Thus, in the South Atlantic War of 1982 the British used forces diverted from their more usual NATO tasks.

The differences, and the problems, have come with revolutionary war where advanced technology has in many ways proved to be more of a hindrance than a help. This certainly would appear to be one of the major lessons available for the United States' forces from their experiences in Southeast Asia and one which, astonishingly, is being re-learned by the Soviet forces currently in Afghanistan. The possession of these advanced equipments, procured at great expense, seems to bring with it a virtually irresistible desire to use them – sometimes, indeed, to engineer circumstances in which they can be used – to satisfy some outside pressure, particularly a political one.

Many of the weapons and much of the technology used in these types of conflicts are the same as those used in general war. Indeed, much of the technology is a by-product of that used and developed for general war, especially since the American withdrawal from Vietnam.

This chapter looks at advanced technology in these areas, but only where it produces systems or weapons which differ significantly from those described in the preceding chapters.

**Left: In stark contrast to the large, modern, sophisticated forces needed for nuclear warfare, unconventional warfare forces must be prepared to operate in small groups using age-old techniques.**

**Above: Special forces must be prepared to operate in every environment and in every part of the world, such as this US Navy SEAL en route to an underwater demolition task in Pacific waters.**

**Above: The terrorist threat has often proved too much for the police, and military special forces have had to be called in, such as these British SAS in the London Iranian Embassy siege in 1981.**

these forces themselves, is very seldom seen in public. It was only due to the fact that the Iranian Embassy siege took place in the centre of London and under the full glare of media attention, for example, that the all-black fighting rig of the SAS and their use of the Heckler and Koch MP5 sub-machine-gun became known. (This is now their "public" image, of course; what their weapons and equipment would be in a general war situation – even the Falklands, for example, is a different matter.)

The United States sees the Special Operations Forces as having a particular role as the repository of knowledge and skills of low-intensity warfare, as well as having a significant role in general war. In peace, SOF are capable of direct action in overseas crises where the overt use of military force would be unacceptable or counter-productive, including counter-action against terrorists attacking US targets. One spectacular example of such action was the unsuccessful attempt to rescue the American hostages held in the embassy in Tehran, although this failed through mischance allied to some planning errors, rather than to deficiencies in men or equipment.

The current composition of US Army SOF is one Special Operations Command, consisting of eight Special Forces (SF) groups with 28 SF battalions; four PSYOP groups with 12 PSYOP battalions; two special operations battalions; one Ranger group with three Ranger battalions; and one civil affairs battalion. The Navy has two special warfare groups comprising three warfare units; six SEAL teams; two SEAL delivery teams; two special boat squadrons, six special boat units, and (at present) one special operations submarine. The United States Air Force has an air division (part of 23rd Air Force), consisting of one special operations wing, five special operations squadrons, one helicopter detachment, and two special operations groups.

These US SOF will be increased in the near future by a ninth Army SF group, another PSYOP battalion (the thirteenth) and a seventh naval SEAL team, while all three services will add much new equipment. Under central defence control is Special Forces Operational Detachment – Delta (known simply as "Delta"). Founded by the redoubtable Colonel Charles Beckwith, Delta carried out the Teheran rescue attempt, and has since been strengthened in equipment, organisation and personnel.

One of the most significant developments is that the now rather elderly SOF transport submarine – the USS *Grayback* – which can carry 67 SF troopers is to be replaced by no fewer than six nuclear-powered submarines, three in each of the two major fleets. The first two are the former SSBNs, *Sam Houston* and *John Marshall*, which have been undergoing refits since retiring from the missile fleet; this has included having their missile tubes plugged with concrete to comply with the provisions of SALT II. These transport submarines will have a global range, will offer comfortable conditions to both crews and passengers, and will give an important enhancement to the USA's "black" (top secret clandestine) capabilities.

The 1st Special Operations Wing of 23rd Air Force has five squadrons equipped with highly specialised versions of the ubiquitous Lockheed C-130 Hercules. The MC-130H, code-named Combat Talon, is an improved MC-130E with better range and payload, improved command and communication fits, enhanced ECM and better avionics. These aircraft can fly at extremely low levels for protracted periods, and are flown like fighters even though built as long-range transports. Also used is the AC-130H, a gunship version, equipped with a miscellany of high-power armament with very precise fire control devices, which enabled them to bring fire to bear very close to their own troops in the Grenada operation in 1983.

The HH-60D Night Hawk is a special operations version of the UH-60 Blackhawk helicopter and is optimised for the combat air rescue mission. It is equipped with terrain-following radar, forward-looking infra-red (FLIR), armoured protection, and an armament mix of guns and missiles. A new helicopter – the AH-X – is due to reach IOC in 1986 to improve 1st SOW's medium-lift capability.

### The Soviet Spetsnaz

Spetsnaz are the Special Forces of the Main Intelligence Directorate (GRU) of the Soviet General Staff, and are formed into units known as "diversionary brigades". Unlike the United States, in which each service has its own special operations forces, the Soviet Union has one force, the Spetsnaz, covering the Army, the Navy and the Air Force. Their wartime tasks would be to conduct sensitive, covert missions abroad and include murdering enemy political and military leaders, attacking enemy nuclear bases and command posts, and the destruction or disruption of economic targets.

**Above: US Special Forces on realistic training. Technology is very valuable for such forces, but must never be allowed to replace the basic military skills; one of the lessons from the Vietnam War.**

**Above: Soldiers of the Soviet Union's Spetsnaz, known as "Diversionary Brigades". These men are very highly trained and are considered in the West to be a major threat in future conflicts.**

**Above: US SOF use several versions of the ubiquitous C-130 for close fire support. The aircraft have heavy firepower, including 20mm Gatlings and 105mm cannon, and excellent fire control systems.**

**Below: A Soviet submersible photographed in the Sea of Japan during the search for the wreckage of Korean Air Lines flight KAL 007. The Spetsnaz bottom-crawling submersible is very similar to this type.**

At a peacetime level of some 30,000 personnel the Spetsnaz are the largest special forces group in the world. A ground forces Spetsnaz brigade is some 3,000 strong and consists of four battalions, an anti-VIP company, and signals and other supporting units. The naval Spetsnaz brigade has an anti-VIP company, a group of midget submarines, three battalions of combat swimmers, a parachute battalion and supporting units. However, both ground and naval Spetsnaz troops generally work in small groups of three men, with a brigade producing 100 or more such groups, but these can come together to form larger groups of any size to perform specific missions.

Their employment is not confined to war, however, and a Spetsnaz brigade was the first Soviet unit into Prague in 1968, while a Spetsnaz "hit-squad" was responsible for the December 1979 murder of Afghanistan President Hafizullah Amin, in a joint KGB/GRU operation.

Spetsnaz training is arduous and concentrates on military and physical skills. As one of the prime targets will be NATO nuclear weapons sites, large training areas exist where life-size mock-ups have been built of equipment such as GLCM and Pershing launchers, M110 nuclear-capable self-propelled howitzers and France's Mirage IV bombers. Clear evidence of such training areas has been obtained by United States reconnaissance satellites and published in open documents.

## Spetsnaz tracked mini subs

Obviously, Spetsnaz units receive high priority for equipment and the vast majority of this is kept very secret in the usual Soviet way. However, one very important item of equipment that has come to light is the tracked miniature submarines, which suddenly came to public attention when they were used by naval Spetsnaz units in a blatant reconnaissance of the Swedish naval base in the Stockholm archipelago in October 1982. These vessels have twin tracks and a large single propeller, and appear to be taken to the vicinity of their targets by a mother-ship. In fact, it is highly probable that these are the mini-submarines known to operate from the Soviet Navy's India-class submarines, and which had been thought to be intended for underwater rescue missions. Incredibly, it has since transpired that a photograph of these tracked mini-subs had been published in *Pravda* some ten years previously, when they were said by the Soviet authorities to be in use to search for the lost city of Atlantis – a somewhat fanciful and improbable story – but nobody in the West seems to have appreciated their military significance. They could be put to a number of uses, apart from the known one of reconnaissance, such as laying nuclear mines at the exit points from NATO nuclear submarine bases. The question that inevitably arises is just what other reconnaissances have been carried out by these machines – could they, for example, have reconnoitred the British nuclear submarine base at Faslane in Scotland or the French base at Brest, or the many American naval bases around the world?

**Spetsnaz targets**
The Soviet Union's Spetsnaz troops are among the most highly trained and best equipped in the world. Their primary tasks in war are to operate against enemy command and control systems, especially those involved in the nuclear warfare process. They receive every assistance in preparing for this task and are given every weapon that Soviet technology can provide. Among their targets will be NATO nuclear weapons systems and an extremely realistic training area has been photographed by US satellites (below). With a peacetime strength of some 30,000 men the Spetsnaz are by far the largest special forces group in the world; they serve in the land and the naval forces and they have their own supporting troops (logistics, signals, etc). In peacetime they are used whenever their special skills are required, for example, in Czechoslovakia in 1968 and throughout the current and very protracted conflict in Afghanistan.

**Spetsnaz soldier**
Like many other special forces, Spetsnaz soldiers have numerous roles and a variety of uniforms to go with them, from NATO combat outfits, through civilian clothes, to Soviet national uniforms. This Spetsnaz soldier is wearing Soviet Army combat uniform and is armed with an SA-7 Grail surface-to-air missile launcher. Like his Western counterparts (US Special Operations Forces, British SAS, etc) the Spetsnaz soldier is trained in basic military skills, deception techniques and in the use of the most modern weapons to be found on the battlefield.

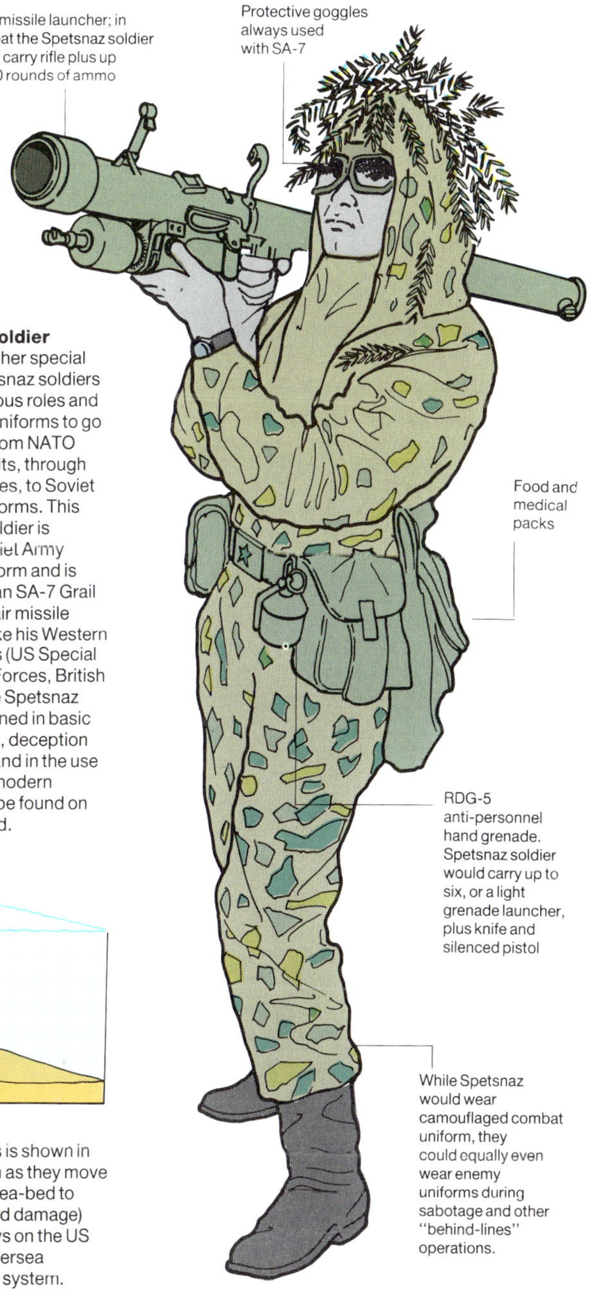

**Undersea operations**
While investigating the numerous Soviet incursions into its territorial waters the Swedish Navy positively identified the use of small submersibles equipped with caterpillar tracks. Such craft were first revealed by Pravda in 1973 when they were stated to be intended for use in "research", including looking for the lost civilization of Atlantis! Two of these craft are carried in an open-topped well on the India-class submarine and one of their more probable applications is shown in this diagram as they move across the sea-bed to examine (and damage) sensor arrays on the US SOSUS undersea surveillance system.

The problem of terrorism has existed for many years, although for a long time terrorist campaigns were linked to a domestic political objective and tended, with but few exceptions, to be confined to the national territory of the antagonists. Today, however, terrorists conduct their operations on a global basis and frequently strike at targets on the territory of a totally uninvolved third party. Thus, for example, in April-May 1980 a group of anti-Khomeni terrorists from Arabistan in southern Iran attacked and occupied the Iranian Embassy in London, England, forcing the British – who had absolutely no connections with the dispute – to try to bring it to some sort of a conclusion.

Training camps for terrorists exist in several countries including the USSR, the German Democratic Republic, Bulgaria, Hungary, Czechoslovakia, Libya and South Yemen. The current major terrorist groups include the Italian Red Brigades, the Irish Republican Army (IRA) and its more extreme splinter-group, the Provisional Wing of the IRA (PIRA), the West German Baader-Meinhof Gang, the Seikgu (Red Guard) in Japan, some elements of the Palestine Liberation Organization, militant Shi'ite groups in the Middle East, and Basque nationalists in Spain.

According to Risk International, Inc., quoted in the New York *Times*, 24 April 1984, during the calendar year 1983 authenticated terrorist activities around the world included 119 kidnappings (7 killed), 1,120 bombing incidents (813 killed), 1,231 other attacks on installations (8,610 killed), 362 assassination incidents (725 killed) and 5 hijackings (4 killed). The following year was no better, although no figures are yet available, but the killing and maiming for "political" ends did not stop. The year included, amongst many other incidents, the murder of a British policewoman by shots from the Libyan People's Bureau (embassy) in London, and the mass murder of many passengers in an explosion in a train travelling through a tunnel in Italy.

**Above: A graphic re-enactment of the flightdeck scene as hijackers strike. The spate of hijackings in June 1985 showed how easily such events can still occur.**

A major difficulty for the security forces is that terrorists generally have the tactical initiative in the sense that they can choose when and where to strike next, frequently lying dormant for months (sometimes even years) and then striking suddenly at a totally unexpected target. Some Western European groups, for example, kidnap or murder important businessmen, carefully selected by them, but in an apparently haphazard pattern. The security forces cannot, in any case, protect every major industrialist and businessman. In Italy the rather odd kidnapping of a US Army logistician (Brigadier-General Dozier) by the Red Brigade was similarly a totally unpredicted (and unpredictable) event. In an effort to obviate such events security forces try to identify and monitor terrorist groups and to anticipate their actions, but without creating the right-wing, "fascist" conditions the terrorists claim they are trying to combat.

Terrorists have never been slow to utilise advanced technology weapons and equipment for their own ends, although they are normally forced to adapt available technology rather than having sophisticated equipment developed specifically for their needs. Thus, for instance, the Provisional

**Below: An Arab hijacker proudly waves from a door of an Iran Air Boeing 747. The troubled Middle East has given rise to more hijackings than any other single area of the world and continues to do so.**

**Irish mortar attack**
The hit-or-miss nature of many terrorist attacks was clearly shown in March 1985 when a nine-barrelled lorry-mounted mortar was used to attack the RUC police station at Newry. Nine steel pipes were secured to an angle-iron frame on the flatbed behind the vehicle cab at a fixed angle. The bombs were made from commercial oxygen cylinders and filled with explosives. Of the nine fired, two exploded prematurely in flight and two failed to detonate when they hit the ground. Three of the bombs overshot the target, exploding in a street behind the police station and one was unaccounted for. That left just one bomb, which hit the target – the police canteen – killing nine officers.

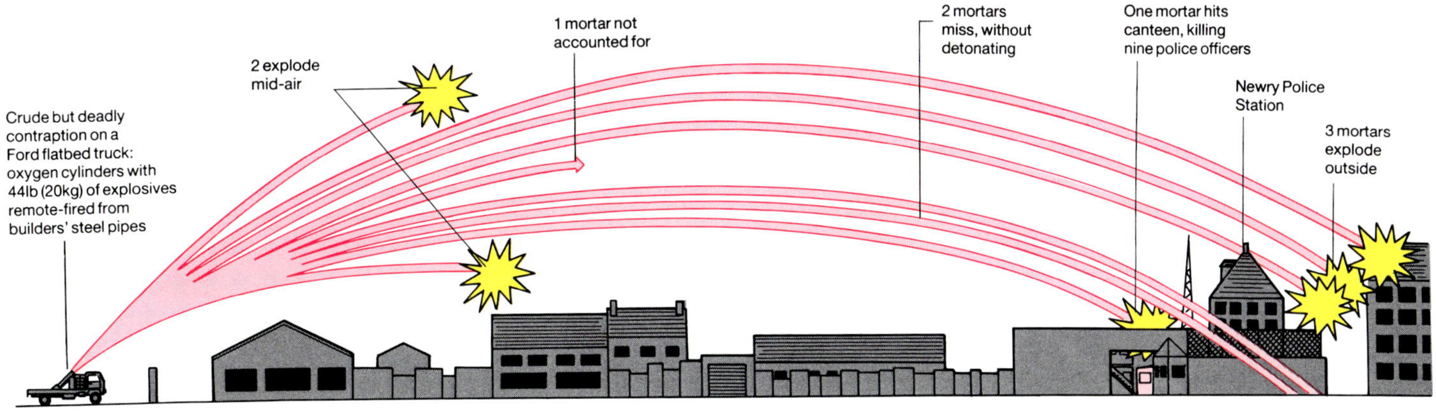


Wing of the Irish Republican Army has used commercially available explosives and electronic components to develop a series of remote-controlled devices. One major example of this was the mortar attack during the night of 28 February 1985 on the police station at Newry in County Down, in which nine off-duty police officers were killed in an incident regarded as one of the worst terrorist outrages in the Province since the present "Troubles" began in 1969. The weapon used was a nine-barrelled mortar, the latest in a series developed since the first IRA mortar attack in 1970. Nine steel pipes were secured by wires to a girder and mounted on a frame made of angle-iron. The projectiles were made from adapted oxygen cylinders, each packed with some 44lb (20kg) of explosives and fitted with rudimentary fins. With everything prepared for firing the lorry was driven to a disused yard about 220 yards (200m) from the police station and the mortars fired by an electronically-operated delay device which gave the terrorists time to escape. Of the nine bombs fired, three were duds, two blew up in mid-air, three overshot the target by a considerable margin (exploding on impact, in a street, fortunately, injuring nobody), but the ninth, apparently falling shorter than the others, went straight through the roof of the canteen, killing the policemen and women. All the components, apart from the explosives, were commercially available in Northern Ireland, and the use of such an extemporary mortar, while inaccurate, meant that the IRA had not had to smuggle proper military mortars and their bombs across the border into Northern Ireland, thus running all the risks associated with such an undertaking.

In somewhat similar, but even more bloody, incidents Lebanese terrorists attacked and killed large numbers of US Marines and French soldiers in Beirut in 1983. They used devastating explosions of exceptional power, in which large charges were strapped around steel cylinders of explosive gas. The resulting implosion/explosions applied at critical points inside the buildings were sufficient to collapse the entire structures, with horrific death tolls, including, in both cases, the drivers of the kamikaze lorries which had rammed their way into the ground-floor entrances of the buildings.

**Below: The consequences of just one of the many hundreds of bombs which have gone off in strife-torn Beirut. The extraordinary vulnerability of modern, rapid building methods is obvious.**

**Below: Member of the Provisional IRA (PIRA) armed, like so many terrorists, with a Soviet weapon. In this case it is an RPG-7V anti-tank rocket launcher; light, accurate and very effective.**

Other "advances" such as small, readily concealable weapons and optical and electronic surveillance devices are all available to terrorists if they know the right sources – and they have shown that they do.

The Israeli Military Industries' Mini-Uzi 9mm sub-machine-gun (SMG), a favourite terrorist weapon, weighs only 6.9lb (3.11kg) with a magazine of 20 rounds and is just 14.2in (360mm) long with its stock folded. If some of the latest handguns reach production they will be even more deadly in terrorist hands. The Heckler and Koch G11 uses caseless ammunition, weighs 9.5lb (4.3kg) with 100 rounds and is 29.5in (750mm) long; a short-barrel SMG version with a smaller number of rounds would be extremely potent in close-quarter battle situations. Although designed for military and police use such powerful weapons are obviously ideal for the terrorists' purposes: powerful, yet small and relatively easy to conceal until needed.

**Anti-riot gear**
Technology has increasingly come to the aid of the soldier engaged in the task of dealing with civil disturbance. This soldier is in typical modern anti-riot gear. He is wearing a specially-made helmet with an inbuilt clear visor to protect his eyes. His flame-proofed combat uniform hides a "flak-jacket", and he is carrying a clear shield and a heavy wooden baton. Thus equipped he is able to deal with most non-military situations.

Left: The British Army's famed "Wheelbarrow" remote-handling equipment, developed as an answer to the PIRA's wide-scale use of bombs in Northern Ireland. Many different devices can be fitted to deal with explosives at safe distances.

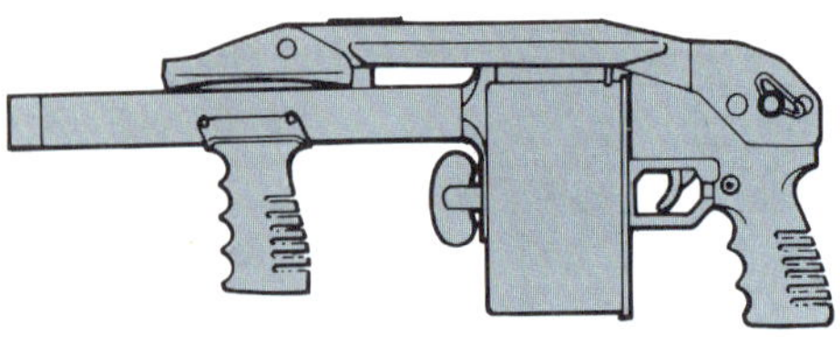

Above: The Striker automatic shotgun is a South African design for closequarter combat by police or military, and for private use against terrorists. Made of aluminium, the gun can fire up to twelve 12-bore rounds in 2.6 seconds.

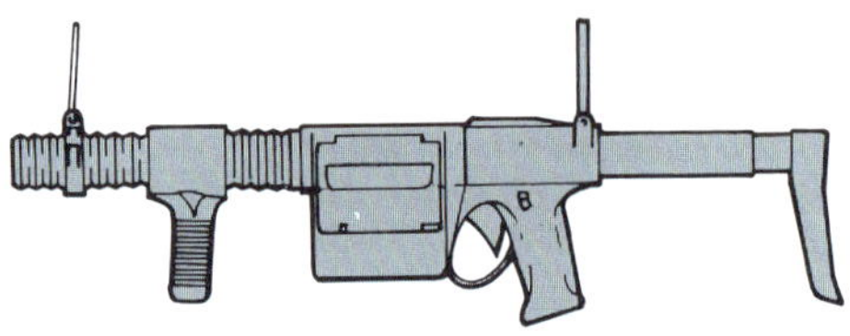

Above: The Arwen 37 anti-riot weapon system is a product of the British Royal Ordnance arms firm. It is a multi-shot, semi-automatic anti-riot weapon, and can fire baton and CS rounds from a five-round, revolver-type magazine.

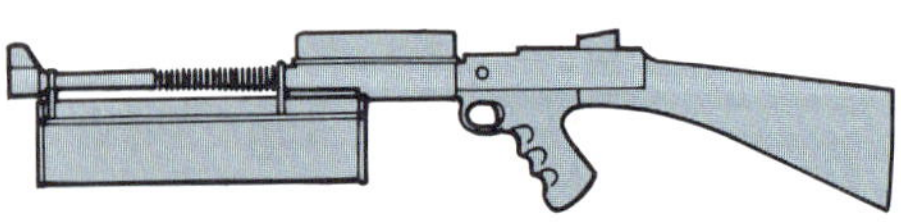

Above: The AM-180 0.22in anti-riot sub-machine gun is specifically designed to use the Laser Lok sight which projects a clear red dot on the target. As the barrel is zeroed on to the dot any round then fired must hit the target.

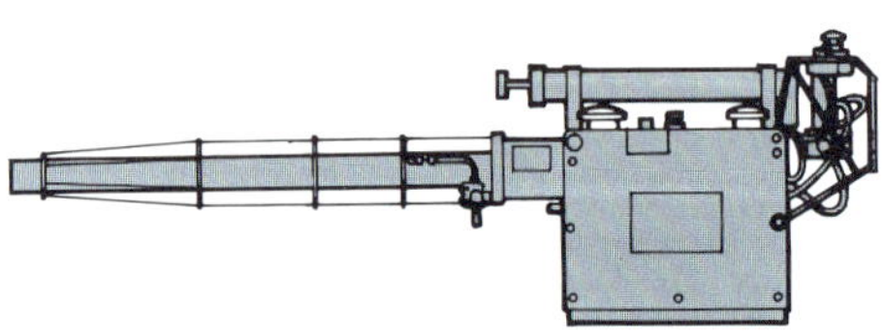

Above: The Smith and Wesson Mark XII-C Pepper Fog Smoke Generator is used to produce 100,000 cu ft (2,831 cu.m) of CN, CS or special CS tear-gas in just 26 seconds. Such use of gas can be effective, but must be carefully controlled.

## Countering terrorism

The counter-terrorist response has been, to a large degree, to use technology in the fight against terrorism. Some of these devices are used overtly: for example, electronic baggage surveillance has become a familiar and accepted part of every journey by air. Others, widely known about but usually unseen by the general public, are electro-optical equipment for examining mail (to counter letter bombs), electronic "sniffers" for detecting explosives (for example, in cars), and electronic "bugs" for listening to telephone and radio conversations.

Information on other equipment is now becoming public. For example, fibreoptic technology has enabled many advances to be made in surveillance, especially in siege operations. Fibreoptic borescopes enable security forces to see "through" apparently solid walls and doors. A very small hole, drilled by a special bit, is required through which the borescope is inserted; viewing can then be done by the naked eye or with a TV monitor. It was presumably a device such as this that enabled the British SAS troopers to know so precisely the internal layout and situation in the Iranian Embassy in the 1980 rescue operation. Such techniques can also be used to "see" into suspect devices such as barrels and vehicle petrol tanks.

**Left: Men of West Germany's highly trained and efficient GSG 9. Unlike many Special Forces this unit is part of the Federal Border Police (Bundesgrenzschutz) and is not a military force.**

One of the problems for counter-terrorist forces is that absolutely guaranteed safety (if, indeed, such a thing is attainable) imposes so many restrictions upon "normal" life that the terrorists' aims are partly realised, anyway. Politicians lose their effectiveness if they are so surrounded by security guards that their public cannot see them, while businessmen (now also on some terrorist target lists) must move about to conduct their affairs.

Special security cars abound, protected by steel armour-plate or Kevlar, and body protection devices are readily available. One company in the USA (Ballistic Shelters Corp) even markets a "safe haven", a bomb- and bullet-proof cabinet, about the size of a linen wardrobe and fitted with communications and air-conditioning; the more expensive models even have a gun-port.

One application of technology in the anti-terrorist battle is the remotely-controlled robot for locating, defusing or destroying explosive devices. The progenitor of the line was the British "Wheelbarrow", developed to deal with the increasingly sophisticated PIRA bombs and the escalating death-toll of bomb-disposal experts. Most such robots run on tracks, which enables them to surmount small obstacles as well as giving the operator very fine and precise control over its movement; a Canadian model, however, runs on six wheels.

Most of these robots are connected to their control-box by cable, although some

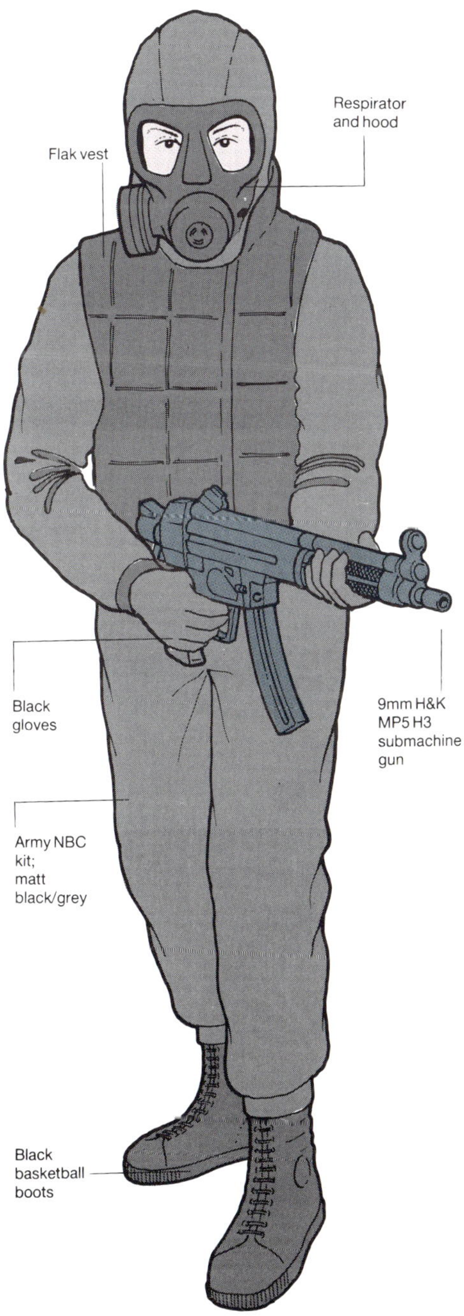


**British SAS soldier**
The SAS smock, flak jacket and trousers, with basketball boots, knitted woolen balaclava helmet and a respirator – all in matt black. The reason for this garb is not only for concealment, but also to create a feeling of awe and dread in the terrorists in the crucial first few seconds (calculated down to 3) of an attack, as was shown in the Iranian Embassy siege in London in 1981. The weapon is a Heckler and Koch MP5A3 9mm SMG.

Right: A large variety of anti-riot weapons are now in use around the world. One widely used item is CS (tear) gas, which can be dispensed using a hand-thrown grenade or projected from a "riot gun", usually of 1.5in (38mm) calibre. This Haley & Weller round has a range of about 70 to 140 yards (65 to 130m), which enables it to be used out of hand-thrown projectile range of the rioters, a valuable tactical advantage in military riot control.

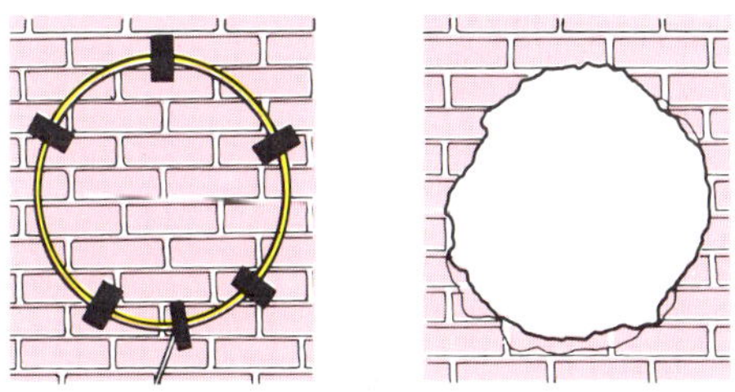

Above: Rapid and precise explosive devices are frequently required by counter-terrorist units. These diagrams show the use of "Dartcord", a linear cutting charge.

Right: A linear cut: "Dartcord" being used to sever a stone pillar during the demolition of an old mill, showing the exceptional power of this simple device.

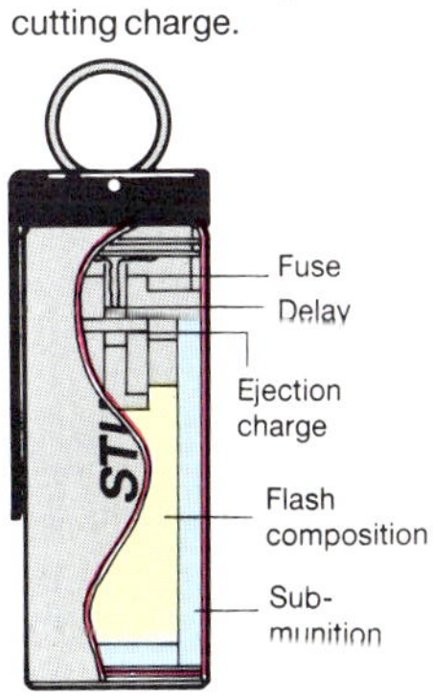


Above: Stun grenades are used to disorientate terrorists without causing any injuries to innocent hostages.

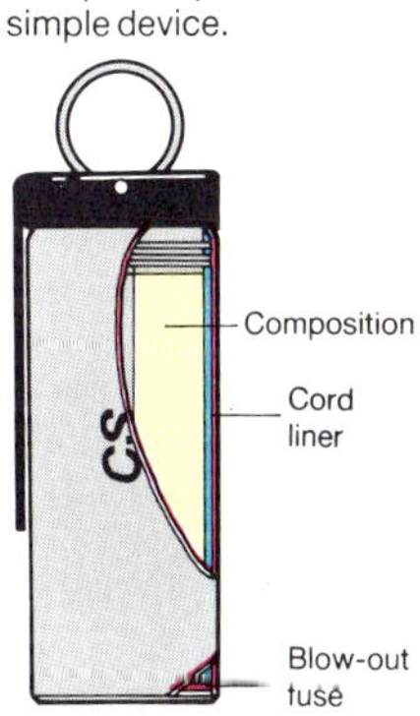


Above: CS grenades are essential in certain circumstances, but if wrongly used can be counter-productive.

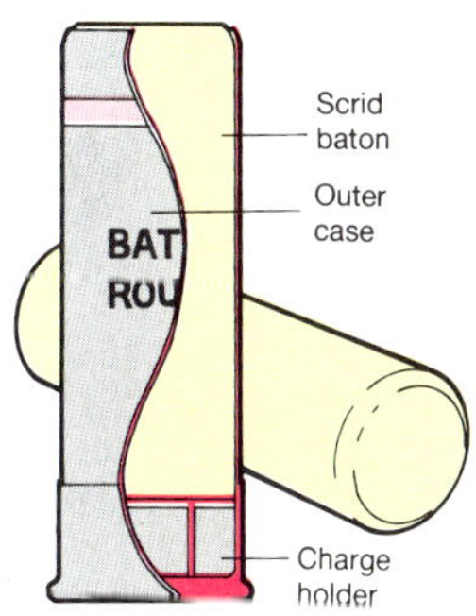


Above: Baton rounds, made of special plastic, are intended to discourage rioters without causing serious wounds.

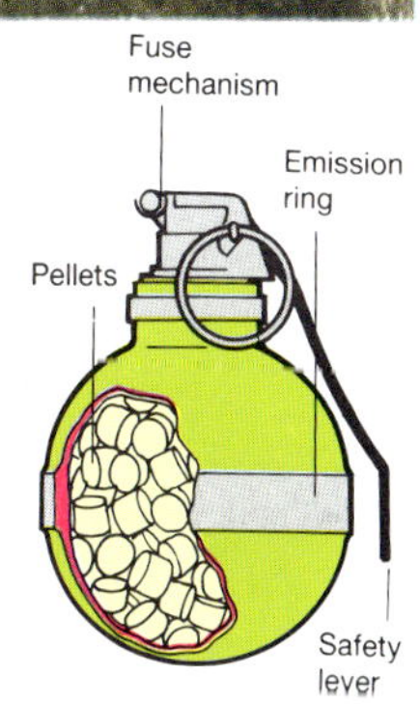


Above: Smith and Wesson No 15 grenade is easy to throw and burns rapidly, giving a rioter little chance to throw it back.

**Left: The Haley & Weller multi-purpose riot gun. There are four separate barrels: for 38mm anti-riot rounds, 12-bore rounds, 38mm signal rounds and 25mm signal rounds, respectively.**

**Above: The Uzi 9mm sub-machine-gun has proved to be one of the finest and most reliable guns of its type in the world. Three separate safety systems overcome one of the major weaknesses of the SMG.**

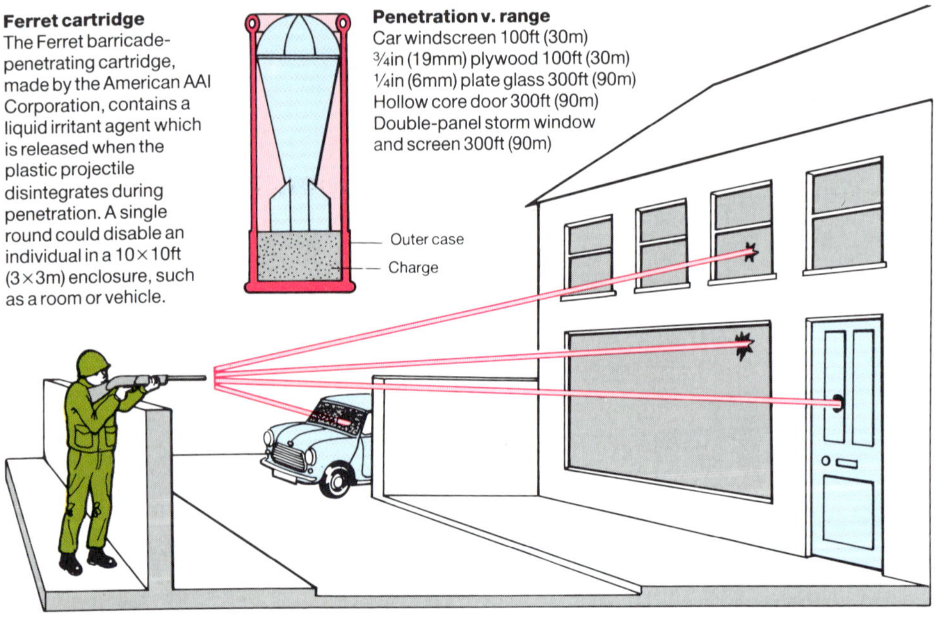


**Ferret cartridge**
The Ferret barricade-penetrating cartridge, made by the American AAI Corporation, contains a liquid irritant agent which is released when the plastic projectile disintegrates during penetration. A single round could disable an individual in a 10×10ft (3×3m) enclosure, such as a room or vehicle.

**Penetration v. range**
Car windscreen 100ft (30m)
¾in (19mm) plywood 100ft (30m)
¼in (6mm) plate glass 300ft (90m)
Hollow core door 300ft (90m)
Double-panel storm window and screen 300ft (90m)

**Barricade penetrators**
Major problems arise for security forces when terrorists gain limited protection in buildings or cars. The Ferret penetrating cartridge seeks to overcome this. Fired from a standard 12-bore shotgun it can penetrate barriers to release 3cc of irritant agent (eg, CS), which, inside small enclosures, is enough to distract an unprotected target.

have appeared with radio control; cable is more awkward, but has the advantage of being unjammable. The robots are equipped with extendable arms with interchangeable handling mechanisms, guns, TV cameras and various types of detectors. These relatively inexpensive machines (the "Wheelbarrow" is reported to sell for about £60,000) have already saved many lives and will continue to do so; indeed, their roles seem likely to extend. There seems to be no reason why robots such as these could not be used by military and police forces to assist in dealing with armed terrorists holed up in siege situations.

A major tool in the hands of counter-terrorist forces is the computer. Terrorists, in the words of Chairman Mao Tse-tung, must swim, like little fish, in the sea of the people. The majority of terrorists thus hide themselves in the larger cities, achieving a relative degree of security among the millions in a modern metropolis. Many security forces have therefore developed sophisticated on-the-street communications systems which enable soldiers or policemen to communicate with computer operators who can check out within seconds a vehicle registration number or a wanted person's description against central files.

Anti-terrorist operations have given birth to a technology all their own, especially in the area of non-lethal weaponry. The "stun-grenade", for example, has been developed to disorientate the target for a brief, but critical, period, such as the moment of entry of rescuers into an aircraft, as was done in the assault by the West German Border Police anti-terrorist squad (GSG 9) in their assault on the hijacked Lufthansa airliner at Mogadishu in October 1977. The British XFS Grenade has a cardboard body and gives off a very loud report (approximately 180dB at a distance of 3ft/1m) and a peak light intensity of 50 million candelas. So marked is the desire to minimise fatalities (among hostages) that a small preliminary charge is used to blow off the metal firing mechanism and aluminium cup one second before the main charge is detonated.

While terrorists have developed the use of explosives to produce bombs capable of killing large numbers of people at little or no cost to the terrorists themselves, then the security forces have also developed explosives, but in a somewhat different way. When the SAS blasted their way into the Iranian Embassy in London they used specially shaped charges to blow in windows and doors with the greatest of precision. So, too, have explosive charges been developed for use in gaining entry to aircraft, ships or trains with the maximum speed and minimum warning to the terrorists inside.

The plastic anti-riot round (widely and inaccurately known as the "rubber bullet") was also developed as a non-lethal weapon, with the purpose of inflicting, at worst, severe bruising and shock. Great controversy has been generated over these rounds, much of it ill-founded, but the search continues to find a weapon which can be used by security forces at such a distance from a crowd of rioters as to

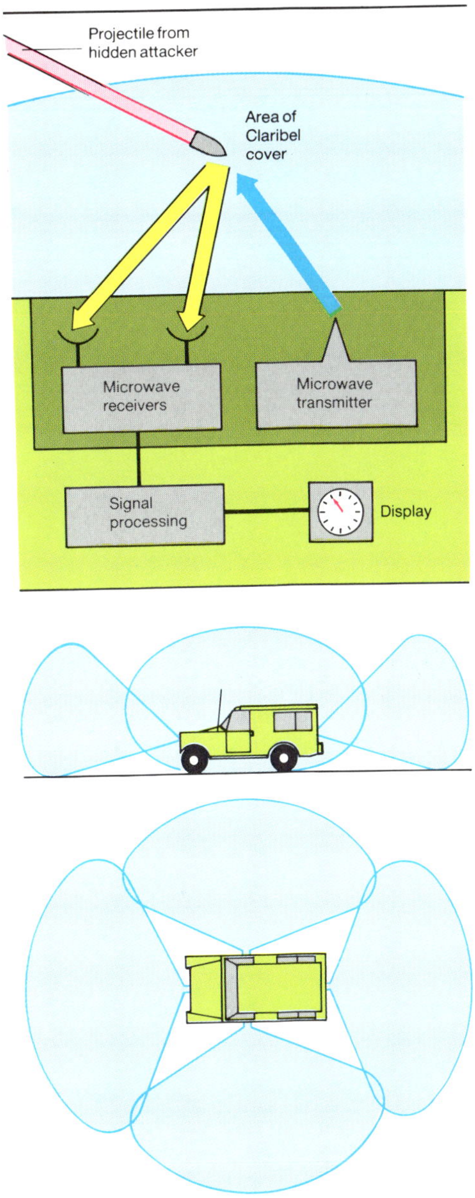


**Claribel system**
In an urban environment it is frequently difficult, if not impossible, for the Security Forces to tell from whence a shot has come, and the British Claribel system was designed to overcome this. A block diagram (top) shows that the signals from a microwave transmitter are reflected from the incoming projectile and are picked up by two (or four) receivers. Using Doppler techniques signal processing circuits operate on the phases between the reflected signals and establish an angle for the incoming projectile, which is displayed on a simple indicator, telling the crew the direction of the fire. The system can be installed on vehicles (above), helicopters or static observation posts.

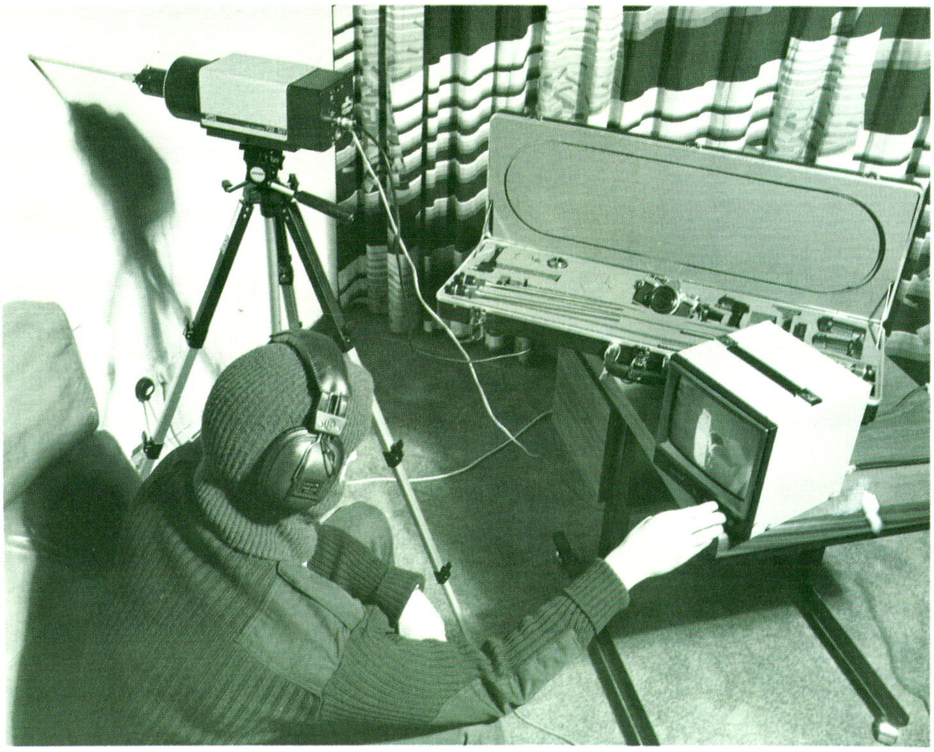

**Above: An operator using a rigid borescope, special camera and CCTV for remote and covert surveillance. Such devices have eased the counter-terrorist problem, especially in siege situations.**

**Below: The Arwen 37 anti-riot weapons system is now in service with a number of US State Police forces, as part of a growing armoury of devices to deal with civil disturbances.**

prevent physical contact between the two, but which will not seriously harm a target and yet still be sufficiently effective to deter troublemakers and make them disperse.

Considerable thought, and R&D effort, has been put into the more esoteric areas of anti-riot operations. The dress of some modern police forces in such duties is primarily meant to protect the wearer, but there can be no doubt that it is also intended, quite deliberately, to be intimidating. The large helmets with their clear-plastic visors, protective clothing (which naturally makes the wearer look larger), heavy boots, riot shield, long truncheon and – not infrequently – weapons, all combine to make the policeman look extremely threatening, like a figure out of a science-fiction film. Some fairly crude psychological techniques are used by some police and military forces to heighten further such an impression; these can include the rhythmic beating of truncheons on shields, the liberal spreading of smoke to create a murky and hostile environment, and the use of specially devised "dazzle" beams to create confusion and disorientation.

Gas has been used for many years in an anti-riot and anti-terrorist role. The old CN (tear-gas) has now been almost universally superseded by CS, an irritant agent. The gas is delivered primarily by grenades, either hand-thrown or by guns and special projectors. In disturbances of the late 1960s rioters rapidly discovered that, if quick, they could either kick or throw the grenades back into the squads of police or soldiers. New devices have had to be designed to make this more difficult. The Brazilians have produced a grenade in which the gas-emission slots are canted outwards to make the grenade spin around on the ground, while others simply have a much shorter time-delay fuse.

The British Royal Small Arms Factory (RSAF) at Enfield has developed a special anti-riot gun the 37mm ARWEN 37. This semi-automatic, multi-shot system can fire a variety of baton and CS rounds out to a maximum range of 110 yards (100m) and weighs only 8.3lb (3.8kg) when loaded with five rounds.

Of the well over 50 military campaigns since World War II, all of them have qualified for the term low intensity, rather than general war. Even the Second Indochina War, which involved the USA as a direct participant on one side supporting South Vietnam, opposing North Vietnam with the USSR and China in a surrogate role on the other, never approached the scale of general war, nor did it ever exhibit any serious signs of spreading beyond the confines of the three Indochinese states of Vietnam, Cambodia and Laos.

It is certainly true that during the course of the Vietnam War the USA used virtually every weapon system available to them, even to the extent of using Iowa class battleships for shore bombardment and B-52 bombers to devastate areas of North Vietnam, but they never took the war beyond the geographic limits of Indochina, nor did they threaten the use of nuclear weapons. It is also the case that in Afghanistan the Soviet Union appears to be using the weapons normally available to their tank and motor rifle divisions and intended for a conflict in Central Europe.

It is now becoming apparent, however, that although this may seem to be a cost-effective answer, especially to politicians who want to see the expensive equipment they have voted for used (and to the military who want to see new systems tested), there is still a good case for developing and producing at least some weapon systems especially for this type of warfare. However, few of these would be radical examples of advanced technology systems (as with, say, green laser communications to submerged submarines), but rather would be examples of the concerted application of technology to produce equipment which is lighter, more effective, longer ranged and more mobile than its predecessor.

**Above: Of equal utility in either high or low intensity warfare are weapons systems such as this US AH-64A attack helicopter armed with Hellfire missiles.**

## Counter-insurgency aircraft

Many of the aircraft in a standard air force inventory have the obvious roles in low-intensity warfare, although many appear over-sophisticated for the task. The American experience in Vietnam, therefore, appeared to suggest the need for a specialised counter-insurgency (CO-IN) aircraft. This form of aircraft was deemed to require a good weapon load, good range, together with sufficient protection to ensure survival in a low-level environment, and lengthy loiter time to await the call-forward from ground forces.

To a large extent this requirement seems to have arisen as a result of the undoubted success of the Douglas A-1 Skyraider, a sturdy and reliable piston-engined strike aircraft, which had some notable achievements in Vietnam. This led to designs such as the Piper Enforcer, an updated, turboprop-powered version of the World War II North American F-51 Mustang, and encouraged foreign industries to follow suit with other counter-insurgency types, such as the Argentine Pucara. Most dramatic of the resulting aircraft, however, was the Fairchild A-10 Thunderbolt II, a truly specialised ground-attack/tank-killer aircraft in the tradition of the Skyraider and the earlier Soviet Schturmovik. The Soviet Air Force has developed a similar type – the Sukhoi Su-25 "Frogfoot" – which bears a remarkable resemblance to the A-9, the losing design in the competition won by the A-10.

**Below: The US Army's new Fast Attack Vehicle (FAV), here armed with the TOW anti-tank missile, is designed for rapid movement in a low-threat environment.**

The other major type to emerge from the Vietnam experience, and for somewhat similar reasons, was the heavily armed and well armoured attack helicopter. The USA has spent a vast amount of money on a series of projects to meet this requirement, including the Lockheed AH-56A, which never attained service status. The latest type, however, the Hughes AH-64 Apache, is now in service, albeit as an attack aircraft for general warfare, but with obvious applications in low intensity warfare. As with the fixed-wing type, the Soviets watched the American Vietnam experience and developed their own type, the highly successful Mil Mi-24 "Hind", which is now being used to great effect in the USSR's "Vietnam" in Afghanistan, and are following up with a new "Apache-type" designated Mil Mi-28 "Havoc".

## Naval forces

Naval forces are, in one sense, constantly engaged in low intensity warfare in peace time in that they are always on patrol and exercise, and come into regular contact with potential opponents. This activity, however, involves all the naval forces required for general war. Naval forces also support ground troops in actual low intensity warfare, by giving sea transport where required and by offshore patrols. Some low intensity wars have even involved naval engagements, but only on a very limited scale, such as those between the Israeli and Egyptian navies in the various Arab-Israeli wars.

In Vietnam the US Navy was spurred to take an interest in special craft, including "monitor-type" vessels and air-cushion vehicles (ACVs), using various British types for combat patrol duties, especially in the Mekong delta.

Generally speaking, however, the true low-intensity wars have had much less effect upon the navies than they have had on the other two services.

## Ground forces

Low intensity warfare predominantly involves ground battle, and it is therefore not surprising that most of the development of dedicated systems has been for

armies, as opposed to navies or air forces. The various threats to ground troops on the potential nuclear battlefield have resulted in ever-more sophisticated, heavier and complex weapons systems, which means that many of them are unsuitable for use in low intensity warfare. The design of self-propelled artillery, for example, has led to weapons such as the SP70, with a 155mm gun and a combat weight of 95,950lb (43,525kg): mammoth equipments, and difficult to maintain and deploy in unsophisticated environments. Similarly, contemporary main battle tanks, most of which now weigh in the region of 134,400lb (60,960kg), are just too large and too heavy for such an environment.

The Americans have twice tried to apply their massive technological resources to designing a tank for low-intensity warfare. The first was the M551 Sheridan, which was intended to be a reconnaissance vehicle; small, well-armoured and agile, but nevertheless with a powerful armament. This led to the multi-purpose 152mm gun/missile launcher which, despite vast expenditure, could never be made to work, and the M551 was an almost total failure.

Current development work is being concentrated on the Rapid Deployment Force Light Tank (RDF/LT), based on the High Survivability Test Vehicle (Lightweight) already under test for the USA Tank Automotive Command. Weighing just 29,600lb (13,425kg) in combat order, the RDF/LT has a 75mm ARES cannon and a crew of 3 (another version with a 76mm M32 gun has also been tested). The 75mm ARES cannon is mounted in a turret of revolutionary design and is fed by an automatic magazine containing 60 rounds of mixed APFSDS and multi-purpose ammunition. Highly mobile, hard hitting, and yet small and compact, eight RDF/LTs can be fitted into a C-5 Galaxy, and two into a C-141 StarLifter or C-130 Hercules; one can be carried underneath a CH-53E helicopter. The other US armoured vehicle programme in this field is the Light Armored Vehicle (LAV), an eight-wheeled vehicle armed with a 25mm Bushmaster cannon, and based on the Swiss Mowag "Piranha".

The British Scorpion/Scimitar light tanks proved their value once again in the South Atlantic War of 1982. The Soviets, too, have a good fire-support/reconnaissance vehicle in their ASU-85.

**Above: In low intensity operations less sophisticated vehicles can be used, such as this Mowag Piranha, rather than the complex machines for general war.**

**Above: Air-Delivered Seismic Intruder Devices (ADSID) on a wing-pylon of a US Navy aircraft, an example of technology helping to solve tactical problems.**

Self-propelled artillery has become so large and heavy that it is no longer air-portable, nor is it readily mobile in some Third World environments. Also, the firing conditions differ, because an SP in a West European environment will seldom have problems of firing out of clearings in forest, in which the tree height will not be very great, while in Southeast Asia jungle clearings will be frequently used for gun positions, and the angle of firing must be great in order to clear the trees, which can be up to 140ft (43m) in height.

Most nations, therefore, retain towed artillery for such conditions, and a surprising number of projects are under way, particularly at 155mm calibre. Here the aim, in virtually every case, is to produce a more readily air-portable gun, especially under a helicopter, with greater range and a more effective shell. In many of these cases the advanced technology is as much in the manufacturing processes as it is in the actual gun or its ballistics. In the British FH-70, for example, very considerable attention was paid to the trail and the barrel to ensure that they were very strong, and yet as light as possible. Many of these guns now also have rocket-assisted projectiles (RAP) which can increase the range considerably; in the case of the FH-70, for example, by 25 per cent.

Another area where low-intensity warfare has differing requirements from general war is in surveillance devices. Although various sensors have been around for some years it was the American requirements in Vietnam which really gave the major urge to development, in particular, for surveillance of the Ho Chi Minh Trail – the Communists' major north-south route for resupply of guerrilla and regular forces operating in South Vietnam. A whole host of devices were produced, most of which depended upon either seismic effects or upon the interruption of an electronic beam. Data was often passed back to aircraft, which relayed the information back to fusion centres.

A contemporary device is the AN-TRS-2(V)PEWS early warning system, in which remotely emplaced sensors can detect movement within about 30ft (10m) and, by sampling the magnetic and seismic patterns, classify it as either human or vehicle. The sensors can be placed up to 1,640 yards (1,500m) from the central control, with communications being either by radio or cable.

## Conclusion

There can be no doubt that in general war advanced technology has a major place; indeed, there are many who would say that it is the central issue. In low-intensity warfare, however, and more particularly in anti-terrorist and anti-riot operations, its place is more controversial. Some advanced, and carefully designed, weapons undoubtedly are beneficial in that they enable the security forces to achieve successful results with the minimum use of force. A good example is the stun grenade and its use during the critical break-in period to overcome terrorists in a siege or a hijacking, where everything is optimised to achieve maximum surprise with the least possible harm to hostages in the vicinity.

There are those who claim, and with some justification, that in other ways advanced technology is being counter-productive. For example, the great range of equipment now available to anti-riot squads may enable them to disperse individual riots (usually at some cost in injuries if not lives), but it also tends to make each riot more serious, and can in the longer term be counter-productive by creating deep resentment and ill-will against both police and the government that control them. This can, in turn, give excellent opportunities for anti-government propaganda. Thus, in these areas advanced technology should not be applied blindly, but must be brought to bear on the problems with care, and with a deliberate consideration of the possible consequences.

# INDEX

Page numbers in **bold** type refer to subjects mentioned in captions to illustrations.

## A



## B



## C



## D



## E



## F



## G



## H

## T



## U



## V



## W



## X



## Y



## Z



# Picture Credits

The British Ministry of Defence has been abbreviated to MoD, the US Department of Defense to DoD

**Back cover:** Top right Scicon, remainder DoD. **6:** Top McDonnell Douglas, bottom Grumman Aerospace. **7:** Hughes. **8:** MoD. **9:** DoD. **10:** Top USAF, bottom DoD. **11:** Top Hughes, bottom DoD. **12:** Top US Army, bottom Grumman. **13:** Scicon. **14:** DoD. **15:** Top US Defense Nuclear Agency, bottom Wimpey. **17:** Top United Technologies Norden Systems, bottom US Navy. **18:** USAF. **20:** Top left Kurass Maffei, right Hollandse Signaalapparaten. **21:** Top Ericcson, centre FFV, bottom Bridport Aviation Products. **22:** Hollandse Signaalapparaten. **23:** Plessey Marine. **24:** Top USAF, bottom Hughes. **25:** Top US Navy, bottom Matra. **26:** Top Matra, bottom left USAF, right Hughes. **27:** Top Rank Pullin, bottom DoD. **28:** Top left Texas Instruments; top right US Army, bottom US Army. **29:** Top and bottom US Army, top right Bofors. **30:** Top Emerson Electric, bottom Plessey. **31:** Racal. **32:** Plessey. **33:** Top Doug Richardson, centre Hughes, bottom Martin Maurietta. **34:** Hughes. **35:** Top USAF, bottom Collins. **38:** DoD. **39:** Top Xerox, centre Northrop, bottom Sanders. **40:** Schermuly. **41:** British Aerospace. **42:** Top British Aerospace, bottom Thomson-CSF. **43:** Left US Army, right USAF. **44-45:** DoD. **46-47:** DoD. **48:** Top ECPA, bottom DoD. **50-51:** DoD. **52:** US Navy. **53:** Top and centre USAF; bottom Boeing. **54:** Top Salamander, bottom DoD. **55:** Top USAF; bottom Salamander. **56:** DoD. **57:** Top McDonnell Douglas, bottom DoD. **58:** DoD. **59:** Top left & right DoD, bottom CEL. **60-61:** DoD. **62:** Top DoD; bottom MoD. **63:** Left and centre ECPA, right DoD. **64:** Top DoD, bottom left USAF, right US Navy. **65:** DoD. **66:** USAF. **67:** Top and centre DoD, bottom AMD-BA/MARS. **68-69:** USAF. **70:** Top Rockwell, bottom DoD. **71:** Royal Swedish Air Force. **72:** NASA. **73:** Top Salamander; bottom Ford Aerospace. **74:** USAF. **75:** DoD. **76:** USAF. **77:** Left Salamander, right DoD. **78:** Top NASA, bottom DoD. **79:** RCA. **80:** DoD. **81:** Top and centre USAF, bottom CNES. **82-83:** DoD. **84:** Top Austin Research Associates; bottom DoD. **85:** USAF. **86:** DoD. **87:** LTV. **88:** US Army. **89:** DoD. **90-91:** DoD. **92:** Top USAF, bottom McDonnell Douglas. **93:** DoD. **94:** Top US Army, bottom left, Ford Aerospace; right MoD. **95:** Top left Wallop Industries, top right and bottom Contraves. **96:** British Aerospace. **97:** Left MBB; right Hughes. **98:** Top USAF; bottom McDonnell Douglas. **99:** USAF. **100-101:** USAF. **102:** Top and centre Ford Aerospace, bottom British Aerospace. **103:** USAF. **104:** Top Alkan, bottom left DoD; right Hughes. **105:** Top MoD, bottom USAF. **106:** Northrop. **107:** Top British Aerospace, bottom Dassault-Breguet. **108:** USAF. **109:** Top Sikorsky; bottom Hughes. **110:** Hunting Engineering. **112:** USAF. **113:** Top G & W Portarrest, bottom McDonnell Douglas. **114:** Top USAF, bottom Rockwell. **115:** Top British Aerospace; centre and bottom USAF. **116:** Top and centre USAF, bottom Matra. **117:** Top and centre Hunting Engineering, bottom MBB. **118:** Left Bofors; right AVCO Systems Division. **119:** CSD. **120:** AVCO Systems Division. **121:** Top and centre Hunting Engineering, bottom General Electric. **122:** Top McDonnell Douglas; bottom Hughes. **123:** Top United Technologies, bottom Saab-Scania. **124:** Top USAF. **125:** A. Johnson. **126:** USAF. **127:** US Navy. **162:** Top US Navy, bottom NATO. **164-165:** US Navy. **166-167:** US Navy. **168:** Top Thomson CSF; bottom US Navy. **169:** US Navy. **170:** US Navy. **171:** Marconi. **172:** Bottom US Navy. **173:** Top MoD. **174:** US Navy. **175:** Top US Navy, bottom British Aerospace. **178:** Top US Navy; bottom Grumman. **179:** Top McDonnell Douglas, bottom Dassault-Breguet. **180:** Top SIRPA-Marine/MARS; bottom US Navy. **181:** Top MoD; bottom DoD. **182-183:** US Navy. **184:** McDonnell Douglas. **185:** Aerospatiale. **187:** US Navy. **188:** Boeing. **189:** Nordco Ltd. **191:** DCN. **192:** US Navy. **193:** Top left MoD, top right US Navy; bottom RNAS Yeovilton. **194:** US Army. **195:** Left US Navy; right London Express News Service. **196-197:** DoD. **198:** DoD. **199:** Left DoD, right Camera Press. **200:** Top German Ministry of the Interior; bottom MoD. **201:** Hallye Weller. **202:** Left Hallye Weller, right Israel Military Industries. **203:** Top Keymed, bottom: Royal Ordnance Factories. **204:** Top Hughes; bottom US Army. **205:** Top Mowag, bottom USAF.